Android Mobile Development Technology

Android
移动开发技术

张　勇　编著

清华大学出版社
北京

内 容 简 介

Android 系统是运行在智能移动设备上的嵌入式操作系统,包括 Linux 内核、系统库与 Java 运行时、应用程序框架层和应用程序层 4 部分,具有公开源代码和免费使用的特点,是目前深受欢迎且全球用户数量最多的嵌入式操作系统。本书讲述基于 Android 系统的应用程序设计方法,全书分为 9 章,内容包括 Android 系统概述、Java 语言、Android 应用程序框架、单用户界面设计、多用户界面设计、数据访问技术、图形与动画、多媒体技术和通信应用技术。本书的特色在于原理讲解透彻,实例丰富且有代表性。

本书是作者近几年来从事嵌入式教学与研究的成果结晶,重点阐述了 Android 应用程序设计的理论与方法,适合作为普通高等院校软件工程、物联网、电子通信和智能控制等专业讲授 Android 移动开发技术的本科生教材或参考书。

本书封面贴有清华大学出版社防伪标签,无标签者不得销售。
版权所有,侵权必究。侵权举报电话: 010-62782989　13701121933

图书在版编目(CIP)数据

Android 移动开发技术/张勇编著. —北京: 清华大学出版社,2017
ISBN 978-7-302-46659-8

Ⅰ. ①A… Ⅱ. ①张… Ⅲ. ①移动终端－应用程序－程序设计　Ⅳ. ①TN929.53

中国版本图书馆 CIP 数据核字(2017)第 036007 号

责任编辑: 梁　颖　薛　阳
封面设计: 常雪影
责任校对: 白　蕾
责任印制: 李红英

出版发行: 清华大学出版社
网　　址: http://www.tup.com.cn, http://www.wqbook.com
地　　址: 北京清华大学学研大厦 A 座　　　　邮　编: 100084
社 总 机: 010-62770175　　　　　　　　　　邮　购: 010-62786544
投稿与读者服务: 010-62776969, c-service@tup.tsinghua.edu.cn
质量反馈: 010-62772015, zhiliang@tup.tsinghua.edu.cn
课件下载: http://www.tup.com.cn,010-62795954

印 装 者: 清华大学印刷厂
经　　销: 全国新华书店
开　　本: 185mm×260mm　　　印　张: 22.25　　　字　数: 540 千字
版　　次: 2017 年 6 月第 1 版　　　　　　　　　　印　次: 2017 年 6 月第 1 次印刷
印　　数: 1～2000
定　　价: 69.00 元

产品编号: 073219-01

智能手机的普及和移动计算技术的发展,促进了嵌入式操作系统的迅速发展。一般地,某类移动设备被称为"智能"的,就是指其加载了某种嵌入式操作系统,从而使其具有类似于计算机的功能,通过加载用户应用程序可以实现复杂的数据计算、控制处理和友好的人机交互工作。在智能移动设备中,嵌入式操作系统所起的作用与桌面视窗 Windows 系统类似,是用来管理设备系统软硬件资源的,而且嵌入式操作系统还具有实时性强、体积小和可裁剪的特点。目前,流行的嵌入式操作系统有 Google 的 Android、Apple 的 iOS、Microsoft 的 Windows CE(Windows 移动版)、嵌入式 Linux、WindRiver 的 VxWorks 和 Micrium 的 μC/OS-III 等。其中,Android 和嵌入式 Linux 是免费且公开源代码的,μC/OS-III 是公开源代码的。

Android 系统是基于 Linux 内核的嵌入式操作系统,严格地说是借用了 Linux 内核硬件驱动和线程调度功能,增加了与用户界面相关的应用程序设计框架和系统库。有些专家认为 Android 不过是类似于 QT 的用户界面程序。的确,Android 在用户界面方面具有强大的优势,其特色在于优化了图形显示技术并专门设计了图标。由于从程序员的角度出发,Android 系统向应用程序提供了完备的系统调用、进程管理与进程通信以及应用程序开发接口等,因此,Android 系统普遍被认可为嵌入式操作系统,并且在全球范围内得到用户和程序员的青睐。它的最大优势在于公开了源代码且可免费使用,并且其应用程序开发环境也是免费的。

本书基于 Google 推出的 Android Studio 集成开发软件,根据 Android 系统和应用软件(APP)设计技术的进展情况,阐述 Android 应用程序设计的最新原理与技术。全书共 9 章。

第 1 章为"概述",介绍 Android 系统的发展历程和系统结构,并详细讲述 Java 语言程序设计的集成开发环境 Eclipse 和 Android 应用程序设计的集成开发环境 Android Studio。

第 2 章为"Java 语言",介绍 Java 语言的语法和数据结构,深入讲解 Java 类、内部类和事件方法等概念。由于 Android 应用程序采用 Java 语言编写,因此,这一章的内容可使没有接触过 Java 语言的读者快速入门。

第 3 章为"Android 应用程序框架",介绍 Hello World 工程及其结构,详细讲述 Hello World 工程的工作原理及应用程序框架的基本组成,并分析 Activity(活动界面)的生命周期。

第 4 章为"单用户界面应用设计",讲述 Activity 的概念和使用方法,详细讲解应用程序布局方法和 Android 系统常用控件的种类及其使用方法。通过"计算器"工程深入分析单用户界面程序设计的特点。

第 5 章为"多用户界面应用设计",介绍 Intent 的概念和不同界面间的数据通信方法。与单用户界面程序相比,多用户界面程序需要进行界面间的数据通信,包括对话框与 Activity 之间、菜单与 Activity 之间以及两个 Activity 之间的数据通信,借助于内部类或 Intent 对象可实现这些通信方法。

第 6 章为"数据访问技术",介绍 Android 系统的 4 种数据存储与访问方式,即 SharedPreferences 文件访问、流文件操作、SQLite 关系数据库和 Content Provider(内容提供者)等,并通过实例对比这 4 种方式的异同点。其中,Content Provider 可实现不同应用程序间的数据共享与通信。

第 7 章为"图形与动画",详细讲述借助于 View 类和 SurfaceView 类进行图形绘制和动画设计的方法,介绍图形绘制的应用程序框架,阐述三种动画方式,即定时器动画、渐变动画和帧切换动画。

第 8 章为"多媒体技术",介绍借助于 Media Player 类播放音频文件和视频文件的方法,并介绍 Service(服务)的程序设计方法。

第 9 章为"通信应用技术",详细介绍基于 Android 智能手机进行短信通信的程序设计方法,本章工程的执行需要借助于真实的 Android 智能手机而非 Android 模拟器。

本书用于课内教学,建议理论学时为 48,课内与课外实验学时为 96。本书的特色在于 Android 应用程序设计原理讲解透彻,语言通俗易懂,程序实例丰富且具有代表性。书中全部工程实例的代码是完整的,可在学习和录入代码的过程中,体会到 Android 移动开发的乐趣,并试着对已有工程进行功能扩展和创新。本书实例代码约 2.5GB,可以在百度云盘下载(链接:http://pan.baidu.com/s/1ciM0s2 密码:di67)。

特别强调的是:本书中的一些图像和音视频素材均来自于互联网,其版权归相关公司所有,书中引用的这些素材仅用于教学和研究。同时,作者保留其余内容的所有权利。

最后,感谢江西财经大学软件与通信工程学院唐颖军博士、陈滨博士和廖汉程博士等专家学者对本书提出的建设性意见,感谢研究生侯文刚、李雪倩和张琼阅读本书的初稿并订正了一些错误。由于作者水平有限,书中难免有纰漏之处,敬请同行专家和读者批评指正。

张 勇

江西财经大学枫林园

2017 年 2 月

目录

第1章　概述 ……………………………………………………………………………… 1
  1.1　Android 操作系统 …………………………………………………………………… 2
  1.2　Android 系统结构 …………………………………………………………………… 4
  1.3　Java 开发环境 ……………………………………………………………………… 7
  1.4　Android 开发环境 …………………………………………………………………… 16
  1.5　本章小结 …………………………………………………………………………… 34

第2章　Java 语言 ………………………………………………………………………… 35
  2.1　Java 程序语法与控制 ……………………………………………………………… 35
    2.1.1　顺序方式 ……………………………………………………………………… 35
    2.1.2　分支方式 ……………………………………………………………………… 40
    2.1.3　循环方式 ……………………………………………………………………… 44
    2.1.4　异常处理 ……………………………………………………………………… 47
  2.2　Java 基本数据类型 ………………………………………………………………… 49
    2.2.1　数值 …………………………………………………………………………… 49
    2.2.2　字符 …………………………………………………………………………… 50
    2.2.3　字符串 ………………………………………………………………………… 51
    2.2.4　布尔数 ………………………………………………………………………… 53
    2.2.5　数组 …………………………………………………………………………… 53
  2.3　Java 类 ……………………………………………………………………………… 55
    2.3.1　类与对象 ……………………………………………………………………… 56
    2.3.2　继承与多态 …………………………………………………………………… 57
    2.3.3　接口 …………………………………………………………………………… 60
  2.4　Java 文件操作 ……………………………………………………………………… 63
  2.5　在命令行窗口中运行 Java 程序 …………………………………………………… 67
  2.6　Java 图形界面 ……………………………………………………………………… 69
    2.6.1　事件响应方法 ………………………………………………………………… 69
    2.6.2　内部类 ………………………………………………………………………… 71

2.6.3　匿名内部类 …………………………………… 72
2.7　本章小结 ………………………………………………… 73

第3章　Android 应用程序框架 …………………………… 75

3.1　Hello World 工程 ………………………………………… 75
3.2　Hello World 应用工作原理 ……………………………… 80
3.3　应用程序框架 …………………………………………… 86
　　3.3.1　应用程序框架基本组成 ………………………… 86
　　3.3.2　Android 配置文件 AndroidManifest.xml …… 87
　　3.3.3　Android 资源文件 ……………………………… 90
　　3.3.4　Android 源程序文件 …………………………… 94
3.4　Activity 生命周期 ………………………………………… 97
3.5　本章小结 ………………………………………………… 101

第4章　单用户界面应用设计 ……………………………… 102

4.1　Activity 概念 ……………………………………………… 102
4.2　布局与控件 ……………………………………………… 103
　　4.2.1　布局软件 DroidDraw ………………………… 103
　　4.2.2　控件事件响应方法 ……………………………… 111
　　4.2.3　Android 常用控件 ……………………………… 118
　　4.2.4　线性布局 LinearLayout ………………………… 120
　　4.2.5　相对布局 RelativeLayout ……………………… 122
　　4.2.6　框架布局 FrameLayout ………………………… 125
　　4.2.7　表格布局 TableLayout 和 TableRow …………… 126
　　4.2.8　约束布局 ConstraintLayout …………………… 129
4.3　"计算器"工程 …………………………………………… 135
4.4　本章小结 ………………………………………………… 145

第5章　多用户界面应用设计 ……………………………… 147

5.1　Intent 概念 ………………………………………………… 147
5.2　对话框 …………………………………………………… 150
　　5.2.1　AlertDialog 对话框 ……………………………… 150
　　5.2.2　自定义对话框 …………………………………… 155
　　5.2.3　Dialog 类 ………………………………………… 161
　　5.2.4　ProgressDialog 对话框 ………………………… 167
5.3　菜单 ……………………………………………………… 171
　　5.3.1　XML 布局菜单 …………………………………… 172
　　5.3.2　动态菜单 ………………………………………… 175
　　5.3.3　上下文菜单 ……………………………………… 177

5.4 多用户界面设计 ····································· 181
    5.4.1 简单多用户界面显示 ······················· 181
    5.4.2 多用户界面数据传递 ······················· 191
    5.4.3 活动界面间双向数据通信 ··················· 195
5.5 本章小结 ··· 207

第 6 章 数据访问技术 ································· 209

6.1 SharedPreferences 文件访问 ························· 209
6.2 流文件操作 ······································· 224
6.3 SQLite 关系数据库 ································ 236
    6.3.1 SQLite 数据库访问方法 ····················· 236
    6.3.2 SQLiteOpenHelper 类 ······················ 243
6.4 内容提供者 ······································· 250
6.5 本章小结 ··· 260

第 7 章 图形与动画 ··································· 261

7.1 绘图 ··· 261
    7.1.1 View 类绘图程序框架 ······················ 262
    7.1.2 SurfaceView 类绘图程序框架 ················ 271
    7.1.3 基本图形与字符串 ························· 275
7.2 动画 ··· 276
    7.2.1 定时器动画 ······························· 276
    7.2.2 渐变动画 ································· 287
    7.2.3 帧切换动画 ······························· 292
7.3 本章小结 ··· 296

第 8 章 多媒体技术 ··································· 297

8.1 音频文件播放 ····································· 297
8.2 服务 ··· 307
8.3 视频文件播放 ····································· 312
8.4 本章小结 ··· 322

第 9 章 通信应用技术 ································· 323

9.1 短信息发送 ······································· 323
9.2 短信息接收 ······································· 328
9.3 短信息加密 ······································· 334
9.4 本章小结 ··· 345

参考文献 ·· 346

第 1 章

概 述

　　Android 音译为"安卓",Android 系统是安装在移动设备,例如智能手机、个人数字助理(PDA)、MP5 播放器、手持终端、平板计算机、上网本、电子书等上的操作系统软件,用于管理和调度移动设备的软硬件资源,其作用相当于个人计算机(PC)上安装的微软 Windows 视窗操作系统,与安装在智能手机上的 Windows Mobile Phone(Windows CE)操作系统相似。Android 系统与桌面 Windows 系统、Windows CE 操作系统的异同点列于表 1-1,从表 1-1 可以对 Android 系统有一个全面直观的认识。

表 1-1　Android 系统与桌面 Windows 系统、Windows CE 操作系统的异同点

比 较 项 目	Android 系统	桌面 Windows 系统	Windows CE 系统
用户界面	有	有	有
标准内存	256MB	2GB	256MB
系统大小	<64MB	约 4GB	<64MB
内核驻留	Flash 芯片	硬盘	Flash 芯片
实时性	强	弱	强
应用设备	嵌入式移动设备	个人计算机	嵌入式移动设备
功耗	低	高	低
应用软件体积	小	大	小
应用软件兼容性	极强	强	强
系统源代码	公开	不公开	核心不公开
内核独立性	基于 Linux	独立	独立
输入设备	触摸屏为主	键盘、鼠标为主	触摸屏为主
输出分辨率	相对低	相对高	相对低
英文直译	科幻机器人	视窗	便携式电子窗口
开发商	Google(谷歌)为主	微软(Microsoft)	微软

Android 系统由 Andy Rubin 首创,最初目的是设计一种新的开放性智能手机操作系统,2003 年美国已经有大量的移动设备嵌入式操作系统,新研发的操作系统想进入市场被用户认可是件很困难的事情。然而当时大部分嵌入式操作系统都不是开源的,维护十分困难,Rubin 等人因此提出这种开源的智能手机操作系统,希望借此挤进竞争激烈、商机无限的嵌入式操作系统市场中,赢利模式主要靠安装、维护和提供专业特色应用软件等技术服务。现在看来,Rubin 的做法成功了。2005 年 8 月 Google 收购 Android 加速了该开源嵌入式操作系统的发展,2007 年以 Google 为首组建了全球性的开放手机联盟(Open Handset Alliance),中国电信、中国移动和中国联通也是其中的成员,在全球范围内推动基于 Android 操作系统的手机开发计划。2008 年 10 月宏达电(HTC)公司推出了第一款 Android 系统的手机,命名为 HTC Dream(G1),如图 1-1(a)所示,这是一款被市场证实成功的手机。随后,几乎在全球范围内形成了研究 Android 操作系统的热潮,Android 操作系统功能和版本逐年提高,目前,已经是第 7.X 版,内部研发版本则更高。图 1-1(b)是基于 Android 2.3.3 版本的 Flyer 智能手机。2011 年初 Android 操作系统已经成为嵌入式操作系统领域最受欢迎的智能操作系统。

图 1-1 HTC Dream (G1)和 HTC Flyer

Android 操作系统是开放源代码的,并且拥有全球最多的研究人员和用户群,源文件中大量的 bug(问题)会被及时发现而纠正,因此,Android 系统版本号更新频繁。但是,基于 Android 系统的应用程序开发技术在各个版本中的方法完全相同,这些正是本书的内容(本书基于目前应用广泛的 Android 系统版本 4.4,同样适用于最新版本的 Android 系统)。

1.1 Android 操作系统

Android 操作系统是基于 Linux 内核的嵌入式操作系统,即其底层(称为第Ⅰ层)为 Linux 操作系统及其驱动,该层源代码是 C 语言编写的;底层上面建构了系统库和 Java 运行时(即 Java 程序运行支持软件包或 Java 虚拟机,"运行时"是由 Runtime 意译而来,在很多书中均采用这一译法),称为第Ⅱ层,这一层是使用 C 或 C++代码写成的;第Ⅲ层为应用程序框架层,为用户开发 Android 程序直接提供 API(应用程序接口)函数,这一层是用 Java 代码实现的;第Ⅳ层为用户应用程序层,由于 Android 操作系统内置了许多用户应用程序,因此,有些专家认为应用程序层可以划分到 Android 操作系统中,当然,用户自己编写的应用程序也属于这一层,这一层的应用程序使用 Java 语言设计。Android 系统结构如图 1-2 所示。

第Ⅳ层 应用程序层
例如欢迎界面、浏览器、通话应用、联系人、日历等
第Ⅲ层 应用框架层
例如活动、窗口、通知、包、通信、资源、本地化管理器以及内容提供者、视图系统等
第Ⅱ层 系统库和Java运行时
例如图形界面管理器、多媒体框架、SQLite 数据库、C语言库、Java虚拟机等
第Ⅰ层 Linux内核
例如文件系统、显示驱动、电源管理、USB 驱动、蓝牙驱动等

图 1-2 Android 系统结构

Android 系统结构将在第 1.2 节中详细介绍。从图 1-2 中可以看出 Android 操作系统采用分层结构,且整个系统建立在 Linux 操作系统内核基础上,借助 Linux 内核硬件驱动进行硬件资源的管理。因此,Android 系统没有独立的硬件底层驱动部分,事实上,Android 系统的软件调度也借助了 Linux 内核进程调度实现,即两个显著的操作系统特征在 Android 系统下没有得到体现,严格意义上讲,Android 系统应该隶属于应用软件系统的范畴。而与 Android 系统竞争市场的 Windows CE 操作系统则完全不同,它包括完整的内核层、驱动层和应用程序层,是真正意义上的嵌入式操作系统。如果从应用程序开发者的角度出发,而不去考虑图 1-2 所示的 Android 系统结构,此时,由于 Android 系统封装了各层间的通信和服务调用,向应用程序开发者提供了完备的系统调用(包括驱动程序开发)、进程管理与进程间通信和应用程序开发接口等,因此,从这个意义上说,Android 系统属于操作系统的范畴。现在,Android 系统研发者和应用程序开发者都普遍认可 Android 系统属于嵌入式操作系统,概念上将它与 Windows CE 等嵌入式操作系统等同。

Android 系统相对于其他嵌入式操作系统而言,具有两个明显的优点,即开放源代码和网络功能强大。前面提到了 Android 系统最初开放源代码的原因,从 2003 年到今天仍然保持着这一独特的优势,除了嵌入式操作系统领域市场竞争激烈外,Android 使用 Linux 作为其底层平台是其开源的另一个重要原因。Google 本身是互联网公司,其下的所有产品都是基于互联网模式发展的,Google 收购的 Android 系统也不会例外,伴随着 Android 系统的诞生和版本升级,Android 系统的网络功能越来越强大,这使得基于 Android 系统编写网络程序比基于其他任何嵌入式操作系统都更加容易。可以说,一部 Android 手机就是一部互联网终端,网上购物、新闻、旅游、导航、智能家居等应用的确给用户生活带来了极大的方便。

Android 系统使用 Java 语言编写应用程序,一定意义上可以说,Android 系统推动了 Java 语言的广泛应用。Java 语言属于面向对象的高级语言,Java 语言程序必须借助于 Java

虚拟机解释执行，比其他高级语言的可移植性都强，在 Android 模拟器上开发运行成功的应用程序，一定能够成功地部署和运行在 Android 系统终端机上，这使得 Android 系统应用程序开发变得异常方便。

Android 系统的图形界面也是它的一个亮点，严格地说，Android 系统不是基于可视化窗口的，而是直接基于图形的，也就是说，Android 系统界面是由一幅幅图画组合在一起的，因此，Android 系统界面比较"炫"。相比于 Windows CE 的视窗而言，人性化更强一些。Android 系统界面美观是其受到用户欢迎的最重要的原因，尽管如此，Google 对现有 Android 系统界面仍然不很满意，据说新版本的 Android 系统在用户界面上还要做大的创新。

目前最新的 Android 系统版本号为 7.X，研发代号为 Nougat（牛轧糖），重点在于扩展 Android 系统的人工智能应用。Android 系统是纯粹的商业性操作系统，在 GPL（General Public License）协议条件下源代码公开和免费使用，这意味着当用户免费使用 Android 系统开发软件产品时，其衍生的软件产品也必须是开源和免费的。需要指出的是，Android 虚拟机不是开源的。

1.2 Android 系统结构

图 1-2 为 Android 系统结构图，由于直接翻译 Android 系统组件的术语并不准确，因此，这里给出了经典的英文 Android 系统结构图，如图 1-3 所示。

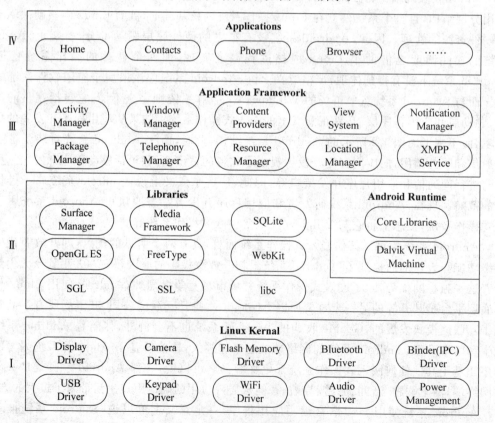

图 1-3 Android 系统结构图

由图 1-3 可知，Android 系统是基于 Linux 内核的操作系统，习惯上把 Linux 内核 (Linux Kernal) 层称为其第Ⅰ层。Linux 是免费和公开源码的实时抢先式多任务操作系统，Linux 内核协助 Android 系统完成进程调度、进程间通信、内存管理、虚文件系统管理、系统安全管理和设备驱动等功能。图 1-3 中仅列举了 Linux 内核实现的 10 种设备驱动功能，即显示驱动、摄像驱动、Flash 存储驱动、蓝牙驱动、Binder IPC 驱动（用于进程间通信管理）、USB 驱动、键盘驱动、WiFi 驱动、音频驱动、功耗管理，事实上，Linux 内核还协助 Android 完成强大的网络管理和驱动等，被视为 Android 系统的一个硬件抽象层。正因为 Android 基于 Linux 内核，很多专家指出，要深入学习 Android 必须加强 Linux 系统的学习。如果重点放在 Android 应用程序设计上，即使不懂 Linux，对学习 Android 程序开发也影响不大。由于 Android 系统是架构在 Linux 系统之上的，Linux 系统不支持的处理器，Android 系统则同样无法支持，即 Android 系统只能移植到 Linux 系统可以运行的处理器上。好在 Linux 系统支持大多数流行的处理器，例如 x86 结构、ARM、MIPS 等，Linux 系统这种广泛的可移植性决定了 Android 系统具有广泛的可移植性。

第Ⅱ层是 Android 系统库 (Libraries) 和 Android 运行时 (Runtime)，这一层是使用 C/C++ 代码编写的。Android 系统库包含了大量的类，通过第Ⅲ层（即应用程序框架层）被应用程序开发者调用，应用程序开发者使用的大量 API (应用程序接口) 函数来自于这些类。API 函数越丰富，则用户开发应用程序的工作量越小，随着 Android 系统版本的升级，API 函数级别也随之上升，如表 1-2 所示。

表 1-2 Android 系统与 API 级别的关系

Android 系统版本号	代号	API 级别 (level)
1.1	—	2
1.5	Cupcake	3
1.6	Donut	4
2.0	Eclair	5
2.0.1	Eclair	6
2.1	Eclair	7
2.2	Froyo	8
2.3.1	Gingerbread	9
2.3.3	Gingerbread	10
2.4	Gingerbread	未公开
3.0	Honeycomb	11
3.1	Honeycomb	12
3.2	Honeycomb	13
4.0	IceCreamSandwich	14
4.0.3	IceCreamSandwich	15
4.1	Jelly Bean	16
4.2	Jelly Bean	17
4.3	Jelly Bean	18
4.4	KitKat	19

续表

Android系统版本号	代　　号	API级别(level)
4.4W	KitKat Wear	20
5.0	Lollipop	21
5.1	Lollipop	22
6.0	Marshmallow	23
6.X	Nougat	24

表1-2显示最新的API级别为24，从表1-2中可以看出Android系统每个版本的研发代号都是一种小食品的名称，依次译为"杯形蛋糕"、"炸面圈"、"指形小饼"、"冻酸奶"、"姜饼"、"蜂巢"、"冰淇淋三明治"、"果冻豆"、"奇巧巧克力"、"棒棒糖"、"棉花糖"和"牛轧糖"，估计发起人也是一位美食家，可以预见下一个版本一定也是一种流行的受人喜爱的小食品名称，这正是Android系统研发的用意所在，希望得到系统研发者和应用程序开发者们的喜爱。

图1-3中第Ⅱ层Android系统库给出了9个组件，即Surface Manager(界面管理器)、Media Framework(多媒体框架)、SQLite、OpenGL ES、FreeType、WebKit、SGL、SSL和libc。这9个组件都十分复杂，下面简要说明一下各个组件的作用。

Surface Manager(界面管理器)负责显示相关的操作，Android系统界面是基于图形系统的，这种图形系统采用客户端/服务器的方式进行工作，"客户端"就是用户的应用程序界面，而"服务器"负责与这个应用程序界面相关的数据管理，由一个称为Surface Flinger的组件管理，"服务器"与"客户端"的通信需要借助于Binder类。界面管理器就是要管理这种工作方式，以实现对二维或三维图形的显示。

Media Framework(多媒体框架)是个非常实用的类库，封装了大量处理多媒体数据的API函数，支持流行的绝大部分多媒体格式，使得应用程序开发者开发多媒体软件异常轻松，只需要调用多媒体框架中的API函数就行了。

SQLite是Android集成的关系型数据库，SQLite是公开源代码的嵌入式数据库，支持ANSI SQL92标准的大部分SQL(结构化查询语言)语句，速度快，体积小(约250KB)，最大支持数据库文件大小为4TB。

OpenGL ES是Android系统中二维和三维图形处理与加速的API函数集。

SGL是Skia Graphics Library的首字母简写，是Android用来处理二维图形的向量图形引擎。所谓的"引擎"术语是从汽车借用过来的，在计算机软件中的引擎是指软件处理中的最核心部分，就像汽车发动机是汽车的核心一样，一个软件功能的升级主要取决于其"引擎"部分的升级。SGL就是Android系统的二维图形引擎。

FreeType是Android系统使用的字体引擎，FreeType是公开源码和免费的，其优点在于提供简单、统一的API函数访问多种字体格式文件，例如位图字体和矢量字体，这使得Android系统处理字体非常方便。

SSL是Secure Sockets Layer的首字母简称，即安全套接层，这里说明Android系统支持SSL，即安全套接层协议。SSL对发送的网络数据进行加密，防止数据在传送至合法目的地网络终端的过程中被非法用户使用或修改，这在电子商务和网上银行的数据交换中尤为重要，SSL 通过对合法用户的认证和数据的加密，确保用户的信息安全，Android 系统自诞生以来，就支持 SSL 协议。事实上，所有的嵌入式操作系统浏览器都支持 SSL 协议。

WebKit 是公开源码的网络浏览器引擎，这个引擎稳定性好、兼容性好、效率高，不仅 Android 系统浏览器基于 WebKit 引擎，苹果的 iPhone 浏览器也基于此引擎。

libc 是通用的 C 语言库，供 Android 系统库调用。

Android 运行时（Runtime）包括核心库（Core Libraries）和 Dalvik 虚拟机（Virtual Machine）。核心库集成了绝大多数 Java 语言核心库的功能，供 Java 语言程序运行时调用。Dalvik 虚拟机解释并运行格式为 dex 的 Java 程序，dex 术语是 Dalvik eXecutable 的缩写，常规的 Java 语言程序（class 字节文件）通过 Android 系统内置的 dx 工具转化为 dex 格式，这种格式被优化为代码内存占用最小，因此，Android 可执行文件扩展名为 .dex。每个 Android 应用程序启动后都对应着一个进程，该进程属于它自己的 Dalvik 虚拟机实例，Dalvik 虚拟机可以同时高效地运行多个虚拟机实例，从而实现多任务处理。

第Ⅲ层为 Android 应用程序框架，进行 Android 应用程序开发必须熟练掌握其 4 种基本组件，即活动（Activity）、服务（Service）、广播接收器、内容提供者（Content Provider）的使用技术。应用程序框架层为开发 Android 应用程序提供了各种 API 函数，这些函数属于不同的类。图 1-3 中列出 10 类组件，即活动管理器（Activity Manager）、窗口管理器（Window Manager）、内容提供者（Content Providers）、视图系统（View System）、通知管理器（Notification Manager）、包管理器（Package Manager）、电话管理器（Telephony Manager）、资源管理器（Resource Manager）、XMPP 服务（XMPP Service）。其中，XMPP 是 Google Talk 的通信协议，简称 GTalk，是 Google 的即时通信方式，就是通常所说的文字或语音聊天，此外，GTalk 还支持 E-mail 功能。第Ⅲ层是进行应用程序开发的基础，也是本书涉及的主要内容，这一层的组件将在后续的章节中详细介绍。

第Ⅳ层为用户应用程序层，这一层的软件包括欢迎界面（Home）、联系人（Contacts）、电话（Phone）、浏览器（Browser）等用户直接使用的程序，当然，也包括用户自己开发的应用程序。本书将详细介绍该层应用程序的设计方法。

1.3 Java 开发环境

Android 程序的开发语言为 Java。Android 应用程序选择 Java 作为其开发语言，而不选用 C 或 C++，其原因在于通过 Java 虚拟机可使 Android 应用程序运行于任意平台上，即 Java 程序具有比 C 和 C++ 语言程序更强的可移植性。

微软公司的 Microsoft Visual Studio（MVS）和 Borland 公司的 Borland Developer Studio（BDS）均集成了优秀的 Java 开发环境。除了 MVS 和 BDS 之外，开源的 Eclipse 是倍受欢迎的集成开发环境，其中集成了 Java 开发环境和 C/C++ 开发环境等。本书将使用 Eclipse 作为 Java 开发环境。

首先从 www.eclipse.org 上下载最新的 Java 开发环境，截至 2016 年 7 月，最新的 Java 开发环境代号为 Neon，建议读者使用最新版本的 Eclipse 软件。如果使用了 64 位的 Windows 视窗系统，则下载其相应的 Windows 64 位版本，压缩包为 eclipse-java-neon-win32-x86_64.zip（该压缩包文件名对应 Neon 版本号，不同的版本号将对应不同的文件名)。现将 eclipse-java-neon-win32-x86_64.zip 解压缩到 D:\Program Files 目录下,如图 1-4 所示。

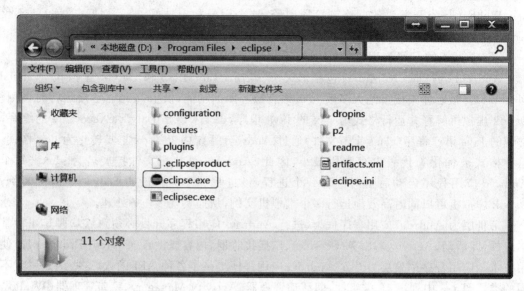

图 1-4　Eclipse Neon 文件目录结构

将图 1-4 中的 eclipse.exe 发送到桌面快捷方式,修改快捷方式的名称为 Eclipse Neon,如图 1-5 所示。

在图 1-5 中双击 Eclipse Neon 图标可进入 Java 开发环境。但在运行 Eclipse 软件之前,需要先安装 Java 运行支持环境,即通常所说的 Java 虚拟机,Java 虚拟机是解释并运行 Java 程序的软件包,通过 Java 虚拟机可使 Java 程序运行在 Windows 系统上,如图 1-6 所示。

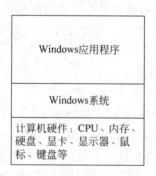

图 1-5　Eclipse Neon 桌面快捷方式　　　　图 1-6　Java 虚拟机的桥梁作用

在 www.oracle.com 上下载 Java 虚拟机,Java 虚拟机更新非常频繁,这里使用了截至 2016 年 7 月的最新版本 Java SE Development Kit 8u102,其 Windows 64 位版本的下载文件为 jdk-8u102-windows-x64.exe,文件名中的 8u102 为版本号。

双击 jdk-8u102-windows-x64.exe 文件(对于 Windows 7 以上系统,尽量使用"以管理员身份运行"),进入图 1-7 所示的安装界面。

在图 1-7 中,单击"下一步"按钮,进入图 1-8 所示的界面。

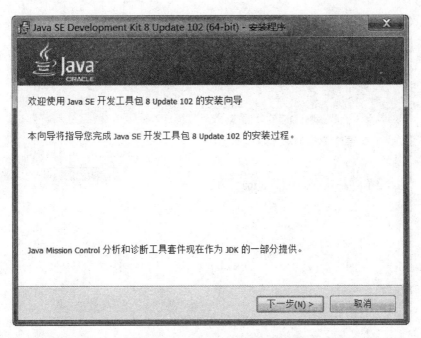

图 1-7 Java 虚拟机安装向导

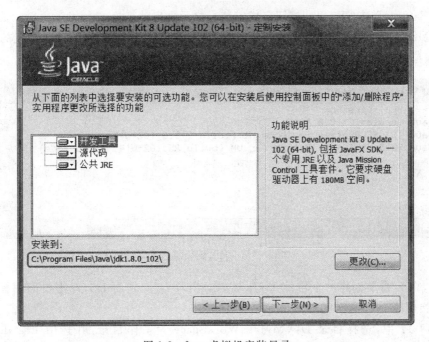

图 1-8 Java 虚拟机安装目录

在图 1-8 中，设定好安装目录，单击"下一步"按钮，然后按照后续的提示，将 Java 虚拟机安装到计算机上。

安装过程中，会提示安装 Java JRE(Java 运行时，即 Java 运行支持库，是严格意义上的 Java 虚拟机)，如图 1-9 所示。

图 1-9　Java JRE 安装提示界面

在图 1-9 中，单击"下一步"按钮完成安装。之后重启计算机。

在"命令提示符"下，输入"java -version"，如图 1-10 所示，将显示 Java 虚拟机的版本号，则 Java 虚拟机安装完成。

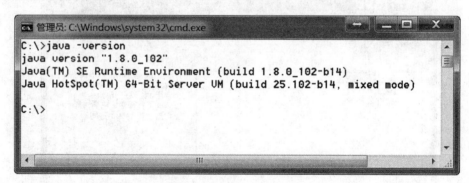

图 1-10　Java 版本信息

现在，在 D 盘上创建一个目录 MyJavaWorkSpace。回到图 1-5，双击桌面图标 Eclipse Neon，将弹出图 1-11 所示的界面。

在图 1-11 中，选择目录 D：\MyJavaWorkSpace 为工作目录，并选中 Use this as the default and do not ask again 复选框，则以后默认使用该工作目录。单击 OK 按钮进入图 1-12 所示的集成开发环境。

图 1-12 为首次运行 Eclipse Neon 的欢迎界面，其中，Create a Hello World application 向导向初学者介绍创建 HelloWorld 工程的详细方法。关闭 Welcome 页面，进入图 1-13 所示的窗口。

在图 1-13 中，选择菜单 File|New|Java Project，进入图 1-14 所示的窗口。

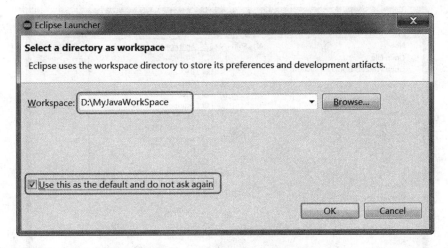

图 1-11　配置 Eclipse 工作目录

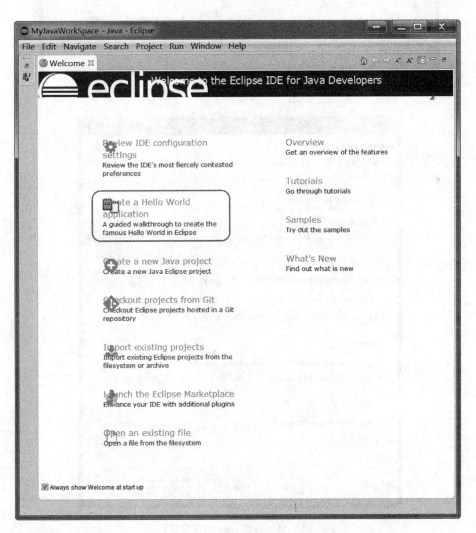

图 1-12　Eclipse Neon 集成开发环境

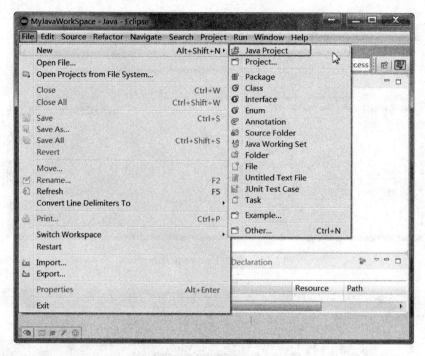

图 1-13　Eclipse Neon 工作窗口

图 1-14　新建 Java 工程窗口

在图 1-14 中的 Project name(即工程名)中输入"MyHelloWorld",单击 Finish 按钮进入图 1-15(注意:在图 1-14 中,单击 Next 按钮可查看工程配置选项,例如,可执行文件保存在目录 D:\MyJavaWorkSpace\MyHelloWorld\bin 中)。

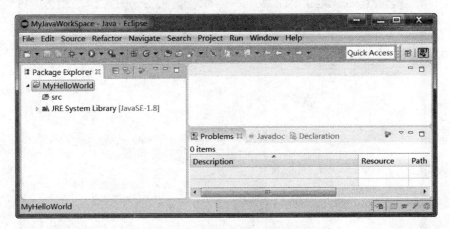

图 1-15　MyHelloWorld 工程窗口

在图 1-15 中,单击菜单 File|New|Class(参考图 1-13),进入图 1-16 所示的界面。

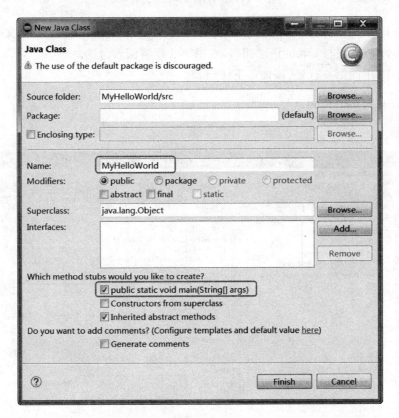

图 1-16　新建类窗口

在图 1-16 中,输入类名为"MyHelloWorld",并选中复选框 public static void main (String[]args),然后,单击 Finish 按钮进入图 1-17。

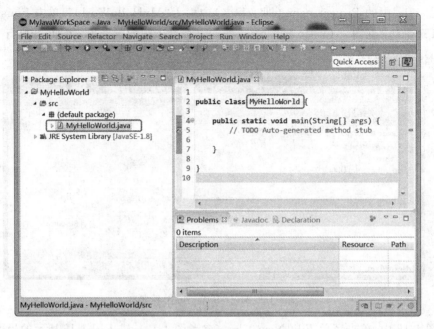

图 1-17　MyHelloWorld 工作区

在图 1-17 中,创建的类名与文件名相同,均为 MyHelloWorld,Java 源文件的扩展名为.java。Java 的语法与 C#的语法相似,本书第 2 章将深入讨论。

在图 1-17 中的第 6 行处,添加如下代码,即

```
System.out.println("Hello World!");
```

如图 1-18 所示。

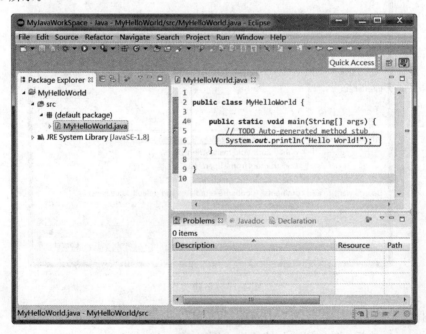

图 1-18　完整的 MyHelloWorld 工程

在图 1-18 中，菜单 Project 包含的子菜单如图 1-19 所示。菜单 Run 包含的子菜单如图 1-20 所示。

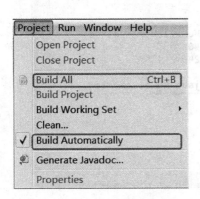

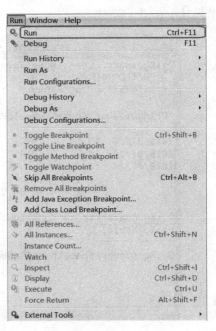

图 1-19　Project 菜单　　　　　　　　　　图 1-20　Run 菜单

在图 1-19 中，如果选中 Build Automatically，表示自动编译，则 Build All（全部编译）和 Build Project（编译工程）子菜单不起作用。

在图 1-20 中，单击菜单 Run|Run，运行结果如图 1-21 所示。

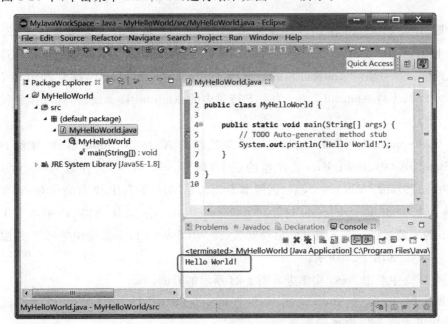

图 1-21　MyHelloWorld 工程运行结果

还可以在"命令提示符"窗口直接运行 Java 类,如图 1-22 所示。编译后的 Java 类文件为 MyHelloWorld.class,扩展名为.class。然后,输入"java MyHelloWorld",按 Enter 键,运行结果显示在图 1-22 中。

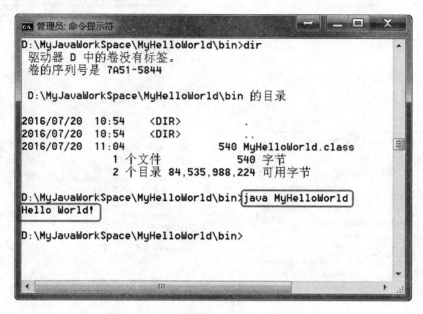

图 1-22 "命令提示符"窗口直接运行 Java 类

至此,第一个 HelloWorld 工程完成了。

1.4 Android 开发环境

早期,开发 Android 移动程序主要借助于 Eclipse 集成开发环境。自 2013 年 5 月 Google 推出 Android Studio(AS)后,使用 AS 集成开发环境开发 Android 移动程序逐步占居主流地位。本书使用 Android Studio 进行 Android 移动程序设计,截至 2016 年 7 月的最新版本为 2.2,支持 Android 7.0。为了支持更多的 Android 移动设备,建议读者使用最新版本的 AS。

从 www.android-studio.org 网站上下载最新的 Android Studio 软件,建议下载包含 SDK(Standard Developer Kit)软件包的版本,即 Bundle 版本,文件名为 android-studio-bundle-143.2915827-windows.exe(该版本为 2.1.1,2016 年 7 月 22 日新发布 AS 版本 2.2 Preview 4,即 android-studio-ide-145.3001415-windows.zip)。下面以 AS 版本 2.1.1 为例介绍 AS 的安装过程。双击该文件启动安装(在 Windows 7 以上系统中尽可能使用"以管理员身份运行"),如图 1-23 所示。

在图 1-23 中单击 Next 按钮进入图 1-24 所示的界面。

在图 1-24 中,勾选 Android SDK 和 Android Virtual Device(Android 虚拟机),单击 Next 按钮进入图 1-25 所示的界面。

图 1-23 AS 安装向导

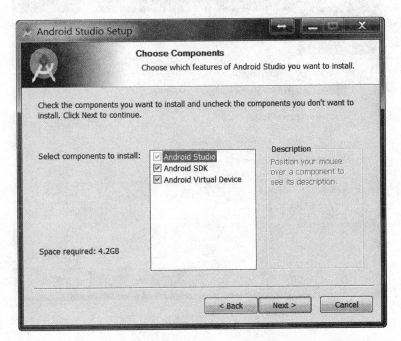

图 1-24 AS 选择组件安装向导

在图 1-25 中,单击 I Agree 按钮同意 Android SDK 使用许可,进入图 1-26 所示的界面。

在图 1-26 中,选择"D:\Android\Android Studio"为 AS 安装目录,选择"D:\Android\sdk"为 Android SDK 安装目录,单击 Next 按钮进入图 1-27 所示的界面。

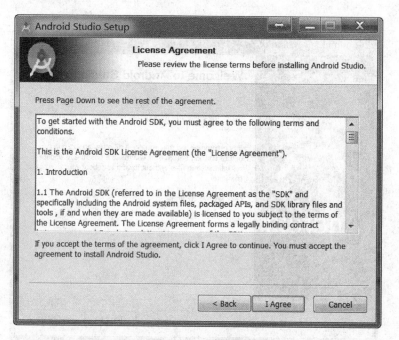

图 1-25　AS 安装许可协议

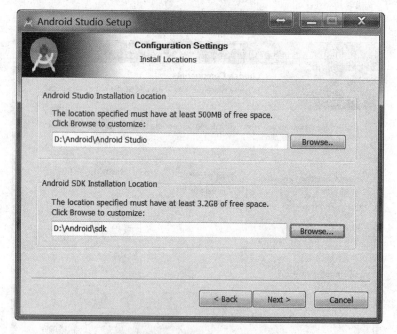

图 1-26　AS 安装目录设定对话框

在图 1-27 中设定 AS 启动菜单名称，单击 Install 按钮启动 AS 安装过程，安装过程需要 5～30 分钟，视所使用的计算机配置而不同。安装完成后，显示图 1-28 所示的界面。

在图 1-28 中，单击 Finish 按钮将启动 AS。此外，在 Windows 启动菜单中，可以找到菜单"所有程序"｜Android Studio｜Android Studio，单击可启动 AS。第一次启动 AS 时，显示

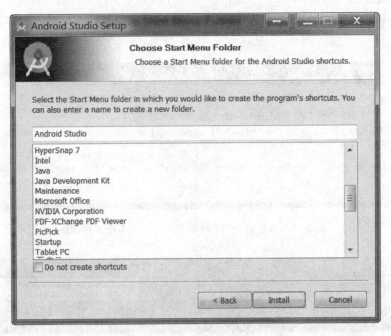

图 1-27　设定 AS 启动菜单名称

图 1-28　AS 安装完成界面

如图 1-29 所示的界面。

　　如果曾使用过旧版本的 AS，可以在图 1-29 中选择 I want to import my settings from a custom location，并指定旧版 AS 的配置文件路径，可导入旧版本的配置信息。如果没有使用过旧版本的 AS,可以在图 1-29 中选择"I do not have a previous version of Studio or I do not want to import my settings.",然后单击 OK 按钮,这里选择了后者,接着进入图 1-30 所示的界面。

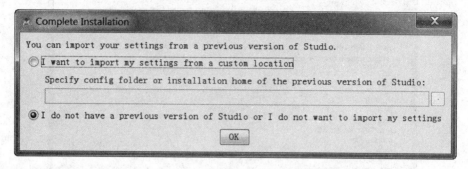

图 1-29　首次启动 AS 时的导入历史版本配置对话框

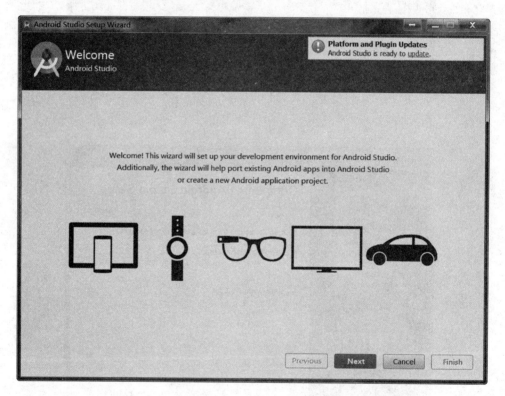

图 1-30　AS 开发环境建设向导

在图 1-30 中,单击 Next 按钮进入图 1-31 所示的界面。

在图 1-31 中,选择 Custom,单击 Next 按钮进入图 1-32 所示的界面。

在图 1-32 中,选择 IntelliJ 界面,单击 Next 按钮进入图 1-33 所示的界面。这里选择 IntelliJ 的目的在于为本书配图更醒目,显然,Darcula 界面风格更有利于保护程序员的眼睛,建议读者选用 Darcula 风格界面。

在图 1-33 中设定安装 Android SDK 的目录为"D:\Android\sdk",单击 Next 按钮进入图 1-34 所示的界面。

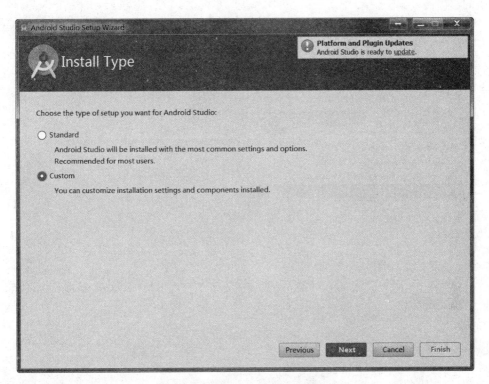

图 1-31　选择 AS 环境风格向导

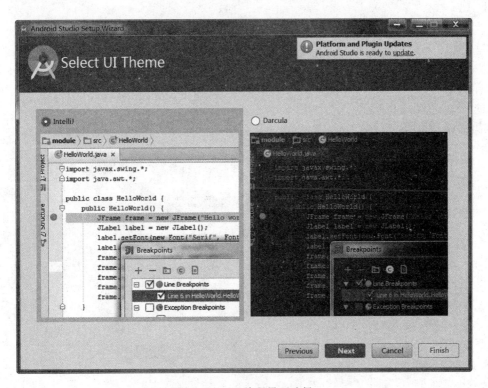

图 1-32　AS 编程界面选择

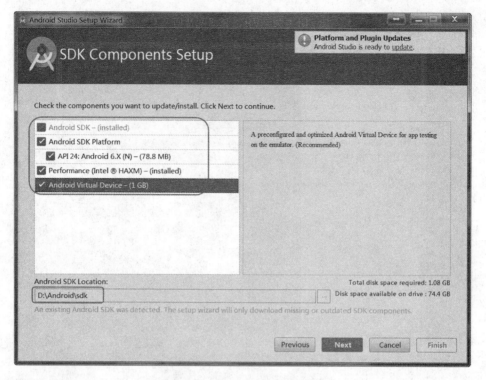

图 1-33　安装 Android SDK 向导

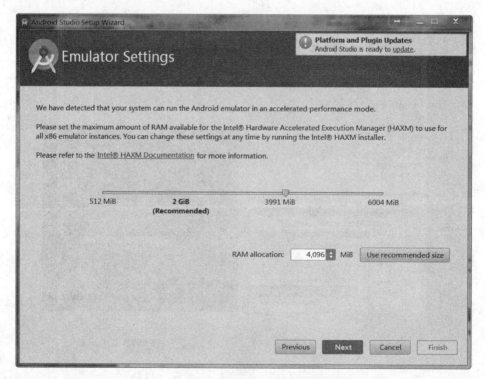

图 1-34　配置 Android 模拟器运行内存为 4GB

在图 1-34 中，单击 Next 按钮进入图 1-35 所示的界面。

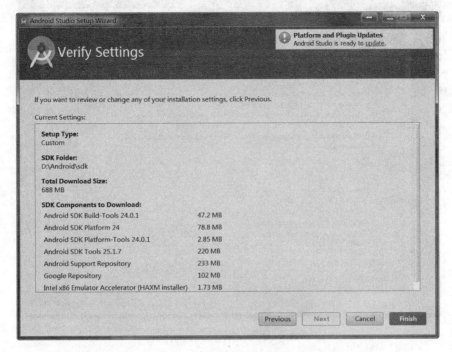

图 1-35　AS 环境配置信息汇总

在图 1-35 中，单击 Finish 按钮启动下载 Android SDK 界面，如图 1-36 所示，下载完成后，界面如图 1-37 所示。

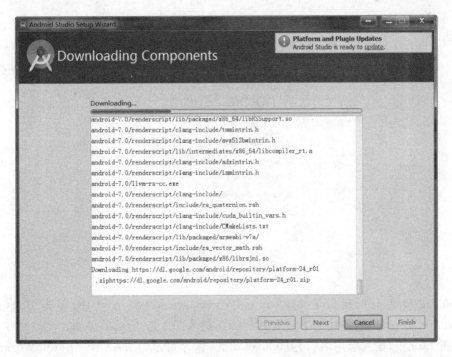

图 1-36　下载 Android SDK 组件

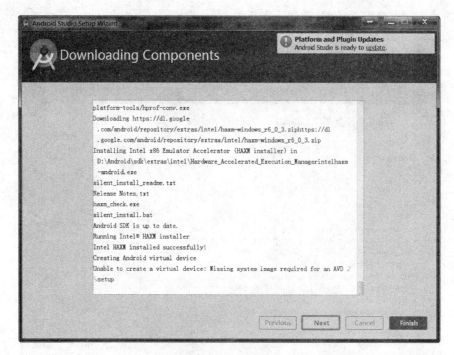

图1-37 下载完 SDK 组件

在图 1-37 中，单击右上角的 Platform and Plugin Updates 将 AS 更新为最新的版本 2.2，此时可能需要网络代理才能完成 AS 更新。

更新完成后的 AS 启动界面如图 1-38 所示。

图1-38 AS 启动界面

现在，在 D 盘上创建目录 MyAndroidWorkSpace。在图 1-38 中，单击 Start a new Android Studio project，创建一个新的 AS 工程，进入图 1-39 所示的界面。

在图 1-39 中，输入应用名"MyHelloWorld"，保存目录为"D:\MyAndroidWorkSpace\MyHelloWorld"，然后单击 Next 按钮进入图 1-40 所示的界面。

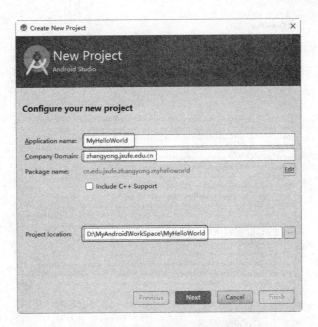

图 1-39 创新 AS 新工程对话框

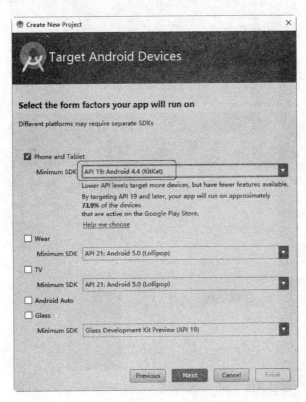

图 1-40 选择 Android 系统 SDK

在图 1-40 中,选择广泛应用的 Android 系统版本 4.4,然后单击 Next 按钮进入图 1-41 所示的界面。

在图 1-41 中,单击 Next 按钮进入图 1-42 所示的界面。

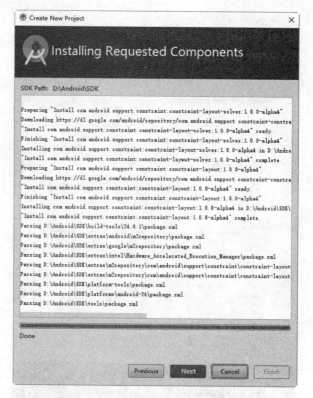

图 1-41　安装需要的 Android SDK 组件

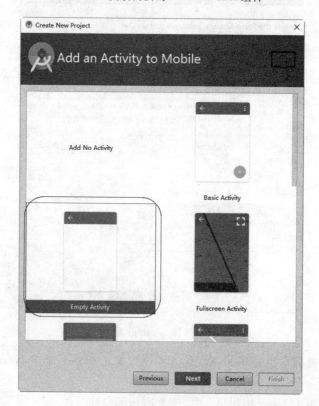

图 1-42　添加 Activity

在图 1-42 中,选择 Empty Activity(空的 Activity),单击 Next 按钮进入图 1-43 所示的界面。

图 1-43　设置 Activity 名称

在图 1-43 中,设置 Activity 名字为 MainActivity,然后单击 Finish 按钮进入图 1-44 所示的界面。

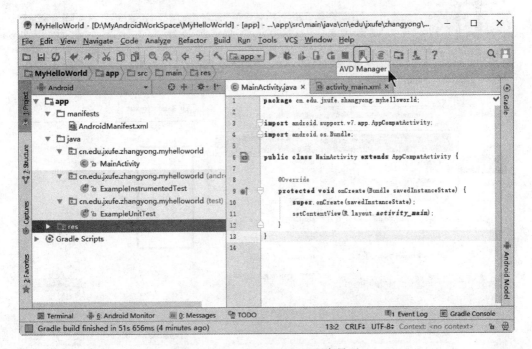

图 1-44　MyHelloWorld 工程主界面

在图 1-44 中，单击快捷按钮 AVD Manager，进入图 1-45 所示的界面，用于创建 Android 模拟器，AS 工程将在 Android 模拟器中运行。

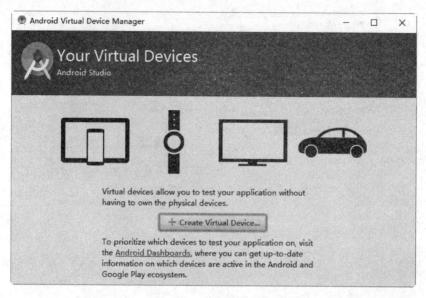

图 1-45　Android 模拟器向导

Android 模拟器是智能设备的软件仿真器，在程序编写和调试阶段，可以借助于 Android 模拟器简化开发平台，当发布工程时，再借助于真实的智能手机设备进行测试，可有效地节省开发成本。在图 1-45 中，单击 Create Virtual Device，进入图 1-46 所示的界面。

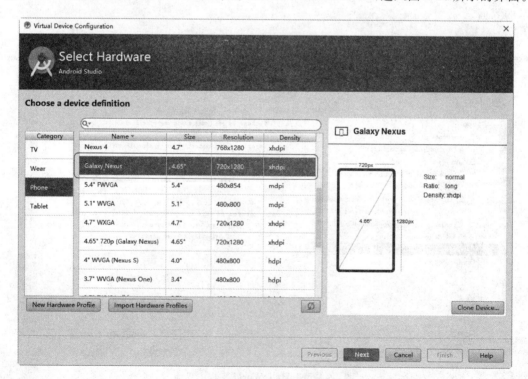

图 1-46　选择 Android 模拟器硬件

在图1-46中,选择Galaxy Nexus作为Android模拟器,屏幕分辩率为720×1280。然后单击Next按钮进入图1-47所示的界面。

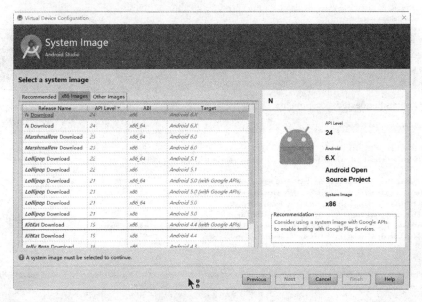

图1-47　选择模拟器上的Android系统

图1-40中设定编写工作在Android系统版本4.4上的应用程序,由于Android系统具有向下兼容性,所以,在图1-47中,必须为模拟器选择4.4以上的Android版本,这里选择"Android 4.4（with Google APIs）",然后单击Download按钮进入图1-48所示的界面。

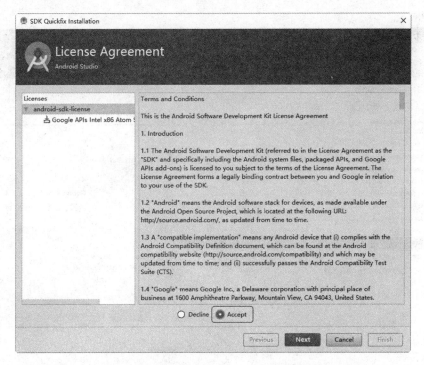

图1-48　模拟器安装Android 4.4许可协议

在图 1-48 中,选择 Accept 接受 SDK 安装许可协议,然后单击 Next 按钮进入 SDK 安装界面,安装完成后进入图 1-49 所示的界面。

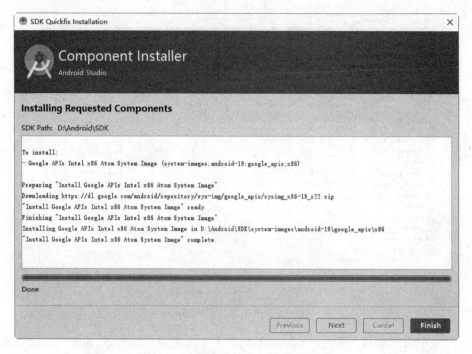

图 1-49　在模拟器安装好 Android 4.4

在图 1-49 中,单击 Finish 按钮进入图 1-50 所示的界面。

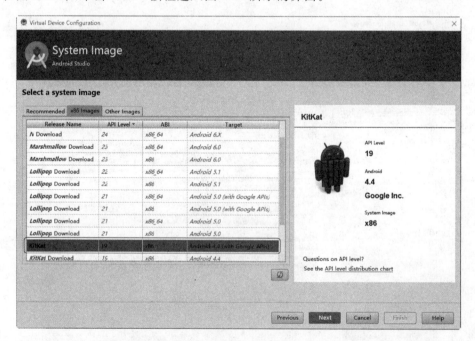

图 1-50　KitKat 下载完成

在图 1-50 中,单击 Next 按钮进入图 1-51 所示的界面。

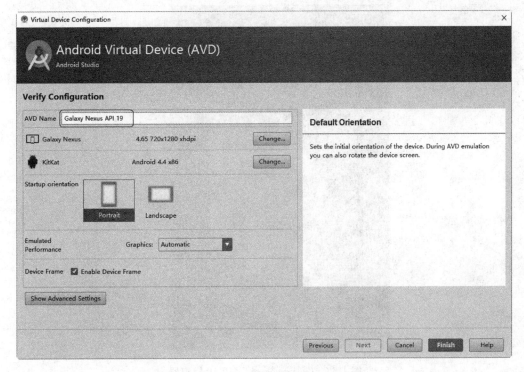

图 1-51　Android 模拟器配置

在图 1-51 中,指定 Android 模拟器的名称为 Galaxy Nexus API 19,可根据自己的需要为 Android 模拟器设定名称。图 1-51 中,Portrait 表示竖屏,Landscape 表示横屏,然后单击 Next 按钮进入图 1-52 所示的界面。

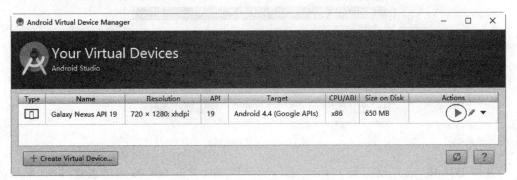

图 1-52　Android 模拟器管理器

在图 1-52 中,单击 Actions 中的启动按钮 ▶ 启动模拟器 Galaxy Nexus API 19,如图 1-53 所示。启动该模拟器后,可关闭图 1-52 所示的窗口。

在图 1-53 中标题栏显示"5554:Galaxy_Nexus_API_19",其中,Galaxy_Nexus_API_19 为模拟器名称,"5554"为模拟器拨号时的电话号码,用于与其他模拟器进行电话和短信通信。图 1-53 左边为 Android 欢迎界面,右边为控制按键区;图 1-53 中鼠标的单击模拟实际智能设备的触屏点击或按键操作。

图 1-53　Galaxy Nexus API 19 模拟器

在图 1-53 中,单击菜单 Run|Run 'app'将弹出图 1-54 所示的界面。

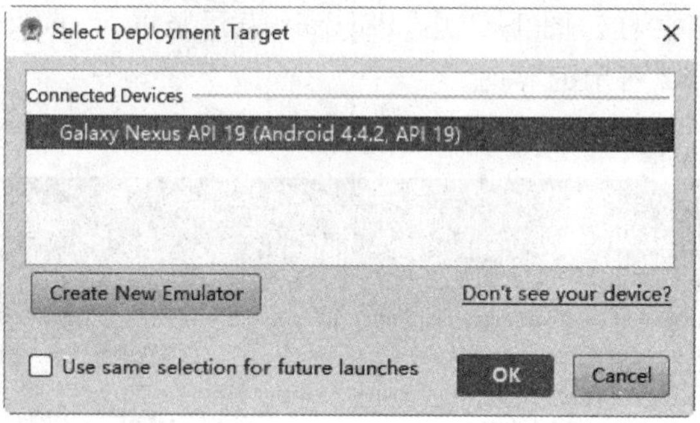

图 1-54　选择模拟器

在图 1-54 中,选择模拟器 Galaxy Nexus API 19,单击 OK 按钮进入图 1-55 所示的界面。

图 1-55 为 MyHelloWorld 工程的执行结果。第 3 章中将详细剖析 MyHelloWorld 工程的组成,这里重点介绍 Android 工程创建、编译和运行的全部开发环境。在图 1-55 中,单击快捷按钮 ⌂ 将进入图 1-56 所示的界面。

在图 1-56 中,MyHelloWorld 图标是 MyHelloWorld 工程对应的 Android 应用程序。可以单击该图标启动 MyHelloWorld 应用程序。Android 系统通常采用黑色背景,这样可以有效地节省移动设备的能量,因为黑色区域的背光灯处于关闭状态。

图 1-55　MyHelloWorld 工程执行结果

图 1-56　Android 模拟器应用程序视图

1.5 本章小结

Android 系统是基于 Linux 内核的嵌入式操作系统,与 Windows CE 等嵌入式操作系统一样,主要针对智能手机和移动设备,具有体积小、实时性强、功耗低和界面人性化等优点。Android 系统采用分层结构,与硬件直接接触的底层为 Linux 内核(第一层),其上为 Android 系统库和 Android 应用程序运行环境(第二层),第三层为 Android 应用程序框架,直接与用户交互的顶层为应用程序层(第四层)。与其他嵌入式操作系统相比,Android 系统的最大优势在于两方面,即公开源代码和免费使用,此外,Android 系统的安全性和网络功能非常强大,因此,Android 系统很适合用于教学和科研。可以借助 Android Studio 集成环境开发 Android 应用程序。Android 系统模拟器功能十分强大,基于模拟器运行良好的应用程序均可以在真实的移动设备上良好地运行。

第 2 章

Java 语 言

Java 语言是具有面向对象特性的高级程序设计语言,其语法与 C♯语言非常相似,Java 语言程序由 Java 虚拟机解释执行,不同操作系统或硬件系统的计算机,只要安装了 Java 虚拟机,则 Java 程序均可运行。因此,Java 语言程序可移植性强,不但在桌面计算机和服务器上流行,而且也广泛应用于移动设备和智能手机上。Java 语言功能强大,本章精选了 Java 语言的基础知识,方便那些没有 Java 语言基础的读者顺利阅读本书,也为了使本书自成体系。想了解 Java 语言的由来和成就,读者可以访问网址 http://www.oracle.com/。

2.1 Java 程序语法与控制

Java 程序文件的扩展名为.java,每个程序文件中只能包含一个 public 类,即公有类,public 是可见性修饰符,用 public 定义的类、方法(或称函数)或数据域(或称变量)可以被任何类访问。Java 程序的入口点是 main 方法,定义在公有类中,main 方法的原型为

public static void main(String[ ]args){ }

与其他高级语言程序设计相同,Java 语言程序具有三种基本的程序控制方式,即顺序、分支(或选择)和循环执行方式,在 main 方法中使用三种程序控制方式进行程序设计。

2.1.1 顺序方式

顺序方式是指 Java 语句按照先后顺序依次执行并得到计算结果的程序执行方式,这是 Java 语言的总体执行方式(由于类只是数据结构,因此类中定义成员数据和方法时不分先后,成员方法中变量必须先定义后使用)。下面例 2-1 输入一个摄氏温度值,将其转化为华氏温度值。例 2-1 详细介绍了借助 Eclipse 编写 Java 程序的步骤,本章中所有实例的创建步骤也都与此类似,因此,后面实例的创建步骤就省略了。

例 2-1 摄氏温度值转化为华氏温度值。

摄氏温度值转化为华氏温度值的关系式为

$$Fah = 1.8 \times Cel + 32$$

式中,Fah 表示华氏温度,Cel 表示摄氏温度。下面分步骤介绍该实例的实现方法。

S1. 创建工程 ex02_01

在 Eclipse 软件主界面(图 1-13)中,选择菜单项 File|New|Java Project,弹出图 2-1 所示的界面。在图 2-1 中输入工程名为"ex02_01",保存在目录"D:\MyJavaWorkSpace\ex02_01"中,然后,直接单击 Finish 按钮完成创建工程向导,如图 2-2 所示。在图 2-2 中,显示了空的工程 ex02_01,其中包含一个 src 标签,该标签下存放 Java 源程序文件,此外,还包含了 JRE System Library,即 Java 程序运行环境库。在图 2-2 界面中,单击菜单项 File|New|Class 即添加新类,进入图 2-3 所示的界面。

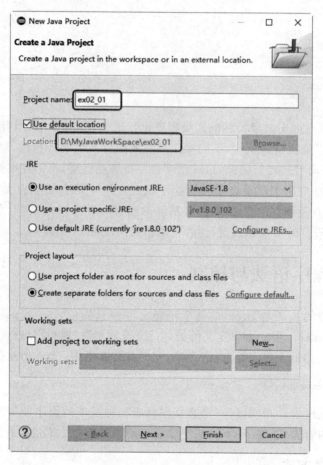

图 2-1 新建 Java 工程

在图 2-3 中输入包名为"cn.edu.jxufe.zhangyong",包(Package)的名称要求具有全球唯一性,包是类的容器,允许位于不同包中的相同名称的类存在。包类似于 C#语言中的命名空间,引入了包的概念后,不同的程序员在命名类名时,更加随意,也不会导致类名冲突。这里的包名"cn.edu.jxufe.zhangyong",表示"中国.教育.江西财经大学.张勇",包名一般

图 2-2 空的工程 ex02_01

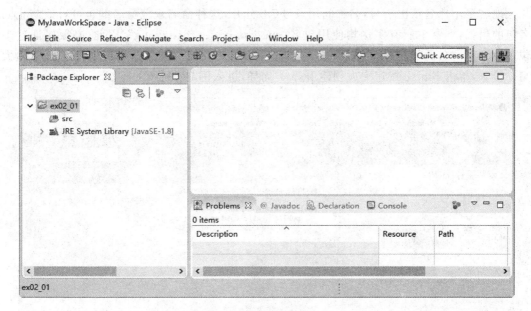

图 2-3 创建新 Java 类

从大地名至小地名依次书写,名称间用点号(.)分开,这种命名规范命名的包可有效防止同名包的出现,本章中所有实例均使用包名"cn. edu. jxufe. zhangyong",因此,必须保证这个包里面没有同名的类出现。然后,输入类名为"MyEx0201",类名命名习惯要求首字母大写。接着,单击 Finish 按钮完成创建 Java 类向导,进入图 2-4 所示的界面。

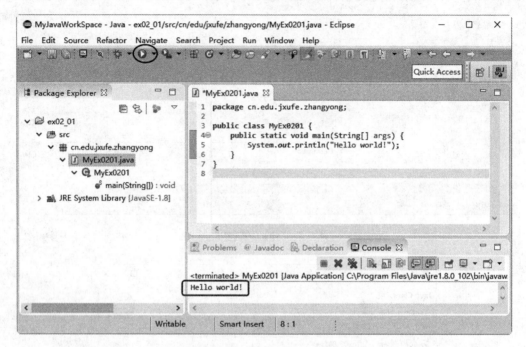

图 2-4　显示"Hello world!"程序代码

S2. 控制台显示 Hello world!

在图 2-4 中输入 Java 程序代码如下:

```
1    package cn. edu. jxufe. zhangyong;
2
3    public class MyEx0201 {
4        public static void main(String[] args){
5            System.out.println("Hello world!");
6        }
7    }
```

上述代码的行号是为了介绍程序方便而添加的。其中,第 1、3、4、6、7 行是创建类向导自动添加的,只有第 4 行和第 6 行代码是因为图 2-3 中选择了 public static void main (String[]args)复选框而自动添加的,第 5 行是程序员输入的代码。第 1 行代码用关键字 package 说明第 3 行的类 MyEx0201 位于包 cn. edu. jxufe. zhangyong 中;第 3 行使用关键字 class 定义了类 MyEx0201,该类为公有类(public),该类中有一个公有、静态且无返回值的方法 main(第 4 行),其唯一的参数 args 为字符串数组类型(String[ ]),该方法中只有一条语名(第 5 行),调用 System. out. println 方法在控制台输出"Hello world!"(System 是一个类,out 为这个类的静态 PrintStream 对象成员变量,println 为 PrintStream 类的方法)。类中的静态方法属于类,因此,使用类名来调用其静态方法;类中的非静态公有方法是类的

对象创建后才有效的（这里的"有效"指为非静态公有方法分配内存并添加访问控制，即通常所说的"注册"），因此，只能使用类的对象调用非静态公有方法。这里的输出终端为控制台，如图 2-4 所示的 Console，其实 Java 支持图形界面功能，然而，输出到控制台只需要一条语句，比较简单。在图 2-4 中单击菜单 Run|Run 或运行快捷按钮（图 2-4 上部圈住的快捷钮）或按 Ctrl+F11 组合键，均可执行工程 ex02_01，运行结果显示在图 2-4 下部。

S3. 摄氏温度转换为华氏温度程序

图 2-4 是一个完整的程序，输出"Hello world!"信息，为了实现摄氏温度转换为华氏温度的功能，改写程序文件 MyEx0201.java 的代码如下：

```
1   package cn.edu.jxufe.zhangyong;
2
3   public class MyEx0201 {
4       public static void main(String[] args){
5           double Cel = 36.5;
6           double Fah;
7           Fah = 1.8 * Cel + 32.0;
8           System.out.println(Cel + " deg. C = "
9                   + Fah + " deg. F.");
10      }
11  }
```

上述代码中，第 5 行定义双精度浮点数 Cel 变量并初始化为 36.5，第 6 行定义双精度浮点数变量 Fah，第 7 行根据公式计算新的 Fah 的值，第 8、9 行调用方法 System.out.println 在控制台输出信息"36.5 deg. C = 97.7 deg. F."，如图 2-5 所示。

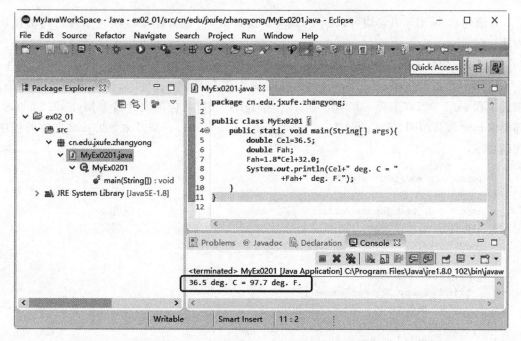

图 2-5　温度转换程序及输出结果

图 2-5 所示工程是一个简单的 Java 顺序执行程序示例,第 5~9 行的语句按先后顺序依次执行,程序的顺序执行方式是最基本的程序控制方式。

2.1.2 分支方式

分支方式是根据条件关系式的逻辑值进行判断,有条件地执行某些语句组。有两种条件语句,即 if-else 语句和 switch 语句,其基本语法分别为:

```
if(布尔表达式){
    布尔表达式为真时执行的语句组;
}
else{
    布尔表达式为假时执行的语句组;
}
```

和

```
switch(表达式){
    case 值 1: 当表达式的值为"值 1"时执行的语句组 1;
        break;
    case 值 2: 当表达式的值为"值 2"时执行的语句组 2;
        break;
    ...
    case 值 N: 当表达式的值为"值 N"时执行的语句组 N;
        break;
    default:
        当表达式的值不等于上述的"值 1"、"值 2"、…、"值 N"时执行的语句组;
}
```

if 语句和 switch 语句均可以嵌套,例 2-2 和例 2-3 分别介绍了这两种条件语句的使用方法。

例 2-2 求解一元二次方程的根。

已知一元二次方程 $ax^2+bx+c=0$,任意输入三个系数 a、b 和 c,计算未知数 x 的值。新建工程 ex02_02,在工程 ex02_02 中新建类 MyEx0202(对应的文件名为 MyEx0202.java,即类名与文件名相同,所在的包为 cn.edu.jxufe.zhangyong)。MyEx0202.java 文件的程序代码如下:

```
1   package cn.edu.jxufe.zhangyong;
2
3   import java.text.DecimalFormat;
4   import java.text.NumberFormat;
5   import java.util.Scanner;
6
7   public class MyEx0202 {
8       private static Scanner scanner;
9       public static void main(String[] args){
10          double a,b,c;
11          double[] x = new double[2];
12          System.out.print("Input 3 coefficients:");
```

```java
13          Scanner scanner = new Scanner(System.in);
14          a = scanner.nextDouble();
15          b = scanner.nextDouble();
16          c = scanner.nextDouble();
17          if(Math.abs(a)<1e-8){
18              System.out.println("Not a quadratic equation!");
19          }
20          else{
21              DecimalFormat df = (DecimalFormat)NumberFormat.getInstance();
22              df.setMaximumFractionDigits(2);
23              if(b*b>=4*a*c){
24                  x[0] = (-b+Math.sqrt(b*b-4*a*c))/(2*a);
25                  x[1] = (-b-Math.sqrt(b*b-4*a*c))/(2*a);
26                  System.out.println("x1 = " + df.format(x[0]) +
27                      "; x2 = " + df.format(x[1]) + ".");
28              }
29              else{
30                  x[0] = -b/(2*a);
31                  x[1] = Math.sqrt(4*a*c-b*b)/(2*a);
32                  System.out.println("x1 = " + df.format(x[0]) + " + " + df.format(x[1])
33                      + "i; x2 = " + df.format(x[0]) + " - " + df.format(x[1]) + "i.");
34              }
35          }
36      }
37 }
```

在 C 语言中使用库函数时需要借助于 include 关键字,在 C♯ 中使用 use 关键字;在 Java 程序中,如果需要调用其他包中的类中的公有方法和数据时,必须借助于 import 关键字导入这些包和类,如上述程序段中的第 3~5 行。第 3 和第 4 行为第 21 和第 22 行的方法服务,第 5 行为第 13 行的方法服务。导入一个包(和它中的某个类)非常方便,例如,在程序中书写了"scanner=new Scanner(System.in);"语句后,会自动提示程序员导入 java.util.Scanner,只需要在提示的右键弹出菜单中单击一下就会在第 5 行添加"import java.util.Scanner;"语句。因此,所有的 import 语句都是用这种方法创建的,不需要程序员书写,比起 C 语言的"♯include"包括头文件语句更加方便。第 8 行定义一个 Scanner 类的静态私有对象 scanner,用于接收控制台的输入。

上述程序段的 main 方法代码为第 10~35 行。第 10 行定义三个 double 型变量 a、b 和 c;第 11 行定义了一个具有两个元素的 double 型数组 x,必须记住 Java 这种定义数组的方法,使用 new 关键字为数组元素开辟存储空间;第 12 行在控制台输出提示信息"Input 3 coefficients:";第 13 行 scanner 对象是第 8 行中类 Scanner 定义一个私有对象,这里用类 Scanner 的构造方法对它进行初始化(在第 2.3 节中将详细介绍构造方法);第 14~16 行调用 scanner 对象的方法 nextDouble 从控制台读入双精度浮点数,并分别赋给变量 a、b 和 c;第 17 行判断 a 的绝对值是否小于 0.000 000 01,即判断 a 是否为 0,如果表达式 Math.abs(a)< 1e-8 为真,则认为 a=0,于是第 18 行输出"Not a quadratic equation!",说明不是一元二次方程;否则,程序执行第 20~35 行代码。第 21 和第 22 行用类 DecimalFormat 创建一个对象 df,第 22 行调用对象 df 的方法 setMaximumFractionDigits 设置小数显示时保留两位小

数,即当调用对象df的方法format时,将方法format的输入参数设置为只显示2位小数,见第26行。第23行判断b*b>=4*a*c是否为真,如果为真,则方程的两个根x[0]和x[1]均为实数(数组的下标从0开始索引);第24和第25行计算两个根的值;第26和第27行在控制台输出两个根的值,为了输出的根只保留两位小数,使用方法df.format格式化输出。如果b*b>=4*a*c为假,则执行第30～33行代码,此时两个根为复数,用x[0]存放根的实部,x[1]存放根的一个虚部,第32和第33行在控制台输出这两个复根。

图2-6和图2-7给出了例2-2的两次执行结果,说明程序工作正常。例2-2中使用if-else语句的两级嵌套实现了一元二次方程的求根运算。

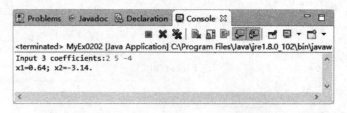

图2-6 例2-2运行结果a=2、b=5和c=-4

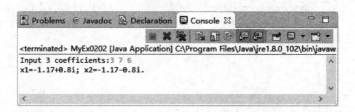

图2-7 例2-2运行结果a=3、b=7和c=6

例2-3 人民币兑换外币计算。

假设已知英镑(pound)、港币(HKD)、美元(dollar)、日元(yen)、欧元(euro)和新加坡元(SGD)对人民币的汇率分别为10.62、0.84、6.53、0.078、9.42和5.22,输入人民币可计算得到其兑换的外币值。新建工程ex02_03,在工程ex02_03中新建类MyEx0203(对应的文件名为MyEx0203.java,所在的包名为cn.edu.jxufe.zhangyong)。MyEx0203.java文件的程序代码如下:

```
1   package cn.edu.jxufe.zhangyong;
2   import java.text.DecimalFormat;
3   import java.text.NumberFormat;
4   import java.util.Scanner;
5
6   public class MyEx0203 {
7       private static Scanner scanner;
8       public static void main(String[] args){
9           final double pound2yuan = 10.62;
10          final double HKD2yuan = 0.84;
11          final double dollar2yuan = 6.53;
12          final double yen2yuan = 0.078;
13          final double euro2yuan = 9.42;
14          final double SGD2yuan = 5.22;
```

```
15          System.out.print("Select(1 - pound,2 - HKD,3 - dollar,4 - yen," +
16                          "5 - euro,6 - SGD):");
17          scanner = new Scanner(System.in);
18          int moneytype = scanner.nextInt();
19          System.out.print("Input the amount of money:");
20          double nativemoney = scanner.nextDouble();
21          double foreignmoney = 0;
22          DecimalFormat df = (DecimalFormat)NumberFormat.getInstance();
23          df.setMaximumFractionDigits(2);
24          switch(moneytype){
25          case 1:
26                  foreignmoney = nativemoney/pound2yuan;
27                  break;
28          case 2:
29                  foreignmoney = nativemoney/HKD2yuan;
30                  break;
31          case 3:
32                  foreignmoney = nativemoney/dollar2yuan;
33                  break;
34          case 4:
35                  foreignmoney = nativemoney/yen2yuan;
36                  break;
37          case 5:
38                  foreignmoney = nativemoney/euro2yuan;
39                  break;
40          case 6:
41                  foreignmoney = nativemoney/SGD2yuan;
42                  break;
43          default:
44                  System.out.println("Input wrong!");
45                  return;
46          }
47          System.out.println("You get:" + df.format(foreignmoney));
48      }
49 }
```

上述代码中,第 1 行指定程序所在的包名为 cn.edu.jxufe.zhangyong,第 6 行定义名为 MyEx0203 的公有类,只要包 cn.edu.jxufe.zhangyong 中没有同名的类,则类名 MyEx0203 就是合法的。第 2~4 行为类 MyEx0203 中的代码将会使用到的包 cn.edu.jxufe.zhangyong 外部的包中的类(同名包中的类可以直接引用)。第 8~48 行为 main 方法,main 方法必须为静态公有方法,且带有一个类型为字符串数组的形式参数 args。第 9~14 行定义了常量 pound2yuan、HDK2yuan、dollar2yuan、yen2yuan、euro2yuan 和 SGD2yuan,Java 中定义常量使用关键字 final(C 语言中使用关键字 const),Java 中的常量与 C 语言中的常量一样,在定义时赋值,且程序执行过程中其值不能被改变,这里将各个外币对人民币的汇率定义为常量(或称常数)。第 15 和第 16 行在控制台输出提示信息"Select(1-pound,2-HKD,3-dollar,4-yen,5-euro,6-SGD):",即输入 1 表示选择人民币兑换为英镑,输入 2 表示选择人民币兑换为港币,等等。第 17 行为类 Scanner 定义的一个控制台输入对象 scanner;第 18 行从控制台读入一个整数,并赋给变量 moneytype。第 19 行在控制台输出提示信息"Input the amount of money:",提示要兑换的人民币数量;第 20 行从控制台读入一个双精度浮点数,赋给变量 nativemoney,即 nativemoney 为要兑换的人民币数量。第 21 行定义 double 型变量 foreignmoney,用来存放兑换的外币数值。第 22 和第 23 行定义了格式化输出对象 df,并设定输出数值保留小数点后两位小数。

第 24~46 行为 switch 语句体,如果 moneytype 的值为 1,执行第 26 和第 27 行;如果 moneytype 的值为 2,执行第 29 和第 30 行;如果为 3,执行第 32 和第 33 行;如果为 4,执行第 35 和第 36 行;如果为 5,执行第 38 和第 39 行;如果为 6,执行第 41 和第 42 行;如果为其他整数值,则执行第 44 和第 45 行。当 moneytype 的值为 1 时,第 26 行计算 foreignmoney 变量的值,第 27 行执行 break 语句跳出 switch 语句体,到第 47 行执行,在控制台输出兑换的外币数量。如果 moneytype 的值不是 1~6 的整数值,则执行第 44 行,在控制台输出"Input wrong!"错误提示,然后,执行第 45 行的 return 语句,结束程序运行,即此时第 47 行得不到执行。

图 2-8 和图 2-9 给出了例 2-3 的两次执行结果,分别是将 1000 元人民币兑换成 153.14 美元和将 2000 元人民币兑换为 383.14 新加坡元,这说明程序运行正常。例 2-3 说明了 switch 语句的用法,从例 2-3 可以体会到,switch 语句实现的程序控制 if-else 语句也能实现。

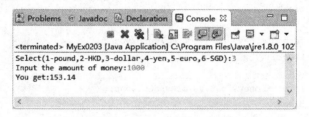

图 2-8 例 2-3 程序运行结果:1000 元兑换美元

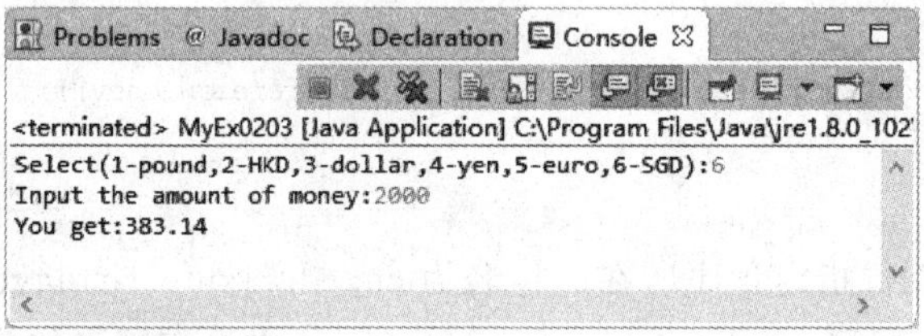

图 2-9 例 2-3 程序运行结果:2000 元兑换新加坡元

2.1.3 循环方式

Java 语言提供了三种循环控制语句,即 while 型、do-while 型和 for 型循环,这三种循环控制语句与 C 语言语法相同,其基本语法如下:

```
while(条件表达式){
    条件表达式为真时执行的语句组(即循环体);
}
Do{
    循环体(先执行一次后判断条件表达式的值,如果为真则再次执行循环体;与 while 型的区别在
```

于当条件表达式为假时,循环体可被执行一次);
}while(条件表达式)
for(初始值；条件表达式；每次循环后执行的语句组){
    循环体
}

以上三种循环控制方式的用法都非常灵活,下面举两个例子说明循环控制的基本用法。

例 2-4 整数的累加运算。

例 2-4 计算小于等于 100 的正整数累加值,实例中用了三种方法计算累加值。新建工程 ex02_04,在工程 ex02_04 中新建类 MyEx0204(对应文件名 MyEx0204.java,所在的包为 cn.edu.jxufe.zhangyong)。MyEx0204.java 文件的程序代码如下:

```java
1   package cn.edu.jxufe.zhangyong;
2
3   public class MyEx0204 {
4       public static void main(String[] args){
5           double sum1 = 0, sum2 = 0, sum3 = 0;
6           int i = 0;
7           while(i <= 100){
8               sum1 = sum1 + i;
9               i++;
10          }
11          i = 0;
12          do{
13              sum2 = sum2 + i;
14              i++;
15          }while(i <= 100);
16          i = -10;
17          while(true){
18              i++;
19              if(i < 0)
20                  continue;
21              if(i > 100)
22                  break;
23              sum3 = sum3 + i;
24          }
25          System.out.println("Sum1 = " + sum1 + "; Sum2 = "
26                  + sum2 + "; Sum3 = " + sum3 + "!");
27      }
28  }
```

上述代码中,第 5 行定义了三个 double 型变量 sum1、sum2 和 sum3,都初始化为 0。第 6 行定义了整型变量 i,初始化为 0；第 7~10 行是一个 while 循环体,当 i 小于等于 100 时,循环执行第 8 和第 9 行,即计算 0+1+2+3+…+100 的值。第 11 行将 i 赋值为 0；第 12~15 行为 do-while 循环体,首先执行第 13 和第 14 行一次,然后判断 i<=100 是否为真,如果为真,则循环执行第 13 和第 14 行,即计算 0+1+2+3+…+100 的值。第 16 行将 i 赋值为-10；第 17~24 行为 while 循环体,此时条件表达式的值为真(true),因此,第 18~

23 行将永远循环执行,但是,第 21 和第 22 行加入了一个 if 语句,当 i 大于 100 时,执行 break 语句,跳出该循环体;第 19 和第 20 行判断 i 的值是否小于 0,如果 i 为负数,则执行第 20 行的 continue 语句,忽略掉第 21~23 行代码,重新回到第 17 行执行循环体。因此,第 17~24 行的循环体仍然是计算 0+1+2+…+100 的值。第 25 和第 26 行向控制台输出运行结果,即信息"Sum1=5050.0;Sum2=5050.0;Sum3=5050.0!"。需要说明的是,break 语句仅跳出包含着它的那一层循环体! continue 语句,用于忽略掉包含它的那一层循环体其后的循环体语句,并从该层循环体开头再次执行循环体! 所以,当循环嵌套时,内层循环体中的 break 和 continue 语句影响不到外层循环体的控制。图 2-10 为例 2-4 的执行结果。

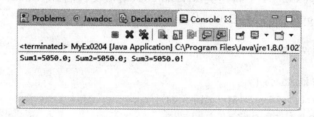

图 2-10　例 2-4 程序运行结果

例 2-5　计算九九乘法表。

例 2-5 使用 for 型循环方式计算九九乘法表。新建工程 ex02_05,在工程 ex02_05 中新建类 MyEx0205(对应文件名为 MyEx0205.java,所在的包为 cn.edu.jxufe.zhangyong)。MyEx0205.java 文件的程序代码如下:

```
1    package cn.edu.jxufe.zhangyong;
2
3    public class MyEx0205 {
4      public static void main(String[] args){
5        for(int i = 1;i<= 9;i++){
6          for(int j = 1;j<= i;j++){
7            System.out.print(String.format("%1d * %1d = %-4d", j,i,i*j));
8          }
9          System.out.println();
10       }
11     }
12   }
```

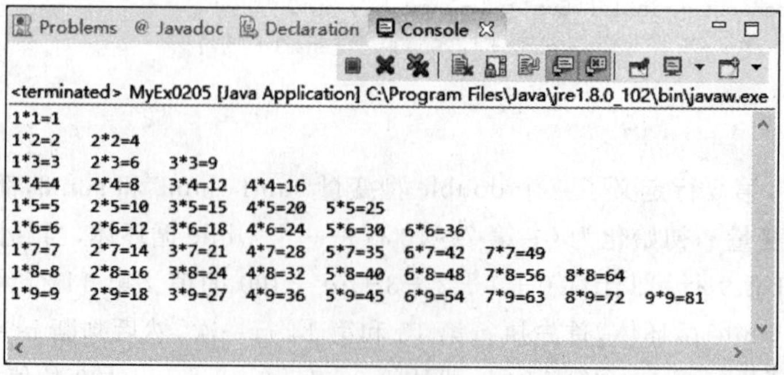

图 2-11　例 2-5 计算的九九乘法表

上述代码的输出结果如图 2-11 所示。第 5～10 行嵌套了两个 for 型循环，外层的 for 型循环中 i 从 1 步进到 9，步长为 1，表示图 2-11 中的 1 至 9 行；第 6～8 行的 for 型循环从 1 步进到 i，步长为 1，表示图 2-11 中的第 1 至 i 列；九九乘法表中第 i 行第 j 列的乘法算式为计算 i 和 j 的乘积，如图 2-11 所示，若 i 等于 6，j 等于 5，即第 6 行第 5 列的式子为"5 * 6 = 30"，于是循环执行第 7 行输出了九九乘法表中的每个乘法式子。第 9 行表示每输出一行后添加一个回车换行。

需要补充的是，第 7 行中使用了语句 String.format("%1d * %1d=%-4d", j,i,i * j)，该语句为字符串格式化输出语句，与 C 语言的 sprintf 语句使用方法相同，%d 将对应的整型变量值转化为字符串，%-4d 表示输出列宽为 4 且左对齐。

2.1.4　异常处理

异常处理也是一种程序控制方式。当程序运行过程中产生了错误（例如数值溢出、输入无效、数组越界等），即所谓的运行错误，往往会导致整个程序运行中突然中断或退出，而程序员无法知道程序退出的原因。异常处理的作用在于当程序运行错误发生时，捕捉运行错误类型（称为异常），并能保证程序仍然正常执行。Java 异常处理功能十分强大，其基本用法为：

```
try{
    被监视的语句组；
}
catch(异常类型 1 异常 1){
    异常 1 发生后的处理语句；
}
catch(异常类型 2 异常 2){
    异常 2 发生后的处理语句；
}
    …
catch(异常类型 N 异常 N){
    异常 N 发生后的处理语句；
}
finally{
    无论是否发生异常都会被执行的语句组；
}
```

当 try"被监视的语句组"中某条语句发生异常时，其后的程序不再执行，而是跳到 catch 语句处，依次判断各个 catch 的异常类型是否与发生的异常相匹配，如果匹配，则将发生的异常（对象）赋给 catch 语句，执行该 catch 块中的语句。无论异常是否发生，都要执行 finally 语句块中的全部语句，这部分语句一般用作内存释放和关闭已打开的文件对象等。

例 2-6 整数除法（除以 0）和输入类型不匹配异常。

从控制台输入两个整数，计算它们的商。新建工程 ex02_06，在工程 ex02_06 中新建类 MyEx0206（对应文件名为 MyEx0206.java，所在的包为 cn.edu.jxufe.zhangyong）。MyEx0206.java 文件的程序代码如下：

```
1   package cn.edu.jxufe.zhangyong;
2   import java.util.Scanner;
3
4   public class MyEx0206 {
5       private static Scanner scanner;
```

```
6   public static void main(String[] args){
7       int nume,deno,quot;
8       scanner = new Scanner(System.in);
9       try{
10          System.out.print("Input 2 integers:");
11          nume = scanner.nextInt();
12          deno = scanner.nextInt();
13          quot = nume/deno;
14          System.out.println("The quotient is:" + quot + ".");
15      }
16      catch(Exception ex){
17          System.out.println(ex.toString());
18      }
19      finally{
20          System.out.println("I always run.");
21      }
22  }
23 }
```

上述代码中，第7行定义了三个整型变量 nume、deno 和 quot，分别用于保存被除数、除数和商；第8行创建了控制台输入对象 scanner。第9～15行为 try 语句块，第10行在控制台输出提示信息"Input 2 integers:"，第11和第12行依次输入两个整数，分别赋给变量 nume 和 deno，此时如果输入的值不是整数，将发生类型不匹配异常；第13行计算 nume 除以 deno 的商值，此时如果输入的除数 deno 值为0，则发生除以0异常；当没有异常发生时，第14行在控制台输出商值。第16～18行为 catch 语句块，这里只有一个 catch 语句，Exception 类是上述两种异常（类）的父类，第17行使用 toString 方法在控制台输出异常信息。第19～21行为 finally 语句块，无论有没有异常发生，第20行总是会执行，在控制台输出"I always run."。

图 2-12～图 2-14 显示了例 2-6 的运行结果，图 2-12 为正常运行，图 2-13 因为输入出现了浮点数 12.2 而发生输入类型不匹配异常，图 2-14 因输入除数为 0 而发生除以 0 异常。

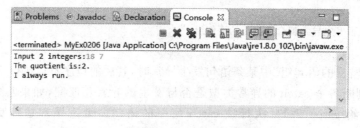

图 2-12　例 2-6 运行结果（无异常）

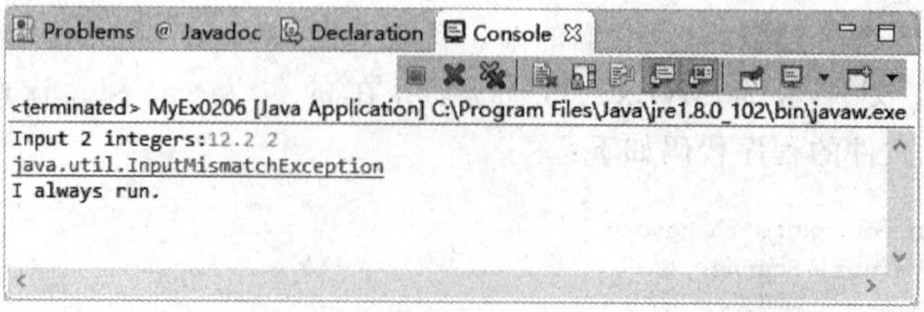

图 2-13　例 2-6 运行结果（输入类型不匹配异常）

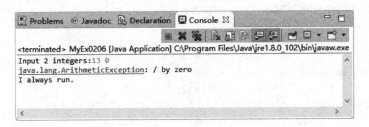

图 2-14 例 2-6 运行结果（除以 0 异常）

2.2 Java 基本数据类型

Java 基本数据类型包括整数、浮点数、字符、字符串、布尔数和数组等。本节将介绍 Java 基本数据类型及其运算符。

2.2.1 数值

数值类型包括整数类型和浮点数类型，整数类型包括 4 类，即字节 byte、短整型 short、整型 int 和长整型 long，依次占有存储字节数为 1、2、4 和 8；浮点数类型包括两类，即单精度浮点数型 float 和双精度浮点数型 double，分别占有存储字节数为 4 和 8。double 和 float 都用来表示小数，为了区分这两种表示，float 型小数后面需要添加"f"或"F"，即 0.53F 被视为 float 型，而 0.53 被视为 double 型。

各种数值类型间可以进行强制类型转换，这一点与 C 语言相同，例如，"int i=(int)15.3;"将 double 型数 15.3 强制转化为整型数 15。强制类型转换也称为显式转换。当一个表达式中出现了 int 型和 double 型，则自动将 int 型转换为 double 型进行计算，这种类型转换称为隐式转换，例如，3/6.0 将得到 0.5。

与数值类型数据相关的 Java 算术运算符包括加、减、乘、除和求余，即"＋"、"－"、"＊"、"/"和"％"；Java 赋值运算符为"＝"；与 C 语言相同，Java 语言支持自增和自减运算符，即"++"和"－－"；Java 语言支持复合赋值运算符，即"+＝"、"－＝"、"＊＝"、"/＝"和"％＝"，注意，复合运算符中两个基本运算符间没有空格，例如"＋ ＝"是不正确的复合运算符。

例 2-7 数值类型演示。

新建工程 ex02_07，在工程 ex02_07 中新建类 MyEx0207（对应文件名为 MyEx0207.java，所在的包为 cn.edu.jxufe.zhangyong）。MyEx0207.java 文件的程序代码如下：

```
1    package cn.edu.jxufe.zhangyong;
2
3    public class MyEx0207 {
4        public static void main(String[] args){
5            byte a1 = (byte)0xF1;
6            short a2 = (short)0xF001;
7            int a3 = 0xF001;
8            long a4 = 0xF001;
9            double b1 = 3/6.0;
```

```
10       float b2 = 15.3000001f;
11       double b3 = 15.3000001;
12       int a5 = (int)b2;
13
14       System.out.println("a1 = " + a1 + "; a2 = " + a2 + "; a3 = " + a3 +
15           "; a4 = " + a4 + "; a5 = " + a5);
16       System.out.println("b1 = " + b1 + "; b2 = " + b2 + "; b3 = " + b3
17           +"; b3 = " + String.format("% - 10.2f", b3));
18   }
19 }
```

上述代码中,第 5 行定义 byte 型变量 a1 并初始化为 0xF1(十六进制数形式),由于 Java 不支持无符号型数,因此 a1 为-15;第 6 行定义 short 型变量 a2 并初始化为 0xF001,此时 a2 等于-4095;第 7 行定义整型变量 a3,由于 a3 占 4 个字节,这里 0xF001 相当于 0x0000F001,因此,a3 为 61441;第 8 行定义长整型变量 a4,这里使用隐式转换,a4 等于 61441。第 9 行定义 double 型变量 b1 并赋初值 3/6.0,即 0.5;第 10 行定义 float 型变量 b2 并赋初值为 15.3000001;第 11 行定义 double 型变量 b3 并赋初值 15.3000001。第 14、15 行输出各个整型数变量的值,第 16、17 行输出各个浮点数变量的值,如图 2-15 所示。从图 2-15 中可以看出,float 型数 b1 的精度比 double 型数 b2 的精度低,b1 无法精确识别 15.3000001。

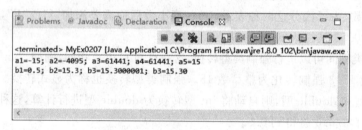

图 2-15 例 2-7 输出结果

2.2.2 字符

Java 支持 Unicode 码,即用两个字节表示的字符,Unicode 码(常被译为统一码)包含了 ASCII 码,几乎可以表示地球上现有的所有人类语言符号。因此,一个 Java 字符占 2 个字节。定义的字符用两个单引号括起来,例如"char ch1='A';"。如果字符不是 ASCII,可以使用该字符的 Unicode 码值来定义字符,例如"char ch2='\u03b1';",这里的 03b1 是十六进制形式表示的码值。与 C 语言相同,字符型数据支持自增和自减运算符。字符型数据与数值型数据可以相互转换,一般借助于显式转换方式。

例 2-8 大写字母转换为小写字母。

例 2-8 执行结果为:输入一个大写字母,输出其对应的小写字母。新建工程 ex02_08,在工程 ex02_08 中新建类 MyEx0208(对应文件名为 MyEx0208.java,所在的包为 cn.edu.jxufe.zhangyong)。MyEx0208.java 文件的程序代码如下:

```
1 package cn.edu.jxufe.zhangyong;
2 import java.util.Scanner;
3
```

```
4   public class MyEx0208 {
5       private static Scanner scanner;
6       public static void main(String[] args){
7           scanner = new Scanner(System.in);
8           System.out.print("Input a letter:");
9           String str = scanner.next();
10          char ch = str.charAt(0);
11          if ((ch >= 'A') && (ch <= 'Z')){
12              ch = (char)(ch + 'a' - 'A');
13          }
14          System.out.println(str.charAt(0) + "\'s lower case:" + ch + ".");
15      }
16  }
```

上述代码中，第7行创建了一个控制台输入对象scanner；第8行在控制台输出提示信息"Input a letter:"；第9行从控制台读入字符串的值并赋给字符串变量str，scanner不支持字符的读入；第10行从字符串str中提取第一个字符，赋给字符变量ch，这里使用了字符串的方法charAt。第11～13行判断字符ch是否为大写字母，如果第11行为真，即字符ch为大写字母，则第12行将其转化为小写字母。第14行输出转化结果，如图2-16所示。

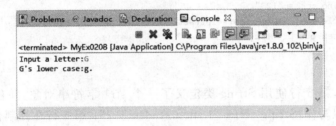

图2-16 例2-8运行结果

Java语言中常用的转义字符如表2-1所示。

表2-1 Java语言常用转义字符

转 义 字 符	Unicode码	含　　义
\\	\u005C	反斜杠
\'	\u0027	单引号
\"	\u0022	双引号
\r	\u000D	回车符
\b	\u0008	退格符
\n	\u000A	换行符
\t	\u0009	Tab键
\f	\u000C	进纸符

2.2.3 字符串

Java中声明字符串变量使用String关键字（实际上是String类），字符串声明时需要初始化，字符串类型为引用类型（即字符串变量是指向字符串的地址），一旦创建后就不能更改字符串的值（注：可以在字符串尾部添加字符串）。如果要创建一个可以改变内容的字符串,需要借助 StringBuffer 类或 StringBuilder 类,这两个类都是通过操作字符缓冲区来更改字符串的值,常用的方法有 append、insert、substring、charAt、setLength、delete 和 toString 等,分别表示追加字符串、插入字符串、定位部分字符串、定位字符串中的字符、设置字符串缓冲区的长度、删除部分字符串和返回字符串。

例 2-9　字符串操作。

新建工程 ex02_09,在工程 ex02_09 中新建类 MyEx0209(对应文件名为 MyEx0209.java,所在的包为 cn.edu.jxufe.zhangyong)。MyEx0209.java 文件的程序代码如下:

```
1    package cn.edu.jxufe.zhangyong;
2
3    public class MyEx0209 {
4        public static void main(String[] args){
5            String str1 = new String("I am a student!");
6            System.out.println(str1);
7            StringBuffer strbuf = new StringBuffer(60);
8            strbuf.append("I student!");
9            strbuf.insert(2,"am a ");
10           strbuf.replace(7, 14, "worker");
11           System.out.println(strbuf.toString());
12           String str2 = "123", str3 = "43.45";
13           double num = Integer.parseInt(str2) + Double.parseDouble(str3);
14           System.out.println(num);
15       }
16   }
```

上述代码中,第 5 行使用 String 类定义了一个 str1 字符串对象,并初始化为"I am a student!",也可以写作"String str1="I am a student!";";第 6 行在控制台输出 str1 字符串。第 7 行使用 StringBuffer 类定义了一个 strbuf 字符缓冲区对象,其长度为 60;第 8 行向 strbuf 中添加字符串"I student!";第 9 行向 strbuf 中插入字符串"am a",插入位置为 2,strbuf 的位置从 0 开始索引,故第 2 个位置为字符's'处,插入操作完成后,strbuf 的值为"I am a student!"。第 10 行将 strbuf 中第 7~14 个字符替换为"worker"字符串,此时 strbuf 的值为"I am a worker!"。第 11 行向控制台输出 strbuf 的值,需要借助 toString 方法。

第 12 行定义了两个字符串对象(或称变量)str2 和 str3,这两个字符串中仅包含数值型字符,可以借助 parseInt 和 parseDouble 方法提取出字符串中的数值,如第 13 行所示,因此,num 等于 123 与 43.45 的和,即 166.45。例 2-9 的运行结果如图 2-17 所示。

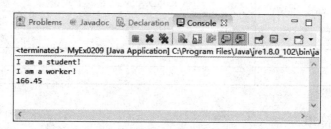

图 2-17　例 2-9 的运行结果

此外,字符串可以转化为字符数组,例如"char[]chs=str1.toCharArray();",这里的 str1 如例 2-9 程序第 5 行所示,那么 chs[0]是字符'I',chs[1]是空格字符,chs[2]是字符'a',等等。

2.2.4 布尔数

Java 语言中,逻辑真为 true,逻辑假为 false,而且只有这两个是布尔数。Java 语言的布尔运算符如表 2-2 所示,参与布尔运算符的只能是布尔数。

表 2-2 布尔运算符

运算符	名称	功能
!	非	逻辑取反,例如 !true 等于 false
&&	与	逻辑与,例如 true && false 等于 false
||	或	逻辑或,例如 true || flase 等于 true
^	异或	逻辑异或,例如 true ^ false 等于 true

Java 语言的关系运算符如表 2-3 所示,关系运算符的结果为布尔数。

表 2-3 关系运算符

运算符	名称	举例	结果
==	等于	4==4	true
!=	不等于	4!=4	false
<	小于	4<5	true
<=	小于等于	4<=5	true
>	大于	4>5	false
>=	大于等于	4>=4	true

布尔数主要用于分支(或选择)控制语句中,定义布尔型变量的方法为"boolean bon=true;"。

2.2.5 数组

Java 语言中定义数组的方法为"数据类型[] 数组名;",例如定义包含 20 个元素的 double 型一维数组 arr,其定义为"double[]arr=new double[20];"或"double[]arr; arr= new double[20];",这一点与 C++ 相同。可以在定义数组时给数组初始化,例如,"double[] arr1={12.3, 2.6, 7.9, 8.8};",相当于语句组"double[]arr1=new double[4]; arr1[0]= 12.3; arr1[1]=2.6; arr1[2]=7.9; arr1[3]=8.8;"。Java 数组的下标从 0 开始索引。

Java 语言也可以定义高维数组,例如,"double[][]arr2=new double[4][5]"定义了一个 4 行 5 列的二维数组,第一个元素的脚标为[0][0],最后一个元素的脚标为[3][4]。

对于 Java 数组,可以使用 foreach 循环方法遍历数组中的所有元素,可以借助 Arrays 类中的方法处理数组元素,如例 2-10 所示。

例 2-10 数组演示实例。

例 2-10 的功能为:产生 10 个随机数,对它们按从小到大排序,并计算它们的和。新建工程 ex02_10,在工程 ex02_10 中新建类 MyEx0210(对应文件名为 MyEx0210.java,所在的包为 cn. edu. jxufe. zhangyong)。MyEx0210.java 文件的程序代码如下:

```
1   package cn.edu.jxufe.zhangyong;
2
3   import java.util.Arrays;
4   import java.util.Random;
5
6   public class MyEx0210 {
7       public static void main(String[] args){
8           double[] arr = new double[10];
9           Random rnd = new Random(10000);
10          System.out.print("The original arr:");
11          for(int i = 0;i < 10;i++){
12              arr[i] = ((int)(rnd.nextDouble() * 10000))/100.0;
13              System.out.print(arr[i] + " ");
14          }
15          System.out.println();
16          Arrays.sort(arr);
17          System.out.print("The sorted arr:");
18          for(double elem:arr){
19              System.out.print(elem + " ");
20          }
21          System.out.println();
22          double sum = 0.0;
23          for(double elem:arr){
24              sum = sum + elem;
25          }
26          System.out.println("The summary of arr:" + String.format("% - 8.2f",sum));
27      }
28  }
```

上述代码中，第 8 行定义 double 型数组 arr，具有 10 个元素。第 9 行使用类 Random 定义随机数对象 rnd，随机数种子为 10 000（即随机数迭代的起始值），指定随机数种子后，每次运行程序产生的随机数序列是相同的。第 11～14 行，在 for 循环中为数组 arr 的每个元素赋值，同时把该元素输出到控制台上，rnd.nextDouble()方法产生 0～1 的浮点数随机数，因此，((int)(rnd.nextDouble() * 10 000))/100.0 产生 0.00～100.00 的随机数。第 16 行调用 Arrays 类的 sort 方法对数组 arr 排序；第 18～20 行输出排序后的数组，这里是一种新的 for 循环形式，即对于 arr 中的每个一元素（elem），都执行第 19 行的语句，即输出到控制台。同样，第 23～25 行也使用这种 for 循环形式计算数组 arr 中所有元素的和，这种 for 循环称为 foreach 循环。例 2-10 的输出结果如图 2-18 所示。

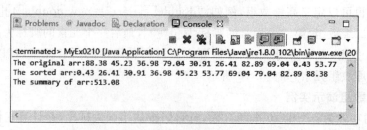

图 2-18　例 2-10 输出结果

2.3 Java 类

Java 是面向对象的高级程序设计语言，类是其基本的数据结构。类也被称为一种程序设计思想，很多学生觉得类难以掌握，很大程度上是受了原来所学的面向过程程序设计思想的影响，而不能完全接受面向对象程序设计的新方法，而这种新方法，大大增强了程序的可扩展性、鲁棒性和可重用性。

类是一种包含数据和方法的集合体数据类型，即所谓的"类封装了数据和方法"。访问类中的数据和方法必须遵循类的"访问控制"要求。一般地，类中的数据和方法有三种访问控制类型，即公有的、保护的和私有的，依次采用关键字 public、protected 和 private 修饰；当不加访问控制修饰符时，相当于私有类型，但是同一个包内可以访问。类中的数据和方法的访问控制权限如表 2-4 所示。

表 2-4 类中数据和方法的访问控制权限

访问控制修饰符	类内访问	包内访问	子类内访问	不同包间访问
public	可以	可以	可以	可以
protected	可以	可以	可以	不可以
无	可以	可以	不可以	不可以
private	可以	不可以	不可以	不可以

使用类时，就是使用类中的数据或方法，主要是调用类的方法完成一些特定的处理。访问类中的公有方法有两种方式：其一，当这种方法为静态方法时，即用 static 定义的方法，这类方法属于类，类创建时静态方法就创建了，因此，直接用"类名.静态方法名"调用；其二，当这种方法不是静态方法时，必须创建类的实例（或称变量），即对象，当创建对象后，类的非静态方法才能被创建，因此，调用非静态方法需要借助于对象，即"对象.非静态方法"。

类中的非静态方法往往只使用类中的数据成员，借助于对象调用类中非静态方法时，必须向类中的数据成员赋值才能正确调用这些方法。请读者再仔细读一遍上一句话。换句话说，类中必须包含初始化它的数据成员的方法，有两种方法可以初始化类中的数据成员：其一，通过构造方法，构造方法是类的同名方法；其二是通过公有的属性设置方法，称为 set 方法。

Java 语言支持类的派生，即一个类可以派生出多个类，前者称为父类（或基类），后者称为子类（或派生类），但是一个子类只能有一个父类，即任何类不能同时继承自两个或两个以上的父类（即 Java 语言不支持多重继承）。子类和父类中可以有同名同参的方法，称为方法覆盖。而一个类中同名不同参的方法，称为方法重载。

当一个类（特别是顶层的父类）使用 abstract 关键字声明时，称为抽象类，抽象类不能定义对象（即不能实例化），抽象类与普通类一样，可以有数据和方法，但是抽象类中还有抽象方法，这类方法只有方法头（即方法声明），没有方法实现。继承该抽象类的子类要全部实现其抽象父类中的抽象方法。普通类中不能有抽象方法。

接口是一种特殊的"抽象类"，它只能包含常量和抽象方法。多个接口可以被一个子类继承，即接口支持多重继承。

2.3.1 类与对象

类使用关键字 class 定义,Java 语言中建议类名的首字母大写。例 2-11 演示了类和对象的定义与实现方法。

例 2-11 类与对象应用演示。

例 2-11 中定义一个表示圆的属性的类,类名为 MyCircle,将圆的半径设为私有数据成员,求圆的周长和面积作为两个公有方法。新建工程 ex02_11,在工程 ex02_11 中新建类 MyCircle(对应文件名为 MyCircle.java,所在的包为 cn.edu.jxufe.zhangyong);还要在工程 ex02_11 中新建类 MyEx0211(对应文件名为 MyEx0211.java,所在的包为 cn.edu.jxufe.zhangyong)。其中,文件 MyCircle.java 的代码如下:

```
1   package cn.edu.jxufe.zhangyong;
2
3   public class MyCircle {
4       private double radius = 1.0;
5       MyCircle(){}
6       MyCircle(double radius){
7           this.radius = radius;
8       }
9       public double circArea(){
10          return Math.PI * radius * radius;
11      }
12      public double circPerimeter(){
13          return 2.0 * Math.PI * radius;
14      }
15  }
```

文件 MyEx0211.java 的代码如下:

```
16  package cn.edu.jxufe.zhangyong;
17
18  public class MyEx0211 {
19      public static void main(String[] args){
20          MyCircle mycircle = new MyCircle(5.0);
21          double area = mycircle.circArea();
22          double peri = mycircle.circPerimeter();
23          System.out.println("The area of the circle:"
24                  + String.format("%-8.2f",area));
25          System.out.println("The perimeter of the circle:"
26                  + String.format("%-8.2f",peri));
27      }
28  }
```

第 3 行代码定义了公有类 MyCircle,即在包 cn.edu.jxufe.zhangyong 外的包也可使用该类。第 4 行定义了类 MyCircle 的私有数据成员 radius,并赋了初值为 1.0,该私有数据成员 radius 只能被类 MyCircle 内部访问。第 5 行为空的构造方法,一般地,每个类都应具有一个空的构造方法,如果没有定义,也会默认生成一个。第 6~8 行为重载的构造方法,这里的参数变量名 radius 与类的私有成员 radius 同名,所以,在构造方法内部使用 this 关键字,即第7行中this.radius表示类的私有成员radius。需要特别注意的是,构造方法一般不使用任何修饰符;但也可以使用"public"修饰符,例如,"public MyCircle()"这种写法是正确的。

第9～11行为类MyCircle的公有方法circArea,可以被类定义的对象调用,返回圆的面积(即第10行)。第12～14行为类MyCircle的公有方法circPerimeter,可以被类定义的对象访问,返回圆的周长。

第19～27行为main方法。第20行使用MyCircle类定义了一个对象mycircle,并使用构造方法(即第9、10行)向私有成员radius赋值5.0,此时对象mycircle中的私有数据成员的值为5.0。第21行通过对象mycircle调用其方法circArea计算圆的面积;第22行通过对象mycircle调用其方法circPerimeter计算圆的周长。第23～26行在控制台输出计算结果,如图2-19所示。

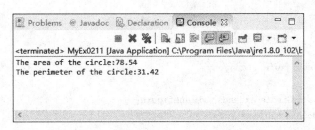

图2-19 例2-11运行结果

通过例2-11,可以体会到MyCircle类的优点,该类中的数据成员为私有成员,外部无法访问,该类的公有方法被外部调用时,仅使用对象内的私有数据成员,保证了公有方法执行时的数据合法性,不会发生方法使用错误的数据源而导致的程序出错。这是设计类的一个基本原则,即类中的公有方法必须仅使用类内的私有数据,不能使用外部数据(类中的静态方法除外,静态方法本质上是函数)。

2.3.2 继承与多态

Java语言中,每个子类只能从一个父类继承,当子类与父类建立了继承关系后,子类将拥有父类所有可以访问的数据和方法(除父类的私有数据和私有方法外的成员),即父类的这些方法可以被子类的实例(对象)调用;但是子类新添加的方法,父类的实例(对象)则不能调用。可以认为子类是父类添加了新特性后的特例,每个子类的实例都是父类的实例,即参数中以子类实例出现的地方,均可以输入父类实例,称之为多态。

例2-12演示了继承与多态的情况,实例中建立了一个平面图形类MyPlanefigure,从该类中派生出类MyCircle和类MyRectangle。

例2-12 继承与多态演示。

新建工程ex02_12,在工程ex02_12中新建三个类,依次为MyPlanefigure、MyCircle和MyRectangle,对应的文件名为MyPlanefigure.java、MyCircle.java和MyRectangle.java,所在的包为cn.edu.jxufe.zhangyong;还要在工程ex02_12中新建类MyEx0212,对应文件名为MyEx0212.java,所在的包为cn.edu.jxufe.zhangyong。

其中,MyPlanefigure.java文件的代码如下:

```
1  package cn.edu.jxufe.zhangyong;
2
3  public class MyPlanefigure {
4      private boolean isFilled = false;
5      MyPlanefigure(){}
6      MyPlanefigure(boolean isFilled){
7          this.isFilled = isFilled;
8      }
9      public String toString(){
10         if(isFilled)
11             return "This is a filled plane figure!";
12         else
13             return "This is an unfilled plane figure!";
14     }
15 }
```

文件 MyCircle.java 的代码如下:

```
16 package cn.edu.jxufe.zhangyong;
17
18 public class MyCircle extends MyPlanefigure{
19     private double radius = 1.0;
20     MyCircle(){}
21     MyCircle(double radius){
22         this.radius = radius;
23     }
24     MyCircle(double radius,boolean isFilled){
25         super(isFilled);
26         this.radius = radius;
27     }
28     public double circArea(){
29         return Math.PI * radius * radius;
30     }
31     public double circPerimeter(){
32         return 2.0 * Math.PI * radius;
33     }
34     public String toString(){
35         return super.toString() + " This is a circle!";
36     }
37 }
```

文件 MyRectangle.java 的代码如下:

```
38 package cn.edu.jxufe.zhangyong;
39
40 public class MyRectangle extends MyPlanefigure{
41     private double width = 1.0;
42     private double height = 1.0;
43     MyRectangle(){}
44     MyRectangle(double width,double height){
```

```java
45        this.width = width;
46        this.height = height;
47    }
48    MyRectangle(double width,double height,boolean isFilled){
49        super(isFilled);
50        this.width = width;
51        this.height = height;
52    }
53    public double rectArea(){
54        return width * height;
55    }
56    public double rectPerimeter(){
57        return 2 * (width + height);
58    }
59 }
```

文件 MyEx0212.java 的代码如下:

```java
60 package cn.edu.jxufe.zhangyong;
61
62 public class MyEx0212 {
63    public static void main(String[] args){
64        MyCircle mycircle = new MyCircle(5.0,false);
65        MyRectangle myrect = new MyRectangle(6.0,8.0,true);
66        calcArea(mycircle);
67        System.out.println(mycircle.toString());
68        calcArea(myrect);
69        System.out.println(myrect.toString());
70    }
71    public static void calcArea(MyPlanefigure myplanefig){
72        if(myplanefig instanceof MyCircle){
73            System.out.print("The area of Circle:");
74            System.out.println(String.format("%-8.2f",
75                ((MyCircle)myplanefig).circArea()));
76        }
77        if(myplanefig instanceof MyRectangle){
78            System.out.print("The area of Rectangle:");
79            System.out.println(String.format("%-8.2f",
80                ((MyRectangle)myplanefig).rectArea()));
81        }
82    }
83 }
```

上述程序代码中,第 3~15 行定义类 MyPlanefigure,该类有一个布尔型私有数据成员 isFilled,用于表示平面图形是否填充;该类有两个构造方法,第 5 行为空构造方法,第 6~8 行的构造方法为私有数据成员赋值。MyPlanefigure 类只有一个公有方法 toString,通过判断 isFilled 私有成员的值,输出提示信息,当 isFilled 为真时输出 "This is a filled plane figure!",否则,输出 "This is an unfilled plane figure!"。

第 18 行定义了类 MyCircle,使用关键字 extends 表示该类继承了类 MyPlanefigure。类 MyCircle 有一个私有数据成员 radius(第 19 行)。该类有三个构造方法,第 20 行为空的构造方法；第 21～23 行的构造方法为私有数据成员 radius 赋值；第 24～27 的构造方法，除了为私有数据成员 radius 赋值外，还使用关键字 super 调用父类的构造方法为父类的私有成员 isFilled 赋值。第 28～30 行的公有方法计算圆面积；第 31～33 行的公有方法计算圆周长。第 34～36 行的方法与父类中的方法同名，这里覆盖了父类的同名方法，第 35 行使用 super 调用了父类的方法 toString。

第 40 行定义了类 MyRectangle，也继承了类 MyPlanefigure。第 41、42 行定义了两个私有数据成员 width 和 height，分别表示长方形的宽和高。类 MyRectangle 有三个构造方法，第 43 行的构造方法为空方法；第 44～47 行的构造方法为私有数据成员 width 和 height 赋值；第 48～52 行的构造方法除了为私有数据成员赋值外，还为父类的私有成员 isFilled 赋值（第 49 行）。第 53～55 行的公有方法 rectArea 计算长方形的面积；第 56～58 的公有方法 rectPerimeter 计算长方形的周长。

第 63～70 行为 main 方法。第 64 行使用类 MyCircle 定义了对象 mycircle；第 65 行使用类 MyRectangle 定义了对象 myrect。第 66 行调用静态方法 calcArea（注：该方法可以设为私有非静态方法！），由于 calcArea 的形式参数是类 MyPlanefigure，而输入的实参为 mycircle，因此，这里是多态应用，即 mycircle 对象传递到了 myplanefig 参数中（第 71 行）；第 72 行使用关键字 instanceof 判定 myplanefig 对象是否属于 MyCircle 类，此时为真，所以，执行第 73～75 行，向控制台输出圆的面积。回到第 67 行，mycircle 对象使用 MyCircle 类中覆盖的 toString 方法向控制台输出信息，这里输出"This is an unfilled plane figure! This is a circle!"，如图 2-20 所示。第 68 行也体现多态应用，calcArea(myrect) 将执行第 78～80 行的代码，即输出长方形的面积。第 69 行中 myrect 对象调用其父类的方法 toString，向控制台输出提示信息。例 2-12 的执行结果如图 2-20 所示。

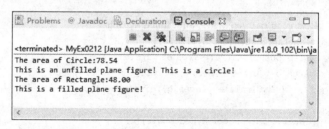

图 2-20　例 2-12 程序运行结果

例 2-12 中几点需要注意的地方：其一，类 MyRectangle 中没有覆盖其父类 MyPlanefigure 的 toString 方法，因此，类 MyRectangle 对象将会调用其父类的方法 toString，即子类实例（对象）可以调用从父类继承的方法；其二，在子类的构造方法中使用"super()"可以调用父类的构造方法；其三，在子类的方法中，使用"super.父类方法"可以调用父类方法；其四，所有子类的对象都是父类的对象（反之不成立），因此子类对象类型的参数可以代入父类对象，由于父类无法调用子类的方法，因此，需要使用显式类型转换为子类对象类型后才能调用子类方法。

2.3.3　接口

接口中只能声明公有的抽象方法和常量，因此，在接口中声明抽象方法时，省略"public

abstract"关键字,在声明常量时,省略"public static final"等关键字。接口可以继承其他接口,而且可以多重继承,即一个接口可继承多个父接口。同样,类继承接口时,也可以多重继承,即一个类可以继承多个接口。当某个类继承了接口后,必须在类中实现该接口的所有抽象方法。

下面在例 2-12 的基础上,添加一个接口 MyComparable,其中声明了一个抽象方法 compareTo,表示比较两个图形的面积;并添加一个新类 MyComparableRectangle,该类继承了类 MyRectangle 和接口 MyComparable,实现了接口 MyComparable 中的抽象方法。

例 2-13 接口应用演示。

新建工程 ex02_13(实际上是在工程 ex02_12 的基础上修改得到),在工程 ex02_13 中新建 4 个类和 1 个接口,即 MyPlanefigure、MyCircle、MyRectangle、MyComparableRectangle 和 MyComparable(使用菜单 File|New|Interface 打开新建接口窗口),对应的文件名分别为 MyPlanefigure.java、MyCircle.java、MyRectangle.java、MyComparableRectangle.java 和 MyComparable.java,所在的包均为 cn.edu.jxufe.zhangyong。然后新建类 MyEx0213,对应的文件名为 MyEx0213.java,所在的包为 cn.edu.jxufe.zhangyong。其中,文件 MyPlanefigure.java、MyCircle.java 和 MyRectangle.java 与例 2-12 中的同名文件内容相同,这里不再列举其源代码,下面罗列了文件 MyComparableRectangle.java、MyComparable.java 和 MyEx0213.java 的代码。

文件 MyComparable.java 的代码如下:

```
1   package cn.edu.jxufe.zhangyong;
2
3   public interface MyComparable {
4       public int compareTo(Object o);
5   }
```

文件 MyComparableRectangle.java 的代码如下:

```
6   package cn.edu.jxufe.zhangyong;
7
8   public class MyComparableRectangle extends MyRectangle
9                                     implements MyComparable{
10      MyComparableRectangle(double width,double height){
11          super(width,height);
12      }
13      public int compareTo(Object o){
14          if(this.rectArea()>((MyComparableRectangle)o).rectArea())
15              return 1;
16
17          else if(this.rectArea()<((MyComparableRectangle)o).rectArea())
18              return -1;
19          else
20              return 0;
21      }
22  }
```

文件 MyEx0213.java 的代码如下：

```
23   package cn.edu.jxufe.zhangyong;
24
25   public class MyEx0213 {
26      public static void main(String[] args){
27         MyComparableRectangle mycomprect1 = new MyComparableRectangle(12.0,8.9);
28         MyComparableRectangle mycomprect2 = new MyComparableRectangle(14.2,7.8);
29         int res = mycomprect1.compareTo(mycomprect2);
30         if(res > 0)
31            System.out.println("Rectangle 1 > Rectangle 2");
32         else if(res < 0)
33            System.out.println("Rectangle 1 < Rectangle 2");
34         else
35            System.out.println("Rectangle 1 = Rectangle 2");
36      }
37   }
```

上述代码中，第3行定义了接口 MyComparable，定义接口需要使用关键字 interface，该接口中只有一个抽象方法，即 compareTo（第4行），该方法返回整型值，这里省略了关键字"abstract"，事实上，第4行的 public 关键字也可以省略。接口中仅包含方法的声明，没有方法的实现（即方法体），接口的方法在继承它的类中实现。

第8~9行定义了类 MyComparableRectangle，该类是类 MyRectangle 的子类，并且也继承了接口 MyComparable。当某个类继承接口时，常称为该类实现了该接口，使用关键字 implements 表示这种关系。注意，接口也可以继承接口，这种继承关系和类间继承关系一样，也使用关键字 extends。由于类 MyComparableRectangle 是类 MyRectangle 的子类，所以，类 MyComparableRectangle 具有（或继承了）类 MyRectangle 所有的公有方法和保护方法，即具有了类 MyRectangle 中的 rectArea 和 rectPerimeter 方法（参考例2-12中程序代码第53~58行）。第10~12行，类 MyComparableRectangle 的构造函数使用 super 关键字调用其父类的构造函数。第13~21行为类 MyComparableRectangle 实现其接口 MyComparable 中的抽象方法 compareTo，该方法中 this.rectArea()调用类 MyComparableRectangle 对象的 rectArea 方法计算长方形面积；而((MyComparableRectangle)o).rectArea()调用参量对象 o 中的 rectArea 方法计算长方形面积，要求参量对象必须也是 MyComparableRectangle 类型。第29行为该方法调用形式，即 mycomprect1.compareTo(mycomprect2)，这样，this.rectArea()是指 mycomprect1.rectArea()，而((MyComparableRectangle)o).rectArea()是指 mycomprect2.rectArea()。回到第14~20行，根据比较结果，返回整数值1、-1或0，分别表示两个长方形间的关系为大于、小于或相等。

第26~36行为 main 方法。第27~28行定义了两个 MyComparableRectangle 类型对象 mycomprect1 和 mycomprect2，并初始化其宽和高分别为12.0、8.9和14.2、7.8。第29行调用 compareTo 方法比较这个长方形的大小（即比较它们的面积大小），返回值存放在整型变量 res 中。第30~35行根据 res 的结果输出比较信息。这里，长方形1的面积小于长方形2的面积，因此输出结果为"Rectangle 1 < Rectangle 2"，如图2-21所示。

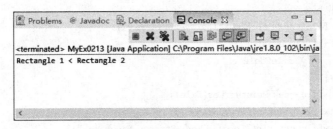

图 2-21　例 2-13 运行结果

从例 2-13 中可以看出,一个设计良好的类在以后的应用中不需要更改其代码,当遇到新的情况时,只需要从其派生出新类并实现新的方法即可,这体现了基于类进行程序设计的优点。

2.4　Java 文件操作

Java 语言中操作文本文件与二进制数据文件的类库不同。如果是操作文本文件,首先使用 File 类定义文件对象,创建文件对象时需要输入文件名,目录分隔符使用"/",这里的文件对象相当于 C 语言的文件指针。创建好文件对象后,如果是文本文件的写操作,则使用类 PrintWriter 创建文件写入对象,使用该对象的方法 print 等向文件中写入数据;如果是文本文件的读操作,则使用类 Scanner 创建文件读出对象,使用该对象的方法 next 等从文件中读出数据。如果是操作二进制文件,创建读或写的二进制文件需要分别借助类 FileInputStream 和 FileOutputStream 创建文件对象;然后,再分别借助类 DataInputStream 和 DataOutputStream 创建文件读/写对象;最后,使用这些对象的方法,例如 readInt、writeInt 等实现二进制文件的读写。此外,类 FileInputStream 和 FileOutputStream 创建的文件对象也具有字节数据的读写方法,即使用类 FileInputStream 和 FileOutputStream 的对象方法实现的读写操作是基于字节数据的,而借助于类 DataInputStream 和 DataOutputStream 的对象方法可实现基本数据类型的读写操作。

例 2-14　文本文件读写演示。

新建工程 ex02_14,在工程 ex02_14 中新建类 MyEx0214,对应的文件名为 MyEx0214.java,所在的包为 cn.edu.jxufe.zhangyong。在工程所在目录(D:\MyJavaWorkSpace\ex02_14)下新建目录 myfiles。文件 MyEx0214.java 的代码如下:

```
1    package cn.edu.jxufe.zhangyong;
2
3    import java.io.File;
4    import java.io.PrintWriter;
5    import java.util.Scanner;
6
7    public class MyEx0214 {
8        public static void main(String[] args){
9            File file;
10           PrintWriter wfile = null;
11
```

```
12      try{
13          file = new File("myfiles/myex0214.txt");
14          if(file.exists()){
15              file.delete();
16          }
17          wfile = new PrintWriter(file);
18          wfile.println("Yong Zhang.");
19          wfile.println("Fenglin West Street, Nanchang.");
20      }
21      catch(Exception ex){
22          System.out.println(ex.toString());
23      }
24      finally{
25          if(wfile!= null)
26              wfile.close();
27      }
28
29      Scanner rfile = null;
30      try{
31          rfile = new Scanner(new File("myfiles/myex0214.txt"));
32          while(rfile.hasNext()){
33              String str1 = rfile.nextLine();
34              System.out.println(str1);
35          }
36      }
37      catch(Exception ex){
38          System.out.println(ex.toString());
39      }
40      finally{
41          if(rfile!= null)
42              rfile.close();
43      }
44  }
45 }
```

上述代码中,第 9 行定义 file 对象;第 10 行定义写方式文件对象 wfile,并赋为 null(空值)。第 12～27 行为 try-catch-finally 结构,第 13 行将对象 file 赋值为文件"myfiles/myex0214.txt",该文件位于工程所在目录下的 myfiles 子目录下,其文件名为 myex0214.txt。第 14～16 行代码判断该文件是否已经存在,如果已经存在,则调用 delete 方法将它删除。第 17 行为对象 wfile 赋值为 file,第 18、19 行调用对象 wfile 的 println 方法写入两行文本。如果程序发生异常,则第 21～23 行得到执行,在控制台输出错误类型。无论有无异常,第 24～27 行的 finally 语句块总会得到执行,如果第 25 行判断文件对象非空,则调用 close 方法将其关闭(或释放)(第 26 行)。

第 29～43 行为读文件操作。第 29 行定义读出方式文件对象 rfile,第 31 行将 rfile 赋为文件 myex0214.txt,即第 10～27 行写入文本数据的文件。第 32～35 行判断文件是否读到文件尾,否则的话反复调用 nextLine 方法从文件中读出一行,并将该行在控制台显示出来。

第41、42行关闭文件对象。

例2-14的执行结果如图2-22所示；打开文件"D:\MyJavaWorkSpace\ex02_14\myfiles\myex02_14.txt"，其内容如图2-23所示。

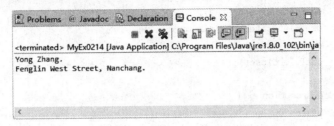

图2-22 例2-14执行结果

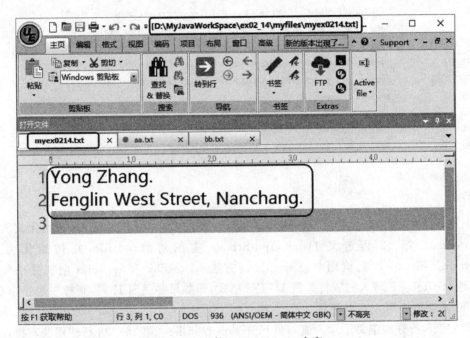

图2-23 文件myex0214.txt内容

例2-15 二进制文件读写演示。

新建工程ex02_15，在工程ex02_15中新建类MyEx0215，对应的文件名为MyEx0215.java，所在的包为cn.edu.jxufe.zhangyong。文件MyEx0215.java的代码如下：

```
1   package cn.edu.jxufe.zhangyong;
2
3   import java.io.DataInputStream;
4   import java.io.DataOutputStream;
5   import java.io.FileInputStream;
6   import java.io.FileOutputStream;
7
8   public class MyEx0215 {
9       public static void main(String[] args){
10          try{
```

```
11        FileOutputStream outStream = new FileOutputStream("myscore.dat");;
12        DataOutputStream outfile = new DataOutputStream(outStream);
13        outfile.writeUTF("Zhang - Enhe:");
14        outfile.writeUTF("Chinese:");
15        outfile.writeInt(100);
16        outfile.writeUTF("Mathematics:");
17        outfile.writeInt(99);
18        outfile.close();
19    }
20    catch(Exception ex){
21        System.out.println(ex.toString());
22    }
23    try{
24        FileInputStream inStream = new FileInputStream("myscore.dat");
25        DataInputStream infile = new DataInputStream(inStream);
26        System.out.println(infile.readUTF());
27        System.out.print(infile.readUTF() + infile.readInt() + "; ");
28        System.out.print(infile.readUTF() + infile.readInt() + ".");
29        infile.close();
30    }
31    catch(Exception ex){
32        System.out.println(ex.toString());
33    }
34  }
35 }
```

上述代码中,第 11 行定义 FileOutputStream 类的对象 outStream,并初始化其文件为 myscore.dat。第 12 行定义 DataOutputStream 类的对象 outfile,并初始化为对象 outStream。第 13~17 行调用对象 outfile 的方法 writeUTF 和 writeInt 向文件中写入数据,其中,writeUTF 写入的字符串按 UTF-8 格式,即如果是 ASCII 码,则按一个字节存储;writeInt 向文件中写入一个整数。

第 24、25 行分别定义了输入流对象 inStream 和 infile。第 26~28 行使用与写操作方法 writeUTF 和 writeInt 对应的读操作方法,即 readUTF 和 readInt,从文件中读出写入的数据并显示在控制台上,如图 2-24 所示。

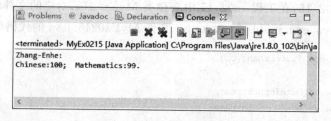

图 2-24 例 2-15 执行结果

采用数据流方法写入到文件中的数据中是二进制格式,文件 myscore.dat 位于目录"D: \MyJavaWorkSpace\ex02_15"下,其内容如图 2-25 所示。

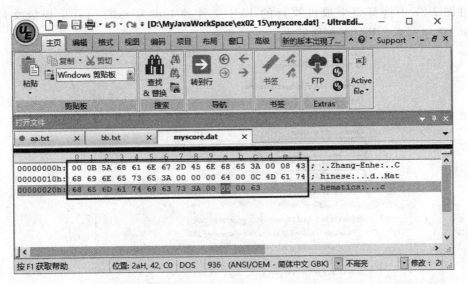

图 2-25 文件 myscore.dat 内容

2.5 在命令行窗口中运行 Java 程序

例 2-1 至例 2-15 的运行,均是在 Eclipse 环境下运行的。事实上,Java 程序可以直接在 Windows 环境下运行(安装了 Java 虚拟机)。在 Eclipse 环境下,选择菜单 File|Export…,在弹出的菜单中选择 Runnable JAR file,如图 2-26 所示,然后单击 Next 按钮进入图 2-27 所示的界面。

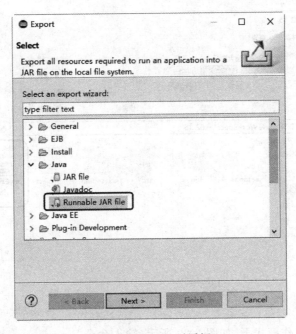

图 2-26 Export 对话框

在图 2-27 中选择要生成可执行的 JAR 文件的工程为 ex02_15，生成的文件名为 myEx0215.jar，单击 Finish 按钮退出图 2-27 界面。

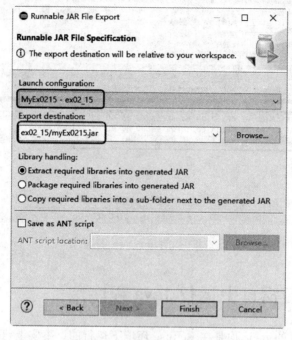

图 2-27 可执行 JAR 文件设置

进入到目录"D:\MyJavaWorkSpace\ex02_15"下，如图 2-28 所示，可执行的 JAR 文件为 myEx0215.jar。

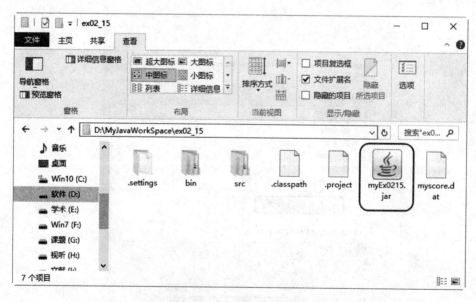

图 2-28 ex02_15 目录文件

在命令提示符工作模式下，如图 2-29 所示，输入命令"java - jar myEx0215.jar"，要求 JAR 文件必须带扩展名.jar，程序执行结果如图 2-29 所示。将图 2-29 的显示结果与图 2-24

的输出结果对比,可见其运行结果完全相同,因此,这两种运行 Java 程序的方式是相同的。一般地,也可以直接在图 2-28 中双击 myEx0215.jar 的图标运行 JAR 程序文件(需要设置 Windows 默认的 JAR 文件运行方式)。

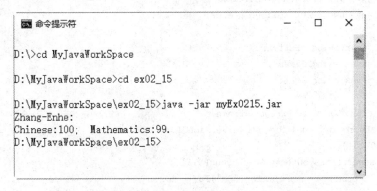

图 2-29 命令提示符下执行 myEx0215.jar 文件

2.6 Java 图形界面

例 2-1 至例 2-15 的 Java 程序均是基于控制台工作的,即程序的输出和输入(如果需要输入时)都从控制台上进行。事实上,Java 语言支持图形用户界面,而且随着 Java 版本的升级,其图形用户界面的功能在不断增强。生成 Java 图形用户界面程序一般需要三个步骤:其一,生成图形界面框架,一般借助于 JFrame 类的实例(对象);其二,在图形框架中使用布局实例(对象)对框架界面进行划分,Java 支持多种布局格式,例如网格布局,这种布局将框架界面分成网格状;其三,按布局的要求在框架中安放控件,控件是 Windows 中的说法,在 Java 中称为 View,即 Java 中的视图就是人们熟悉的 Windows 窗口(或控件)。本书中没有完全按照 Java 语言的规定命名视图,而是采用了人们熟悉的控件名称。

Java 图形用户界面设计好后,还需要为每个控件(即视图)编写事件响应方法,例如,对于命令按钮,需要编写它的单击事件响应方法。Java 语言通过创建监听器对象实现对各个控件的事件响应,例如,对于命令按钮,为其创建一个监听器对象,此时,命令按钮称为事件的源,将监听器对象注册到源上(即通过方法 addActionListener 将监听器对象添加到源控件中),命令按钮的单击事件将调用监听器对象的方法,即所谓的单击事件方法。由于在第 3 章以后,将主要介绍基于 Android 的图形用户界面设计,而且 Java 语言的图形界面与 Android 图形界面区别很大,所以,本节仅简单地介绍一下 Java 图形用户界面设计方法,并借此介绍事件响应方法、内部类和匿名内部类的概念。

2.6.1 事件响应方法

例 2-16 的程序具有基本的图形用户界面,单击命令按钮时将向控制台输出提示信息。

例 2-16 图形界面程序及事件响应演示。

新建工程 ex02_16,在工程 ex02_16 中新建类 MyEx0216,对应的文件名为 MyEx0216.java,所在的包为 cn.edu.jxufe.zhangyong。文件 MyEx0216.java 的代码如下:

```java
1   package cn.edu.jxufe.zhangyong;
2
3   import java.awt.GridLayout;
4   import java.awt.event.ActionEvent;
5   import java.awt.event.ActionListener;
6
7   import javax.swing.JButton;
8   import javax.swing.JFrame;
9   import javax.swing.JLabel;
10
11  public class MyEx0216 extends JFrame{
12      private static final long serialVersionUID = 1L;
13      MyEx0216(){
14          super.setLayout(new GridLayout(3,1));
15          JButton btn = new JButton("OK");
16          add(new JLabel(" Name: Yong."));
17          add(new JLabel(" Last Name: Zhang."));
18          add(btn);
19          ActionListener listener = new OKListener();
20          btn.addActionListener(listener);
21      }
22      public static void main(String[] args){
23          JFrame frame = new MyEx0216();
24          frame.setSize(320,200);
25          frame.setLocationRelativeTo(null);
26          frame.setDefaultCloseOperation(JFrame.EXIT_ON_CLOSE);
27          frame.setVisible(true);
28      }
29  }
30  class OKListener implements ActionListener{
31      @Override
32      public void actionPerformed(ActionEvent arg0) {
33          System.out.println("You pressed OK!");
34      }
35  }
```

上述代码中，第13~21行为类 MyEx0216 的构造方法 MyEx0216，一般地，对于图形用户界面，需要继承类 JFrame（第11行），并在构造方法中设定布局格式（第14行，布局网络为3行1列）；然后，依次创建各个控件并添加到布局中，例如第15、18行创建命令按钮 btn 对象并添加到布局中，或者使用第16行的方法在 add 方法中直接使用语句"new JLabel(" Name: Yong.")"，在创建标签的同时将标签添加到布局中。这里的 JButton 类为命令按钮类，而JLabel 类为标签（即静态文本框）类，GridLayout 类为网格化布局类。

不同控件触发的事件不完全相同，而且，不同事件的监听器接口也不相同。要响应命令按钮的单击事件，必须使得监听器对象的类实现该类事件监听器的接口。命令控件的单击事件为 ActionEvent，其监听器接口为 ActionListener，该接口只有一个抽象方法 actionPerformed（ActionEvent），实现该接口的监听器类必须覆盖该接口的全部抽象方法（这里只有一个抽象方法）。因此，第30~35行的类 OKListener 实现了接口 ActionListener，并在第32~34行覆盖了其唯一的抽象方法 actionPerformed，当有事件发生时，第 33 行向控制台输出信息"You pressed OK!"。

回到第 19 行，声明一个接口对象 listener，第 20 行将 listener 对象注册到 btn 对象。

第 23 行调用构造方法 MyEx0216 得到 JFrame 类定义的 frame 对象，第 24 行指定界面的大小为 320×200，第 25 行将界面显示在窗口中央，第 26 行指定界面关闭后退出应用程序，第 27 行显示界面。

例 2-16 的执行结果如图 2-30 所示。从例 2-16 可以看出，Java 界面程序设计包括三部分，其一，在构造方法中进行界面的设计；其二，在事件监听类中覆盖事件监听器接口的方法，即响应事件的方法；其三，在 main 方法中显示图形用户界面。

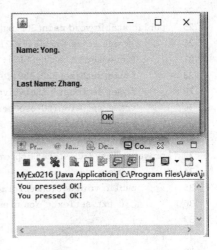

图 2-30　例 2-16 执行结果

2.6.2　内部类

内部类是指定义在类中的类，内部类可以访问定义它的类的私有数据成员和方法。一般地，内部类仅为定义它的类服务。

例 2-17　图形用户界面与内部类演示。

新建工程 ex02_17，在工程 ex02_17 中新建类 MyEx0217，对应的文件名为 MyEx0217.java，所在的包为 cn.edu.jxufe.zhangyong。文件 MyEx0217.java 的代码如下：

```
1   package cn.edu.jxufe.zhangyong;
2
3   import java.awt.GridLayout;
4   import java.awt.event.ActionEvent;
5   import java.awt.event.ActionListener;
6
7   import javax.swing.JButton;
8   import javax.swing.JFrame;
9   import javax.swing.JLabel;
10
11  public class MyEx0217 extends JFrame{
12      private static final long serialVersionUID = 1L;
13      private JLabel txt = new JLabel();
14      private int times = 0;
15      public MyEx0217(){
16          super.setLayout(new GridLayout(4,1));
17          JButton btn = new JButton("OK");
18          add(new JLabel(" Name: Yong."));
19          add(new JLabel(" Last Name: Zhang."));
20          add(txt);
21          add(btn);
22          ActionListener listener = new OKListener();
23          btn.addActionListener(listener);
```

```
24     }
25     public static void main(String[] args){
26         JFrame frame = new MyEx0217();
27         frame.setTitle("MyGUI");
28         frame.setSize(320,200);
29         frame.setLocationRelativeTo(null);
30         frame.setDefaultCloseOperation(JFrame.EXIT_ON_CLOSE);
31         frame.setVisible(true);
32     }
33     private class OKListener implements ActionListener{
34         @Override
35         public void actionPerformed(ActionEvent arg0) {
36             txt.setText("You pressed OK " + (times++) + " times!");
37         }
38     }
39 }
```

对比例 2-16 的程序代码,会发现类 OKListener 移到了类 MyEx0217 内部,而且第 36 行使用类 MyEx0217 的两个私有数据成员,即 txt 对象(第 13 行)和 times 变量(第 14 行),这里 txt 对象的 setText 方法在静态本文框 txt 中输出文本。例 2-17 的执行结果如图 2-31 所示。

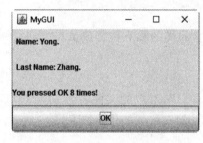

图 2-31 例 2-17 执行结果

在图 2-31 中每单击 OK 按钮一次,则显示的信息 "You pressed OK 8 times!"中的数字值将加 1。由图 2-31 可见增加了一个标签(即静态文本框),所以第 16 行的网格为 4 行 1 列,在第 3 行 1 列的位置处放置对象 txt,这个静态文本框用于显示提示信息。此外,第 27 行添加了 setTitle 方法为程序界面设置标题。

2.6.3 匿名内部类

在例 2-17 的基础上,将内部类 OKListener 移动到方法 addActionListener 中,使用 new 关键字+接口名的形式,便构成匿名内部类,如例 2-18 所示。

例 2-18 图形用户界面与匿名内部类演示。

新建工程 ex02_18,在工程 ex02_18 中新建类 MyEx0218,对应的文件名为 MyEx0218.java,所在的包为 cn.edu.jxufe.zhangyong。文件 MyEx0218.java 的代码如下:

```
1  package cn.edu.jxufe.zhangyong;
2
3  import java.awt.GridLayout;
4  import java.awt.event.ActionEvent;
5  import java.awt.event.ActionListener;
6
7  import javax.swing.JButton;
8  import javax.swing.JFrame;
9  import javax.swing.JLabel;
10
```

```
11  public class MyEx0218 extends JFrame{
12      private static final long serialVersionUID = 1L;
13      private JLabel txt = new JLabel();
14      private int times = 0;
15      public MyEx0218(){
16          super.setLayout(new GridLayout(4,1));
17          JButton btn = new JButton("OK");
18          add(new JLabel(" Name: Yong."));
19          add(new JLabel(" Last Name: Zhang."));
20          add(txt);
21          add(btn);
22
23          btn.addActionListener(new ActionListener(){
24              @Override
25              public void actionPerformed(ActionEvent arg0) {
26                  txt.setText("You pressed OK " + (times++) + " times!");
27              }
28          });
29      }
30      public static void main(String[] args){
31          JFrame frame = new MyEx0218();
32          frame.setSize(320,200);
33          frame.setLocationRelativeTo(null);
34          frame.setDefaultCloseOperation(JFrame.EXIT_ON_CLOSE);
35          frame.setVisible(true);
36      }
37  }
```

对比例 2-17 程序代码和例 2-18 程序代码，可见，例 2-17 中的监听类 OKListener 移动到了第 23～28 行间，即作为了 addActionListener 的对象参数，而且使用 new ActionListener 取代了原来的监听类名 OKListener，此处就是典型的匿名内部类。匿名内部类和内部类（以及普通监听器类）一样，都必须实现接口所有的（抽象）方法，这里接口 ActionListener 只有一个方法，故匿名内部类中只覆盖了一个方法，该方法即事件的响应方法。例 2-18 的运行结果与例 2-17 完全相同，如图 2-31 所示。

2.7 本章小结

Java 语言是面向对象的高级程序设计语言，相比于 C++ 语言而言，其数据结构（类）更简单易学。有趣的是，Java 语言语法在很大程度上与 C/C++/C♯ 相似，每条语句都以";"号结尾，其控制语句几乎完全相同，基本数据类型也大同小异，数组与 C++ 比较相似，关系运算符和逻辑运算符与 C 语言也相同，类的基本用法和概念与 C++ 相似，并且比 C++ 简单。这些都有助于应用程序开发者从 C 语言快速转向 Java 语言的学习。Java 程序界面设计遵循一定的方法，即首先继承界面框架，然后，使用布局类进行界面布局，最后，再按界面布局摆放各种控件。对于图形界面控件事件的响应要求做到：其一，定义监听类实现相应事件监听器接口的所有方法，即事件响应方法；其二，使用监听类对象注册控件对象。这种方法类似于C++/C♯中的委派机制，或者说就是委派机制。Android应用程序基于Java语言开发，如果读者不能完全看懂本章的程序，则需要借助于Java相关的书籍进一步学习本章，笔者建议读者在熟练掌握了本章内容后，才可进入下一章的阅读。关于Java语言的学习，笔者推荐Y. Daniel Liang(梁勇)著的《Java语言程序设计(基础篇)》。

第 3 章

Android 应用程序框架

本章介绍 Android 应用程序的开发过程,分析 Android 工程项目的目录结构,解释 Hello World 工程的工作原理及其源代码含义,并深入剖析 Android 工程的工作原理。

3.1 Hello World 工程

绝大多数程序设计语言的教科书均以显示"Hello World!"字符串的工程作为第一个程序实例,称这个工程为"Hello World 工程",而且,大部分高级程序设计语言几乎只要输入一条代码就可以实现这个功能。在 Android 下实现 Hello World 工程也非常方便,甚至不用输入代码,下面介绍基于 Android Studio 集成开发环境创建 Android 的 Hello World 工程的过程。

例 3-1 显示"Hello World"工程。

在图 1-38 所示界面下单击 Start a new Android Studio project,或者在图 1-44 所示界面下单击菜单 File|New|New Project,弹出图 3-1 所示的界面。

在图 3-1 中,输入应用名称为"MyHelloApp",要求首字母大写,本书中命名应用名称的方式为"My"+"表示应用功能的英文单词"+"App",如上述的"MyHelloApp"。应用名默认为应用程序的标题,如图 1-56 所示,那里 MyHelloWorld 为应用名,除此之外,没有以"应用名"命名的文件,应用名没有特别的用途。

在 Android 中,一个应用的所有类都隶属于一个包,包的名称使用公司名称的反序表示,例如,公司名称的域名为"zhangyong.jxufe.edu.cn",则包名自动设置为"cn.edu.jxufe.zhangyong"+应用名称,包名必须具有全球唯一性。这里的包名中文意思为"中国.教育.江西财经大学.张勇",由范围大的名称到个体的名称,这种包的命名方法是最常用的,可以有效地避免包的同名。包名要求唯一性是指不同开发小组或开发者使用的包名应该不同,如果出现了同名包,将严重影响开发者命名类名的自由度,因为同一个包中不能出现同名的类。本书中使用目录"D:\MyAndroidWorkSpace"作为默认工作路径,每个应用的存储路径为默认工作路径+应用名称。在图 3-1 中单击 Next 按钮进入图 3-2 所示的界面。

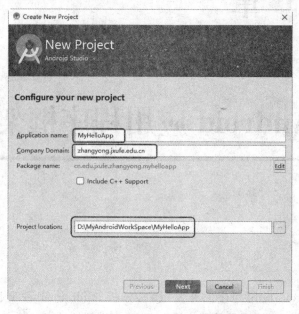

图 3-1　创建新工程对话框

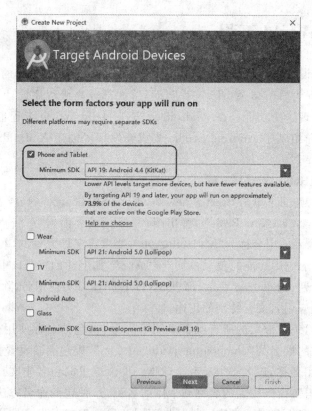

图 3-2　配置应用 SDK

在图 3-2 中为新创建的应用 MyHelloApp 选择其工作的平台和所需要的 SDK 库，这里选择了 Android 4.4（KitKat）。从图 3-2 可知，全球大约有 73.9% 的 Android 智能手机或智能平板上可运行应用 MyHelloApp。在图 3-2 中单击 Next 按钮进入图 3-3 所示的界面。

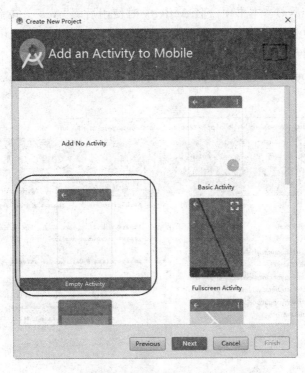

图 3-3　选择 Activity 类型

在图 3-3 中，为应用 MyHelloApp 选择一个空的 Activity。可根据应用需要选择合适的 Activity。在图 3-3 中，单击 Next 按钮进入图 3-4 所示的界面。

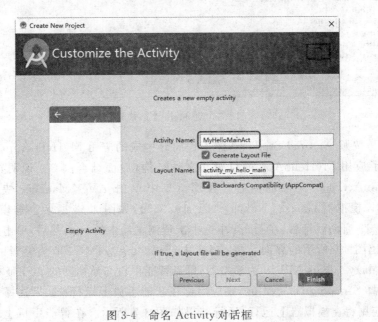

图 3-4　命名 Activity 对话框

在图 3-4 中，输入 Activity 名称为"MyHelloMainAct"，Activity 译为活动，Android 中一个 Activity 就是一个可视界面，建议使用"活动界面"这一译法或者直接使用"Activity"。Activity 名 MyHelloMainAct 将作为应用程序的类名和其对应的 Java 文件名（MyHelloMainAct.java）。本书中命名 Activity 时使用后缀"Act"。在命名 Activity 时，自动为其创建界面布局文件（Layout）名，默认为 activity＋"Activity 名称中的英文单词，各个英文单词间用下画线分开"，例如，Activity 名为 MyHelloMainAct 时，布局文件名为 activity_my_hello_main，布局文件扩展名为.xml。在图 3-4 中单击 Finish 按钮，进入图 3-5 所示的界面。

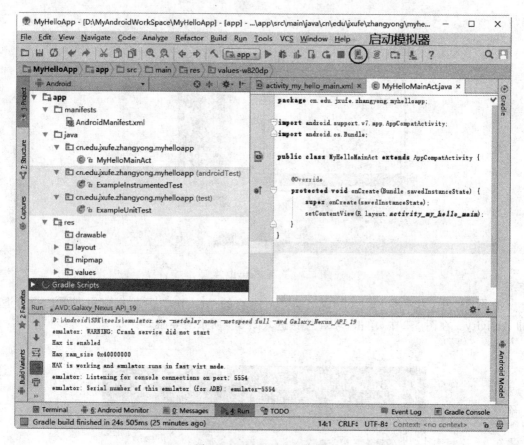

图 3-5　MyHelloApp 应用

图 3-5 是按照 Android Studio 新建工程向导生成的应用 MyHelloApp，左边为工程浏览器，显示了应用 MyHelloApp 的目录和文件结构（后面结合图 3-9 说明）；右边显示了MyHelloMainAct.java 文件的内容。在图 3-5 中单击 AVD Manager 快捷按钮，启动Android SDK 模拟器 Galaxy_Nexus_API_19:5554，如图 1-53 所示。等模拟器 Galaxy_Nexus_API_19 启动就绪后，在图 3-5 中选中工程浏览器中的 app，然后，单击菜单 Run|Run 'app'，将弹出图 3-6 所示的界面。在图 3-6 中，Samsung GT-N7100 为笔者的智能手机（已连接），而 Galaxy Nexus API 19 为模拟器，可见智能手机与模拟器的 API 版本号相同，均为19。如果选择 Samsung GT-N7100，则应用将在智能手机上运行；如果选择 Galaxy Nexus API 19，则应用将在模拟器上运行，运行结果如图 3-7 所示。在智能手机上的运行结果与

图 3-7 相同。

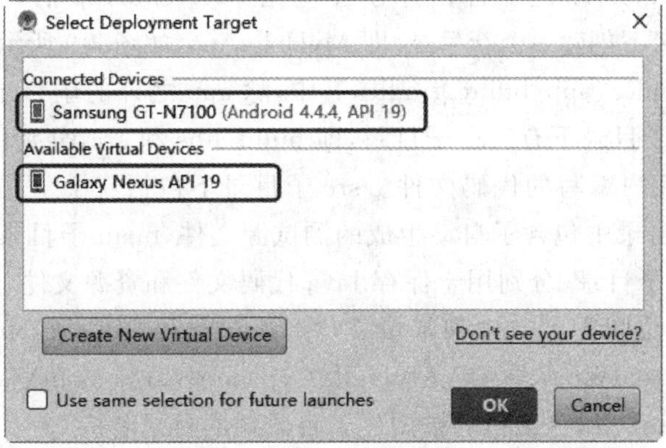

图 3-6 选择目标运行平台

在图 3-7 中,模拟器 Galaxy Nexus API 19 中央显示一行文字"Hello World!",显示窗口标题名为 MyHelloApp,即应用名。因此,只需要按照 Android 新建工程向导就可以生成一个显示"Hello World!"的工程,无须程序员输入一句代码。

在图 3-5 中,Build 菜单如图 3-8 所示。

图 3-7 应用 MyHelloApp 运行结果

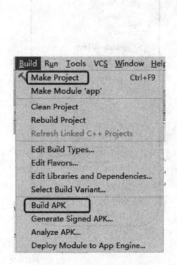

图 3-8 Build 菜单结构

在图 3-8 中,常用 Make Project 编译整个工程,用 Clean Project 清除生成的目标文件和中间目标文件,用 Build APK 生成将应用安装到智能手机上的.apk 文件。应用 MyHelloApp 生成的.apk 文件位于目录"D:\MyAndroidWorkSpace\MyHelloApp\app\build\outputs\apk"下,文件名为 app-debug.apk,这里的 debug 表示调试版本。.apk 文件实际

上是压缩包文件,可以使用 WinRAR 等软件解压,解压后可以看到 classes.dex 可执行文件。

执行 Android Studio 新建工程向导创建应用 MyHelloApp 后,将在工作目录"D:\MyAndroidWorkSpace"下创建一个新目录,即 MyHelloApp,如图 3-9 所示。目录 MyHelloApp 下有子目录.gradle、.idea、app、build、gradle,其中,除 app 之外的子目录与工程编译信息和版本信息有关。app 子目录下有三个子目录,即 build、libs 和 src,分别用于保存应用的生成目标文件信息、库和用户编写的代码文件。src 子目录下包括三个子目录,即 androidTest、test 和 main,前两个目录中包含了自动生成的测试源文件,main 子目录结构如图 3-9 所示,包括 java 和 res 两个子目录,分别用于保存 Java 代码文件和资源文件,其中 res 资源文件包括各种分辨率大小的位图图像文件和常量文件。Java 代码文件 MyHelloMainAct.java 位于目录"D:\MyAndroidWorkSpace\MyHelloApp\app\src\main\java\cn\edu\jxufe\zhangyong\myhelloapp"下,可见包名被用作子目录树的目录名。

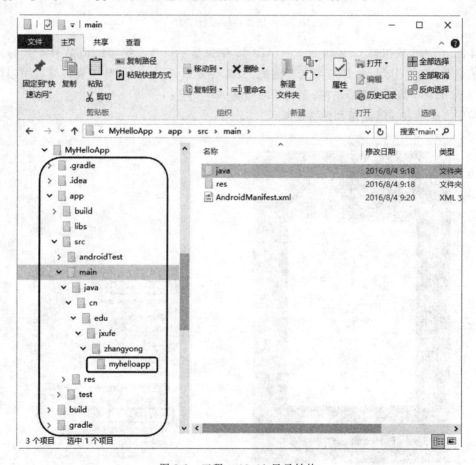

图 3-9 工程 ex03_01 目录结构

3.2 Hello World 应用工作原理

对于传统的面向过程 C 语言程序,其程序执行入口为 main 函数,程序的运行就是 main 函数中各条语句按语法规则依次执行。对汇编语言程序设计熟悉的程序员,还有那些做过

DSP 或单片机开发的程序员,对程序的运行有着更深刻的认识。CPU(中央处理单元)总是执行其 PC(程序计数器寄存器)指针指向的程序空间地址处的指令代码,对于分支和跳转程序,需要修改 PC 的值。对于汇编语言程序设计而言,程序的执行过程非常清晰,程序员负责所有的资源调用,并安排程序指令的执行。汇编语言与程序实现的算法和功能的细节直接相关,开发这类程序需要专业开发人员,难度相对较大。高级语言是接近自然语言的一种程序设计语言,当基于这种高级语言的应用程序设计越简单,那么这类应用程序的执行过程越不容易理解。Android 应用程序基于 Java 高级语言和 Dalvik 虚拟机,它的执行过程比传统的 Java 程序(例如第 2 章的程序)更难理解,这时往往需要按 Android 应用程序设计的规定方法进行程序设计,即按照 Android 框架进行程序设计,并理解其程序的执行过程。事实上,不可能做到像理解汇编语言程序的执行过程一样理解 Android 应用程序的执行过程。

网址 android.xsoftlab.net 是 Android 开发者的镜像网址,可以不用代理服务器直接访问,登录到网址"http://android.xsoftlab.net/reference/android/app/Application.html",如图 3-10 所示(在 Android SDK 安装目录的 docs 子目录下,这里为 D:\Android\SDK\docs 下,打开 index.html,也可进入 Android 开发者参考手册,这些内容需要在 Android Studio 环境下下载 Documentation for Android SDK)。

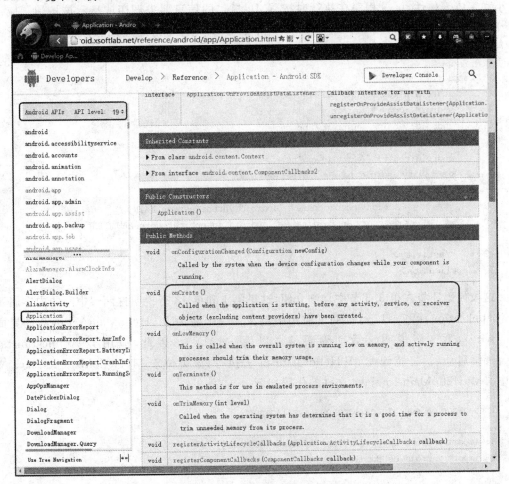

图 3-10 Android 开发者参考手册

从图 3-10 中找到所使用的 API 版本号，这里为 19，找到包 android.app，再定位到 Application 类，最后定位到 Application 类的公有方法 onCreate，该方法是应用程序启动时首先执行的方法，即"Called when the application is starting, before any activity, service, or receiver objects (excluding content providers) have been created."（当应用程序要执行时被调用，其他的应用程序对象都还没有创建）。从第 2 章的学习知道，一个类的公有方法的调用有两种途径，即被该类的实例调用或被该类中的方法调用，因此，可以推断该方法将被 Application 类中的某个静态方法（可能就是 mian 方法）调用。所以，从 Android 开发者手册可知，Android 程序的执行入口是 Application 类的 onCreate 方法。需要强调指出的是，只有熟练地检索 Android 开发者手册，才能成为真正的 Android 应用程序员。

对于显示"Hello World"的 Android 应用程序工程 MyHelloApp 来说，其程序执行入口也是 Application 类的 OnCreate 方法。在 MyHelloApp 应用中有一个文件 AndroidManifest.xml，该文件内容是 XML 语言的程序。XML 是 eXtensible Markup Language 的缩写，译为可扩充标记语言，这类语言不像 C 或 Java 语言等可执行高级语言，XML 语言是不可执行的，它的优点在格式统一且固定，用于为其他高级语言指定（或标记、配置）其数据或资源的位置和相互关系，例如，Java 语言界面中需要使用一个按钮，可以使用 XML 语言为按钮指定名称、大小和字体等属性，Java 程序在执行时按 XML 提供的信息要求调用或显示该按钮。应用 MyHelloApp 的文件 AndroidManifest.xml 中通过以下语句指定应用程序启动后执行的 Activity：

```
1   <application
2       android:allowBackup = "true"
3       android:icon = "@mipmap/ic_launcher"
4       android:label = "@string/app_name"
5       android:supportsRtl = "true"
6       android:theme = "@style/AppTheme">
7       <activity android:name = ".MyHelloMainAct">
8           <intent-filter>
9               <action android:name = "android.intent.action.MAIN" />
10              <category android:name = "android.intent.category.LAUNCHER" />
11          </intent-filter>
12      </activity>
13  </application>
```

这段代码还将在第 3.3 节中详细介绍，这里第 1 和第 13 行说明这两行中间的内容属于应用 application，第 7～12 行说明了应用中包含了一个名为 MyHelloMainAct 的活动界面（Activity），第 9 行说明该活动界面是主界面（MAIN）。因此，应用类 Application 的 onCreate 启动后，将转到类 MyHelloMainAct 去执行，该类位于 MyHelloMainAct.java 文件中。MyHelloMainAct.java 文件的代码如下：

```
1   package cn.edu.jxufe.zhangyong.myhelloapp;
2
3   import android.support.v7.app.AppCompatActivity;
4   import android.os.Bundle;
5
6   public class MyHelloMainAct extends AppCompatActivity {
```

```
 7        @Override
 8        protected void onCreate(Bundle savedInstanceState) {
 9            super.onCreate(savedInstanceState);
10            setContentView(R.layout.activity_my_hello_main);
11        }
12    }
```

上述代码中,第 6 行定义了 MyHelloMainAct 类,该类继承了类 AppCompatActivity,类 MyHelloMainAct 中仅有一个方法,即 onCreate 方法,该方法覆盖了其父类 AppCompatActivity 的同名方法。由于类 MyHelloMainAct 的 onCreate 方法是公有方法,因此,要想调用该方法,必须先创建 MyHelloMainAct 类的实例(对象),但是,这一步工作由 Android 框架帮程序员完成了,即 Android 框架帮程序员定义了类 MyHelloMainAct 的实例(对象)并通过其对象调用了 onCreate 方法。所以,应用 MyHelloApp 从开始执行到执行到类 MyHelloMainAct 的 onCreate 方法之前所做的工作都被 Android 框架隐藏了,程序员会误认为应用 MyHelloApp 最开始执行的是第 8 行的 onCreate 方法。

根据上面的解释可知,应用 MyHelloApp 启动后,Android 操作系统框架层会引导它执行到类 MyHelloMainAct 的 onCreate 方法。假设 Android 框架层为类 MyHelloMainAct 创建的对象名为 myHelloMainObj,整个应用 MyHelloApp 启动后成为 Android 系统的一个进程(Android 系统中每个应用程序对应着一个进程),该进程通过对象 myHelloMainObj 执行其 onCreate 方法,第 9 行执行其父类的方法 onCreate 初始化对象 myHelloMainObj 继承的父类 AppCompatActivity 中的公有和保护数据。第 10 行调用 setContentView 方法使用布局 R.layout.activity_my_hello_main 设置活动界面内容,这里方法 setContentView 是类 AppCompatActivity 的公有方法,可以直接被其子类 MyHelloMainAct 的对象调用。类 AppCompatActivity 中有三种重载形式的 setContentView 方法,这里调用的方法原型为

```
    public void setContentView(int layoutResID);
```

参数 layoutResID 被设置为 R.layout.activity_my_hello_main(第 10 行),这里 R 表示资源类,layout 为 R 的一个内部数,activity_my_hello_main 为 layout 类的公有静态整型常量,指向同名的布局文件 activity_my_hello_main.xml,例如:

```
1    public static final class layout {
2        public static final int activity_my_hello_main = 0x7f030000;
3    }
```

类 layout 为公有静态类,且用 final 修饰,表示该类不能派生新类,该类中有一个公有静态整型常量 activity_my_hello_main(第 2 行),这里的 activity_my_hello_main 对应于图 3-11 中资源 res 中的 activity_my_hello_main.xml 文件,即这里的 R.layout.activity_my_hello_main 是 activity_my_hello_main.xml 的标识符,所以语句 setContentView(R.layout.activity_my_hello_main)将设置活动界面的内容为 activity_my_hello_main.xml。

在图 3-11 中,文件 activity_my_hello_main.xml 的内容如下:

```
1    <?xml version="1.0" encoding="utf-8"?>
2    <android.support.constraint.ConstraintLayout xmlns:android="http://schemas.android.
     com/apk/res/android"
```

```
3      xmlns:app = "http://schemas.android.com/apk/res-auto"
4      xmlns:tools = "http://schemas.android.com/tools"
5      android:id = "@+id/activity_my_hello_main"
6      android:layout_width = "match_parent"
7      android:layout_height = "match_parent"
8      tools:context = "cn.edu.jxufe.zhangyong.myhelloapp.MyHelloMainAct">
9
10     <TextView
11         android:layout_width = "wrap_content"
12         android:layout_height = "wrap_content"
13         android:text = "Hello World!"
14         app:layout_constraintBottom_toBottomOf = "@+id/activity_my_hello_main"
15         app:layout_constraintLeft_toLeftOf = "@+id/activity_my_hello_main"
16         app:layout_constraintRight_toRightOf = "@+id/activity_my_hello_main"
17         app:layout_constraintTop_toTopOf = "@+id/activity_my_hello_main" />
18
19 </android.support.constraint.ConstraintLayout>
```

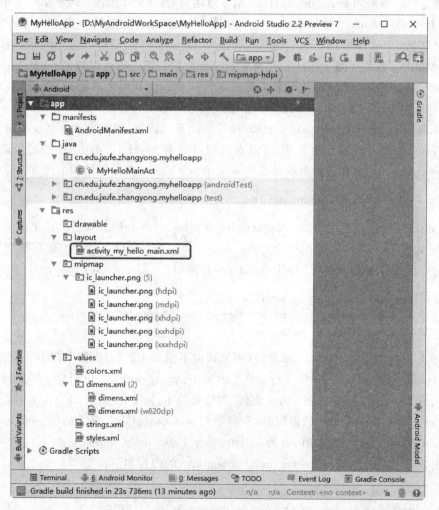

图 3-11　MyHelloApp 应用

上述代码是 XML 格式,第 2 行说明布局为 ConstraintLayout 布局(该布局是 Android 2.3 后续版本才支持的布局),第 4 章将深入介绍 ConstraintLayout 布局,而且将重点使用这种布方式。第 5 行为布局文件名称对应的标识号;第 6、7 行表示布局页面的宽和高均匹配其父视图的大小。第 10~17 行表示这是一个静态文本框(或文本视图,TextView),其宽度与高度由其包含的文本决定,第 13 行表示静态文本框的文本字符串内容为"Hello World!",即显示在图 3-7 中央的文字;第 14~17 行为静态文本框的位置,表示居中显示。需要说明的是,第 5 行指定 layout 的标识符为 activity_my_hello_main,而第 14~17 行中引用该标识符,表示静态文本框的位置居中。

现在通过执行对象 myHelloMainObj 的方法 onCreate 已经把显示内容准备好了,但是该活动界面还没有显示出来,Android 框架还会自动通过对象 myHelloMainObj 调用 onStart 方法,该方法在应用 MyHelloApp 中没有体现,调用 onStart 方法后,应用 MyHelloApp 的界面就显示出来了;然后,Android 框架还要自动通过对象 myHelloMainObj 调用 onResume 方法,该方法调用后,应用 MyHelloApp 的活动界面将得到焦点,成为显示界面的"前台"界面。这些方法 onStart、onResume 还有 onCreate 方法,都是类 Activity 的保护方法,将被其子类(这里是 MyHelloMainAct 类,即 AppCompatActivity 是 Activity 的子类,而 MyHelloMainAct 类是 AppCompatActivity 的子类)继承到,因此,可被其子类对象 myHelloMainObj 调用。

经过上述的过程之后,才能得到图 3-7 所示的界面。一旦该程序运行,将无法被人工关闭,用户可以通过单击模拟器的 Home 按钮回到欢迎界面,或者在图 3-12 中单击 MyHelloApp 应用程序图标重新进入"Hello World"界面。总之,该应用程序无法关闭。Android 系统会自动管理那些没有显示(或没有使用)的应用程序,根据内存和应用程序的使用情况自动关闭那些不再被使用的应用程序,同时释放它们占用的内存空间。

综上所述,对于 Android 程序员来说,只需要明白创建的用户活动界面类继承了类 AppCompatActivity,当应用程序开始执行时,将自动执行其 onCreate 方法,程序员需要在 onCreate 方法中添加界面初始化的代码,然后将应用程序交给 Android 系统管理。对于应用 MyHelloApp 而言,在类 MyHelloMainAct 的 onCreate 方法中启动了包含字符串"Hello World!"的显示界面,即完成了显示"Hello World"的任务。同时,Android 应用程序中需要的各种显示元素均以资源的形式保存,并借助 XML 语言进行管理,而且,整个 Android 应用都需要借助 AndroidManifest.xml 进行配置。

图 3-12 应用程序图标

结合图 3-11,应用 MyHelloApp 中整个工程的配置文件为 AndroidManifest.xml,这个文件只能有一个。位于资源 res 的 layout 下的 XML 文件均为布局文件,应用 MyHelloApp 中只有一个布局文件,即 activity_my_hello_main.xml,如果有多个 Activity,则对应地有多个布局文件。在资源 res 的 values 下的文件为常量定义文件,可以有多个这类文件,例如,应用 MyHelloApp 中含有一个字符串资源文件,即 strings.xml。所有 XML 文件内容均可以图形化地创建,例如,可借助图形化方式创建常量字符串资源 strings.xml 文件。strings.xml 文件的内容如下:

```
1   < resources >
2       < string name = "app_name">MyHelloApp</string>
3   </resources>
```

这里,第 1 行与第 3 行为资源标识符,第 2 行表示字符串常量名为 app_name,字符串内容为 MyHelloApp。

当应用 MyHelloApp 中没有文件 strings.xml 和 activity_my_hello_main.xml 等资源文件时,完全依赖 Java 语言代码也可以实现显示"Hello World"的程序功能。只是因为 Android 系统支持国际化语言,使用资源文件可以方便程序工作在各种语言环境下,所以,建议与显示界面相关的元素全部使用资源描述。实际上,这是这些资源文件的价值所在。

3.3 应用程序框架

Android 应用程序框架是指借助 Android Studio 进行 Android 应用程序设计时,自动生成的应用程序目录及文件结构,大约有 1100 个文件和 500 个子目录,程序员基于应用程序框架,添加文件和代码可实现相应的功能。可以把"Hello World!"工程视为 Android 应用程序框架,更广义的应用程序框架是指当用户应用程序设计完成后,所有 Android 自动生成的代码、文件和目录均被视为 Android 应用程序框架,或者说,除去用户输入的 Java 代码程序文件,其他所有文件都属于 Android 应用程序框架。在第 3.1 节、3.2 节的应用 MyHelloApp 中,阐述了 Hello World 工程的创建过程和执行情况,这里将详细介绍应用程序框架中各个文件的内容及其作用。

3.3.1 应用程序框架基本组成

例 3-2 应用程序框架实例。

新建应用 MyFrameApp,即应用名为 MyFrameApp,公司域名为 zhangyong.jxufe.edu.cn,即包名为 cn.edu.jxufe.zhangyong.myframeapp,保存目录为 D:\MyAndroidWorkSpace\MyFrameApp,使用 SDK 版本号为 KitKat,API 版本号为 19,选择空的 Activity,活动界面(Activity)名为 MyFrameMainAct,对应的布局为 activity_my_frame_main。应用 MyFrameApp 的结构如图 3-13 所示。

图 3-13 中,程序员只需要关注被圈住的内容,而在圈外面的 Gradle Scripts 是调试、优化、库和编译配置信息,位于 java 分组下的带有"(androidTest)"和"(test)"后缀的 cn.edu.jxufe.zhangyong.myframeapp 分组下的文件是测试文件,这些文件均为自动生成的 ASCII 文件,可以打开浏览其内容,但一般不需要程序员干预。AndroidManifest.xml 是 Android

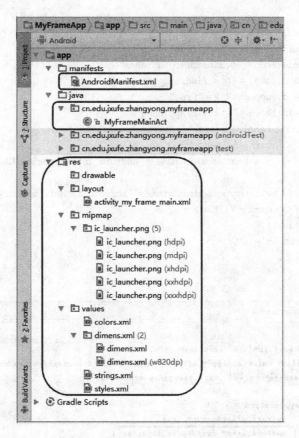

图 3-13 应用 MyFrameApp 工程结构

工程的全局配置文件。java 分组下的 cn.edu.jxufe.zhangyong.myframeapp 用于存放程序员编写的 Java 源程序,目前只有一个源文件 MyFrameMainAct.java。res 分组下包含了项目所有可能使用的资源文件,例如图形文件、布局文件、常量文件等。

图 3-13 中的工程浏览器中,分组树(或称目录树)中的小图标各具含义,例如,包用图标 ▣ 表示,类用图标 ⓒ 表示,方法用图标 ⓜ 表示,资源用图标 ▤ 表示,目录用图标 ▭ 表示等等,认识这些图标对应的文件类型对了解工程结构有很大帮助。

3.3.2 Android 配置文件 AndroidManifest.xml

在图 3-13 中双击 AndroidManifest.xml 文件,弹出图 3-14 所示的界面。AndroidManifest.xml 有两种显示方式,即文本显示方式和合并同类属性的显示方式,如图 3-14 状态栏中所示,其中标签 Text 为文本显示方式。由于应用 MyFrameApp 中只有一个活动界面,故图 3-14 中只有一个 Activity,即 MyFrameMainAct,当需要显示多个界面时,必须将新的 Activity 添加到 AndroidManifest.xml 文件中。

AndroidManifest.xml 文件的内容如下:

```
1  <?xml version = "1.0" encoding = "utf-8"?>
2  <manifest xmlns:android = "http://schemas.android.com/apk/res/android"
3      package = "cn.edu.jxufe.zhangyong.myframeapp">
```

```
 4
 5     <application
 6         android:allowBackup = "true"
 7         android:icon = "@mipmap/ic_launcher"
 8         android:label = "@string/app_name"
 9         android:supportsRtl = "true"
10         android:theme = "@style/AppTheme">
11         <activity android:name = ".MyFrameMainAct">
12             <intent - filter>
13                 <action android:name = "android.intent.action.MAIN" />
14
15                 <category android:name = "android.intent.category.LAUNCHER" />
16             </intent - filter>
17         </activity>
18     </application>
19
20 </manifest>
```

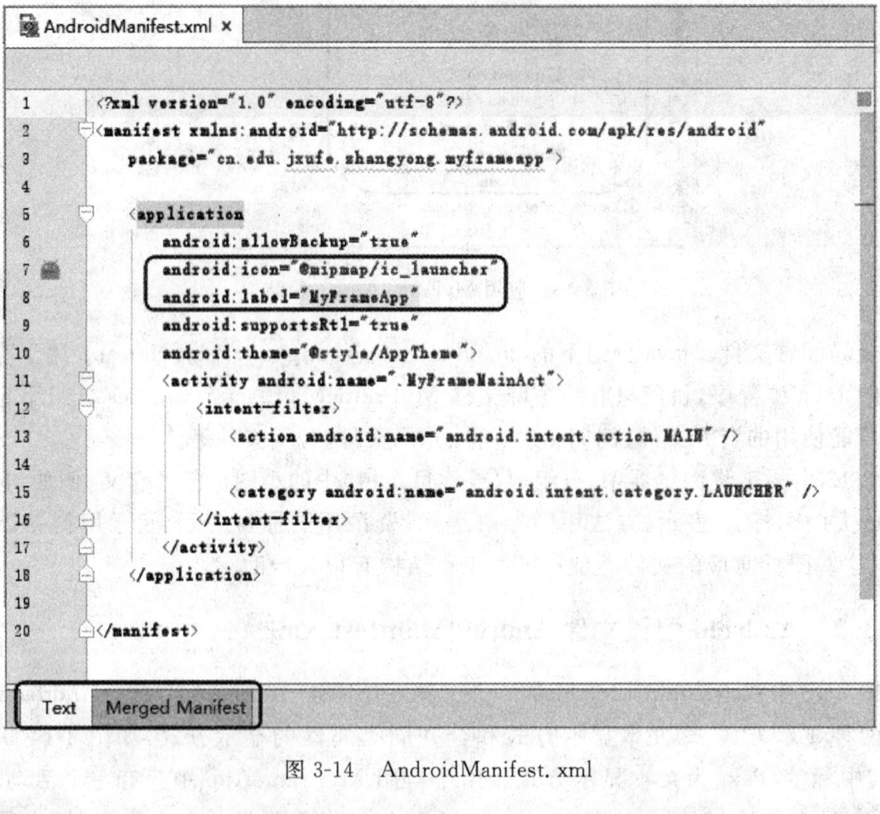

图 3-14 AndroidManifest.xml

上述 XML 代码中,第 1 行说明 XML 版本为 1.0,采用 UTF-8 编码方法,这种编码方法中,如果某个字符在 Unicode 码中的值小于 256,则使用一个字节存储,这样有效地节约了存储空间。第 2 行的 manifest 与第 20 行的/manifest 配对,表示这两行之间为配置内容,配置字段名使用":"连接,例如 xmlns:android 和 android:allowBackup 等。第 2 行的 xmlns:android=http://schemas.android.com/apk/res/android 指定工程中使用的 Android 程序包,这一句不能改动。第 3 行指定应用程序的包名为 cn.edu.jxufe.zhangyong。

myframeapp。第 7 行的 android:icon 指定应用程序图标为资源文件"@mipmap/ic_launcher",即图 3-13 中的 mipmap 分组下的文件 ic_laucher.png,多个同名的.png 文件的分辨率不同,保存在不同的目录下,一般地,mdpi、hdpi、xhdpi、xxhdpi 和 xxxdhdpi 的分辨率依次为 48×48、72×72、96×96、144×144 和 192×192,Android 系统将根据智能终端的显示屏分辨率自动选择合适的图标,因此,这些图标文件名是相同的,保存在硬盘的不同目录下,如图 3-15 所示。默认的 ic_launcher.png 是绿色小机器人,如图 3-12 所示。

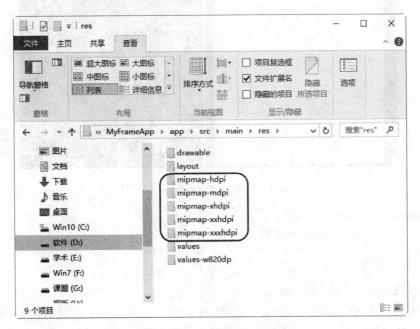

图 3-15　不同分辨率的同名图标的存储目录结构

第 8 行的 android:label 指应用程序的名称(这个名称可作为出现在应用程序图标下面的字符串,见图 3-12),"@string/app_name"表示字符串资源文件 strings.xml 中的名称为 app_name 对应的常量字符串,strings.xml 文件的内容如下:

```
1    <resources>
2        <string name="app_name">框架实例</string>
3    </resources>
```

第 2 行将 app_name 由原来的"MyFrameApp"修改为"框架实例"。上述所说的"@string/app_name"就是字符串"框架实例",将被用作应用的图标名称。现在运行 MyFrameApp 应用,其运行结果如图 3-16 所示。

但是,往往不希望改变 app_name,而是修改文件 AndroidManifest.xml 的第 11 行如下:

```
<activity android:name=".MyFrameMainAct" android:label="@string/mainact_name">
```

表示 activity 的标题为 mainact_name(这个标题同时作为应用程序图标下的字符串,见图 3-16),然后,修改 strings.xml 如下:

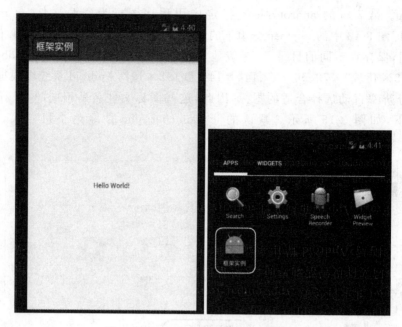

图 3-16 应用 MyFrameApp 及其图标

```
1  < resources >
2      < string name = "app_name">MyFrameApp</string >
3      < string name = "mainact_name">框架实例</string >
4  </resources >
```

运行结果如图 3-16 所示。

现在回到文件 AndroidManifest.xml,第 11~17 行为 Activity 的说明代码,第 11 行说明 Activity 的名称为 MyFrameMainAct,即类名;第 12~16 行为 Intent-filter,Intent 译为意图,用于管理 Activity 的数据,Intent-filter 说明 Activity 的执行方式(或称动作);第 13 行指定动作(action)的名称为"MAIN",第 15 行指定类属名称为"LAUNCHER",表示名称为 MyFrameAct 的 Activity 将作为主窗口启动。

在指定应用程序名或活动界面标题时,除了可以使用"@string/app_name"从字符串资源文件中查找字符串外,还可以直接指定字符串,例如前述 AndroidManifest.xml 文件代码的第 8 行可写为形式"android:label="MyFrameApp ""而不影响应用的执行情况。

3.3.3 Android 资源文件

如图 3-13 所示,新建应用 MyFrameApp 时 Android Studio 自动创建的资源文件分为 4 类,即 drawable(存放图像)、layout(布局)、mipmap(多分辨率图像和图标)和 values(值)。布局将在第 4 章详细介绍,mipmap 在上一小节已经介绍过,这里重点介绍 values 和 drawable 类型的资源文件应用方法。

此时的应用 MyFrameApp 工作界面如图 3-17 所示。

在图 3-17 中,单击菜单 File|New|XML|Values XML File,弹出图 3-18 所示的界面。

在图 3-18 中,单击 Finish 按钮,则文件 strings_hz.xml 自动添加到工程管理器的 res|values 分组下,如图 3-19 所示。

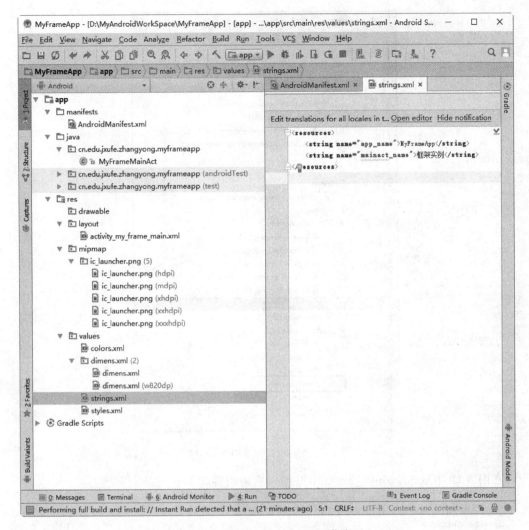

图 3-17 应用 MyFrameApp 工作窗口

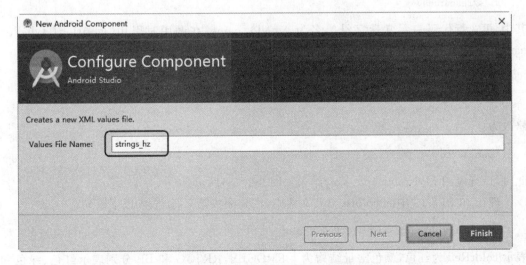

图 3-18 新建常量资源文件

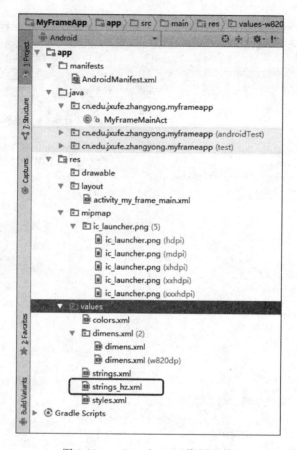

图 3-19　strings_hz.xml 资源文件

在图 3-19 中双击 strings_hz.xml,并编辑它的内容如下:

```
1    <?xml version = "1.0" encoding = "utf-8"?>
2    <resources>
3        <string name = "tvhello_name">爱护我们共同的家园——地球!/string>
4    </resources>
```

其中,第 3 行代码表示常量字符串名为 tvhello_name,在调用时引用"@string/ tvhello_name",其字符串值(即字符串内容)为"爱护我们共同的家园——地球!"。

然后,打开布局文件 activity_my_frame_main.xml,将其中的语句

"android:text = "Hello World!""

修改为:

"android:text = "@string/tvhello_name""

此时,运行应用 MyFrameApp,运行结果如图 3-20(a)所示。

现在,在图 3-19 中的 colors.xml 文件中(在其倒数第 2 行)插入以下语句:

<color name = "colorRed">♯FF0000</color>

表示 colorRed 为红色,颜色常量结构为♯RRGGBB,RR、GG 和 BB 分别表示红色、绿色和

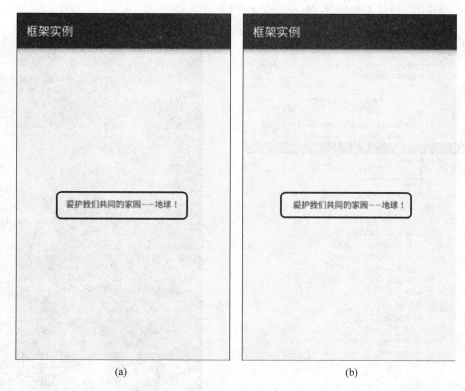

图 3-20 应用 MyFrameApp 运行结果

蓝色分量的大小,用十六进制数表示,取值均为 0x00～0xFF,例如♯FF0000 表示红色,♯00FF00 表示绿色,♯0000FF 表示蓝色。

接着在布局文件 activity_my_frame_main.xml 中的第 14 行插入以下语句:

android:textColor = "@color/colorRed"

然后,执行应用 MyFrameApp,运行结果如图 3-20(b)所示,文本以红色显示。

回到图 3-19,图形文件可以存放在 drawable 下,也可以存放在 mipmap 下,事实上,可以在 drawable 下创建子目录 drawable-hdpi、drawable-mdpi 和 drawable-ldpi,分别用于存放高分辨率、中分辨率和低分辨率图形文件,与 mipmap 类似将文件名相同的不同分辨率的图像文件存放在不同的目录下。由于 mipmap 下可存放各种分辨率的文件,这里建议 drawable 下存放任意分辨率的图像文件,Android 系统会根据显示区域对图像进行自动缩放处理。

从网上下载了一幅地球的图像,命名为 omearth.jpg(这里故意使用了 1280×720 分辨率的图像,其实可以为任意分辨率),将其保存在目录"D:\MyAndroidWorkSpace\MyFrameApp\app\src\main\res\drawable"下,则 Android Studio 自动将 omearth.jpg 添加到工程管理器中,如图 3-21 所示。

现在,修改布局文件 activity_my_frame_main.xml,在其第 8 行插入以下代码:

android:background = "@drawable/omearth"

表示使用 drawable 类型的资源图像 omearth.jpg 作为 activity 的背影。此时执行应用

MyFrameApp，运行结果如图 3-22 所示。

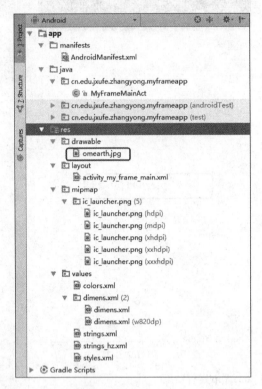

图 3-21　添加了图像 omearth.jpg 的 drawable 资源

图 3-22　应用 MyFrameApp 运行结果

现在回到图 3-21，从图 3-21 中可见工程管理器的资源 res|values 中还有 styles.xml 和 dimens.xml 文件，前者用于指定 activity 或各种控制的样式风格，后者用于指定 activity 显示区域的边界宽度等。样式 styles.xml 文件可由布局文件自动生成，并且样式风格可以继承。layout 分组下为布局文件，应用 MyFrameApp 中只有一个 Activity，即只有一个活动界面，所以，这里只有一个 activity_my_frame_main.xml 文件。当工程中有多个 Activity 时，需要在 layout 下添加多个布局文件。

通过熟练和巧妙地使用资源文件，可有效地减少工程中的 Java 代码量，并且增强程序的层次性和可维护性。新建的资源文件均会自动存放在 res 分组下的相应子分组中。

3.3.4　Android 源程序文件

参考图 3-21，可知应用 MyFrameApp 中只有一个程序文件，即 MyFrameMainAct.java，它包含一个公有类 MyFrameMainAct，该类位于包 cn.edu.jxufe.zhangyong.myframeapp 中，与程序文件同名，其中的方法为 onCreate（返回值为 void）。文件 MyFrameMainAct.java 的代码如下：

```
1    package cn.edu.jxufe.zhangyong.myframeapp;
2
3    import android.support.v7.app.AppCompatActivity;
4    import android.os.Bundle;
5
```

```
6   public class MyFrameMainAct extends AppCompatActivity {
7
8       @Override
9       protected void onCreate(Bundle savedInstanceState) {
10          super.onCreate(savedInstanceState);
11          setContentView(R.layout.activity_my_frame_main);
12      }
13  }
```

上述代码是创建新应用工程向导自动生成的。第 6 行说明类 MyFrameMainAct 继承了类 AppCompatActivity；第 9～12 行的 onCreate 方法只有在该 Activity 第一次执行时被调用，在其执行过程中不会被再次调用，因此，onCreate 方法一般用于初始化应用程序的数据和界面。第 9 行中显示 onCreate 方法具有一个 Bundle 型参量，Bundle 类继承了 java.lang.Oject 类并实现了接口 android.os.Parcelable，因此，Bundle 类的实例（对象）可用于保存和恢复其他类的实例（对象）。如果有一个 MyFramceMainAct 类的实例正处于休眠态中（即由于其他活动界面处于执行状态而被停止运行），那么处于休眠态的实例对象将被保存在参量 savedInstanceState 中；当再次执行类 MyFrameMainAct 的实例时，即调用 onCreate 方法时，原来的对象实例将通过参量 savedInstanceState 传递；如果原来的休眠态实例不存在，则参量 savedInstanceState 为空。显然，第一次创建类 MyFrameMainAct 的实例时，参量 savedInstanceState 为空；当该实例执行过程中，由于其他应用程序的执行，而被迫停止时，则将类 MyFrameMainAct 的实例保存在参量 savedInstanceState 中；当类 MyFrameAct 的实例需要再次执行时，它原来的对象实例将通过参量 savedInstanceState 传递（这部分内容可参考第 3.4 节关于 Activity 的生命周期的详细说明）。

第 10 行借助关键字 super 调用父类 Activity 的 onCreate 方法；第 11 行调用方法 setContentView（继承自父类 AppCompatActivity 的公有方法），根据资源的 ID 号（索引号）调用相应的布局文件设置活动界面，这里是调用 activity_my_frame_main.xml 布局文件设置活动界面。随后 Android 操作系统会管理 MyFrameMainAct 应用程序界面的显示和运行。

应用 MyFrameApp 的布局文件 activity_my_frame_main.xml 的代码如下：

```
1   <?xml version = "1.0" encoding = "utf-8"?>
2   <android.support.constraint.ConstraintLayout xmlns:android = "http://schemas.android.com/apk/res/android"
3       xmlns:app = "http://schemas.android.com/apk/res-auto"
4       xmlns:tools = "http://schemas.android.com/tools"
5       android:id = "@+id/activity_my_frame_main"
6       android:layout_width = "match_parent"
7       android:layout_height = "match_parent"
8       android:background = "@drawable/omearth"
9       tools:context = "cn.edu.jxufe.zhangyong.myframeapp.MyFrameMainAct">
10
11      <TextView
12          android:layout_width = "wrap_content"
13          android:layout_height = "wrap_content"
14          android:text = "@string/tvhello_name"
```

```
15            android:textColor = "@color/colorRed"
16            app:layout_constraintBottom_toBottomOf = "@ + id/activity_my_frame_main"
17            app:layout_constraintLeft_toLeftOf = "@ + id/activity_my_frame_main"
18            app:layout_constraintRight_toRightOf = "@ + id/activity_my_frame_main"
19            app:layout_constraintTop_toTopOf = "@ + id/activity_my_frame_main" />
20
21    </android.support.constraint.ConstraintLayout>
```

该布局文件的内容在第 4 章还将深入介绍,这里第 11~19 行表示一个静态文本框控件 TextView,在第 16 行插入语句:

android:id = "@ + id/tvEarth"

表示该 TextView 的 ID 号为 tvEarth。可在 onCreate 方法中调用该控件的方法 setText 设置其显示的字符串。修改应用 MyFrameApp 的代码文件 MyFrameMainAct.java 代码如下:

```
1    package cn.edu.jxufe.zhangyong.myframeapp;
2
3    import android.graphics.Color;
4    import android.support.v7.app.AppCompatActivity;
5    import android.os.Bundle;
6    import android.widget.TextView;
7
8    public class MyFrameMainAct extends AppCompatActivity {
9        private static TextView textView;
10       @Override
11       protected void onCreate(Bundle savedInstanceState) {
12           super.onCreate(savedInstanceState);
13           setContentView(R.layout.activity_my_frame_main);
14
15           this.setTitle("爱护地球");
16
17           textView = (TextView)findViewById(R.id.tvEarth);
18           textView.setText("我们热爱大自然!");
19           textView.setTextColor(Color.RED);
20       }
21   }
```

上述代码中,在第 9 行中添加类 MyFrameMainAct 的私有静态文本框类对象成员 textView;在第 15 行调用 setTitle 方法设置活动界面的标题为"爱护地球";第 17 行通过 findViewById 方法由控件的资源 ID 号获得控件实例(对象);第 18 行调用 setText 方法设置 textView 对象的显示文字为"我们热爱大自然!";第 19 行调用 setTextColor 方法设置字体颜色为红色。此时执行应用 MyFrameApp,运行结果如图 3-23 所示。

对比图 3-22 和图 3-23,可知标题由原来的"框架实例"变为"爱护地球",视图中央显示的文字由原来的"爱护我们共同的家园——地球"变为"我们热爱大自然!"。

图 3-23　MyFrameApp 应用执行结果

3.4　Activity 生命周期

　　Activity 生命周期是指活动界面实例(对象)从创建到被 Android 操作系统关闭的整个生存周期,在这一过程中,Android 系统将依次自动调用 Activity 的 6 种方法,即 OnCreate、OnStart、OnResume、OnPause、OnStop 和 OnDestroy。这 6 种方法相对于 Android 系统而言,类似于 Activity 的 6 个"钩子"函数(钩子函数是指挂接在某个方法中的空函数,当扩展这个方法的功能时,只需要向其钩子函数中添加特定代码即可,而不用修改这个方法的语句)。Activity 的生命周期如图 3-24 所示,为力求表达准确无误,这里直接引用了 Android 开发者手册上的英文框图。

　　由图 3-24 可知,当某时刻 Activity 启动了,那么其中的方法 onCreate 首先得到执行,程序员可以在该方法中添加初始化 Activity 的代码,此时 Activity 界面仍然不可见;然后方法 onStart 得到执行,一般可在该方法中添加一些服务,该方法执行后,Activity 界面才可见;之后,执行 onResume 方法,使 Activity 获得显示焦点而处于与用户交互的状态,即"Activity is running"。当另一个 Activity 由于系统调度原因在前台显示而获得焦点时,即"Another activity comes in front of the activity",该 Activity 将执行 onPause 方法,暂停当前 Activity 的执行,onPause 方法中不宜添加大量代码,因为只有 onPuase 方法执行完成后,另一个 Activity 的 Resume 方法才能执行。当该 Activity 不可见时,将调用 onStop 方法停止该 Activity。在该 Activity 被彻底关闭(指在内存中也不存在了)前,将调用 onDestroy 方法,之后,该 Activity 生命周期结束,即"Activity is shut down"。

　　因此,onCreate、onStart 和 onResume 方法是主动式执行,而 onPause、onStop 和

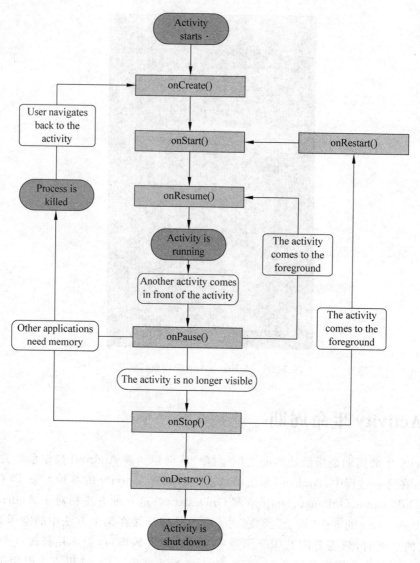

图 3-24　Activity 生命周期

onDestroy 方法是被动式执行，即调用 onCreate 方法后，Activity 在内存中创建；调用 onStart 方法后，Activity 可见（不可用）；调用 onResume 方法后，Activity 可见可用。当 Activity 不可用时，调用 onPause 方法；当 Activity 不可见时，调用 onStop 方法；当 Activity 要从内存中删除时，调用 onDestroy 方法。

Android 应用程序往往具有多个 Activity，当一个 Activity 界面（记为 A）在可用状态下，弹出了对话框 Activity 界面（记为 B），B 界面处于获得焦点状态，但 B 界面没有完全覆盖 A 界面；然后 B 界面关闭，A 界面又获得焦点。对于 A 而言这一处理过程为：A 的 onResume 方法执行使 A 可用可见；A 可见不可用使得 A 的 onPause 方法执行；再次执行 A 的 onResume 方法使 A 可用可见。如图 3-24 中部的循环：onResume→Activity is running→Another activity comes in front of the activity→onPause→The activity comes to the foreground→onResume。

当一个 Activity 界面（记为 A）在可用状态下，弹出了另一个 Activity 界面（记为 B），B 界面处于获得焦点状态，且 B 界面完全覆盖 A 界面；然后 B 界面关闭，A 界面又获得焦点。对于 A 而言这一处理过程为：A 的 onStart 和 onResume 方法执行使 A 可用可见；A 不可用不可见使得 A 的 onPause 和 onStop 方法执行；执行 Restart 方法，然后再次执行 A 的 onStart 和 onResume 方法使 A 可用可见。如图 3-24 右部的循环：onStart→onResume→Activity is running→Another activity comes in front of the activity→onPause→The activity is no longer visible→onStop→The activity comes to the foreground→onRestart→onStart。

当一个 Activity 界面（记为 A）在可用状态下，弹出了另一个 Activity 界面（记为 B），B 界面处于获得焦点状态，但 B 界面没有完全覆盖 A 界面；此时由于 Android 系统需要内存，把 A 运行时占用的内存（即进程）清空了，但是创建 A 的数据结构的内存仍然存在。然后 B 界面关闭，A 界面又再次获得焦点。对于 A 而言这一处理过程为：A 的 onCreate、onStart 和 onResume 方法执行使 A 可用可见；A 不可用使得 A 的 onPause 方法执行；执行 onCreate 方法，然后再次执行 A 的 onCreate、onStart 和 onResume 方法使 A 可用可见。如图 3-24 左部的循环：onCreate→onStart→onResume→Activity is running→Another activity comes in front of the activity→onPause→Other applications need memory→Process is killed→User navigates back to the activity→onCreate。

最后一种情况为：当一个 Activity 界面（记为 A）在可用状态下，弹出了另一个 Activity 界面（记为 B），B 界面处于获得焦点状态，且 B 界面完全覆盖 A 界面；此时由于 Android 系统需要内存，把 A 运行时占用的内存（即进程）清空了，但是创建 A 的数据结构的内存仍然存在。然后 B 界面关闭，A 界面又再次获得焦点。对于 A 而言这一处理过程为：A 的 onCreate、onStart 和 onResume 方法执行使 A 可用可见；A 不可用不可见使得 A 的 onPause 和 onStop 方法执行；执行 onCreate 方法，然后再次执行 A 的 onCreate、onStart 和 onResume 方法使 A 可用可见。如图 3-24 左部的循环：onCreate→onStart→onResume→Activity is running→Another activity comes in front of the activity→onPause→The activity is no longer visible→onStop→Other applications need memory→Process is killed→User navigates back to the activity→onCreate。

下面通过例 3-3 说明图 3-24 中各个方法的调用情况。由于 onPause、onStop 和 onDestroy 方法是被动执行方法，特别是 onDestroy 方法，必须是 Activity 被关闭时才能调用。可以使用事件处理技术（详见第 4 章）调用 Activity 的 finish 方法关闭活动界面，此时会调用 onDestroy 方法，由于目前没有介绍 Android 事件处理技术，因此，例 3-3 没有显示 onDestroy 方法的调用。

例 3-3 Activity 生命周期中各方法的调用顺序。

新建应用 MyOnFrameApp，即应用名为 MyOnFrameApp，包名为 cn. edu. jxufe. zhangyong. myonframeapp，活动界面名为 MyOnFrameAct，SDK 版本为 KitKat。修改 MyOnFrameAct. java 文件内容如下：

```
1   package cn.edu.jxufe.zhangyong.myonframeapp;
2
3   import android.support.v7.app.AppCompatActivity;
```

```
4    import android.os.Bundle;
5    import android.util.Log;
6
7    public class MyOnFrameAct extends AppCompatActivity {
8        private static final String TAG = "MyOnFrameAct";
9        @Override
10       protected void onCreate(Bundle savedInstanceState) {
11           super.onCreate(savedInstanceState);
12           setContentView(R.layout.activity_my_on_frame);
13           Log.i(TAG,"onCreate-Method");
14       }
15       public void onStart(){
16           super.onStart();
17           Log.i(TAG,"onStart-Method");
18       }
19       public void onResume(){
20           super.onResume();
21           Log.i(TAG,"onResume-Method");
22       }
23       public void onRestart(){
24           super.onRestart();
25           Log.i(TAG,"onRestart-Method");
26       }
27       public void onDestroy(){
28           super.onDestroy();
29           Log.i(TAG,"onDestroy-Method");
30       }
31       public void onStop(){
32           super.onStop();
33           Log.i(TAG,"onStop-Method");
34       }
35       public void onPause(){
36           super.onPause();
37           Log.i(TAG,"onPause-Method");
38       }
39   }
```

上述代码中，在第8行添加常量字符串 TAG，其值为"MyOnFrameAct"。在 onCreate 方法中添加"Log.i(TAG,"onCreate-Method");"（第13行），android.util.Log 类直接继承自 java.lang.Object 类，Log 类不能派生子类，用于输出日志信息（log）。这里使用 Log 类的静态公有方法 i(String tag, String msg)，即使用类名 Log 调用，Log.i(TAG,"onCreate-Method") 输出日志信息"onCreate-Method"，日志名（tag）为 TAG 字符串，即"MyOnFrameAct"。

第15～18行为 onStart 方法，首先调用父类的 onStart 方法，然后输出日志信息；第19～22行为 onResume 方法，第23～26行为 onRestart 方法，第27～30行为 onDestroy 方法，第31～34行为 onStop 方法，第35～38行为 onPause 方法，所有这些方法均需要先调用父类的同名方法，然后调用 Log.i 方法输出日志信息。

所谓的日志信息可以在 Android Monitor 观测窗口中显示出来，如图3-25所示。

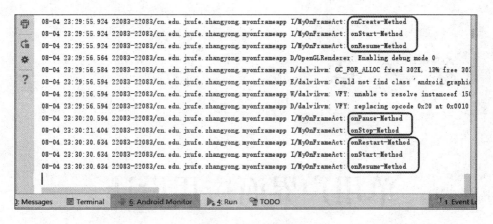

图 3-25 日志信息

执行应用 MyOnFrameApp，接着，在模拟器中单击 Home 按钮进入欢迎界面，然后，在模拟器的应用程序界面中单击 MyOnFrameAct 应用图标再次进入该程序，其 MyOnFrameAct 日志显示结果如图 3-25 所示。由图 3-25 的执行时间栏可知，当程序开始执行时，依次（几乎同时）执行 onCreate、onStart 和 onResume 方法，执行完 onResume 方法约 25 秒之后，操作模拟器进入到欢迎界面，此时依次执行 onPause 和 onStop 方法；然后又过了约 9 秒钟，操作模拟器重新运行 MyOnFrameAct 工程，则依次（几乎同时）执行 onRestart、onStart 和 onResume 方法。

了解 Activity 的生命周期的主要目的在于理解 Android 应用程序的执行过程。由于 Android 是多任务操作系统，其应用程序是面向事件而执行的，与传统的面向过程应用程序相比，其执行过程很难理解。传统的面向过程应用程序在其全部执行过程中独占 CPU 使用权，程序退出时释放 CPU 使用权。而 Android 应用程序一旦执行，就把 CPU 使用权交给 Android 系统，Android 系统综合调度和管理所有运行的应用程序。与其他嵌入式操作系统一样，Android 系统直接管理外部事件，根据外部事件的事件源和事件类型，将事件信息交给相应的应用程序处理。

3.5 本章小结

Android 应用程序框架使得编写 Android 应用程序更加容易，不需要输入代码即可生成一个显示"Hello World"的 Android 应用程序工程。Android 应用程序框架包括工程配置文件、资源文件和 Java 源程序文件等，合理使用资源文件可以增强 Android 工程的层次性，工程配置文件用于定义和管理工程中的活动界面、意图（Intent）、服务和内容提供者等程序元素，Java 源程序文件包含程序执行的动作。Android 应用程序启动后，其活动界面中的 onCreate 方法将首先被调用，因此，该方法一般用于初始化活动界面的数据和显示控件。活动界面的生命周期从执行 onCreate 方法开始，到执行 onDestroy 方法结束。此外，网址 http://android.xsoftlab.net/reference/packages.html 是国内镜像的 Android 开发者手册，在阅读本书过程中应经常查阅该手册，特别是本书中关于一些类和其方法的说明比较简略的地方，在 Android 开发者手册中有详细的介绍。

第 4 章 单用户界面应用设计

本章介绍只含有一个 Activity 的程序设计方法，即单用户界面应用设计。通过单用户界面应用设计，详细介绍 Android 应用程序界面布局的方法以及 Android 各种控件的使用方法。

4.1 Activity 概念

Android 应用程序由 4 个基本的模块组成，即 Activity（活动界面）、Intent（意图）、Content Provider（内容提供者）、Service（服务）。其中，Activity（译为活动或活动界面）被视为应用程序界面的"画布"，在 Activity 中布局并放置各种控件组成与用户交互的界面，Activity 管理可视化界面的所有控件；其余三个组成模块将依次在第 5~8 章介绍。

Activity 的继承关系为 java.lang.Object → android.content.Context → android.content.ContextWrapper → android.view.ContextThemeWrapper → android.app.Activity → android.support.v4.app.FragmentActivity→android.support.v7.app.AppCompatActivity。

上述为 AppCompatActivity 类的父类，Android 应用程序基于 Java 语言，Android 应用程序框架所有的类均直接或间接继承类 java.lang.Object。直接继承 Activity 类的子类有 ActivityGroup、ListActivity、ExpandableListActivity、NativeActivity、AccountAuthenticatorActivity 和 AliasActivity，间接继承 Activity 类的子类有类 LauncherActivity、PreferenceActivity 和 TabActivity。

当应用中只有一个 Activity 时，通过配置 AndroidManifest.xml 使得应用程序启动时自动启动该 Activity，并首先执行它的 onCreate 方法（见第 3.2 节）。当应用中有多个 Activity 时，在当前运行的 Activity 中调用 startActivity 或 startActivityForResult 方法启动一个新的 Activity，方法 startActivityForResult 将借助 onActivityResult 方法在两个

Activity 间传递数据，新的 Activity 必须在 AndroidManifest.xml 中声明。

　　Activity 类有 150 多种公有和保护方法，其中大部分方法属于必须掌握的常用方法，请读者参考 Android 开发者手册。用户定义的活动界面类均需要继承 Activity 类，因此，将继承 Activity 类的全部公有和保护方法，例如，可以使用公有方法 findViewById 通过控件的 ID 号找到控件实例(对象)、调用公有方法 setContentView 设置用户界面布局或控件等。由于 Android 系统管理用户界面类的实例(对象)，程序员只需要针对用户界面类进行界面设计即可，即添加布局(默认为约束布局)和控件(视图)，通过布局文件 activity_my_frame_main.xml 等，这一用户界面设计过程基本是所见即所得的。Android Studio 集成开发环境具有优秀的图形化布局功能，第 4.2.1 节还将介绍 DroidDraw 界面布局软件，可使初学者直观简单地布局用户界面，而且，DroidDraw 也提供了一种学习用 XML 语言编写布局文件的有效途径。

　　用户设计的活动界面类，还要管理界面控件的事件响应处理方法，这些方法是使用 Java 事件响应机制工作的，通过实现监听事件的接口的抽象方法对各种控件事件进行响应，为了增强程序的可读性，后续章节的实例大多使用匿名内部类的方法(见第 2.6.3 节)。

　　最后，需要强调指出的是，类属于数据结构的范畴，由数据成员和方法成员构成，类不是执行单元，类中的方法执行需要创建类的实例(对象)，可以把实例(对象)看作是类在内存中执行的一个实现，当然，因为一个类可以创建多个实例(对象)，所以，一个类可以有多个实现，即类的实例(对象)的执行对应着用户界面，而不是类！这一点在 Android 应用程序中完全得不到体现，好像 Android 用户界面对应着 Activity 类一样，这主要是因为 Android 应用程序框架隐藏了对活动界面对象的实现。总之，程序员必须适应"在类中编写数据和方法，然后就直接运行程序"这种新型的 Android 程序设计理念。

4.2　布局与控件

　　Android 应用程序一般都具有人性化的图形用户界面，有两种设计用户界面的方法，其一是用 Java 语言编写；其二是用 XML 布局文件实现。由于用户界面在程序执行过程中其布局和控件大都保持不变，因此，用第二种方法设计用户界面可以有效地节省 Java 程序代码，增强程序的可读性；更重要的是，第二种方法是所见即所得的界面设计方法，直观方便。基于 XML 语言进行界面设计的基本思路为：首先定义一种布局方式；然后在该布局方式下嵌入各种控件；有时为了满足控件对齐等排列方式的要求，还需要在一种布局方式下嵌入其他布局方式，在新嵌入的布局中摆放控件。用 XML 语言描述布局和控件的格式固定、语句简单，下面首先通过第 4.2.1 节对布局软件 DroidDraw 的介绍，讲述 XML 语言编写布局文件的方法；然后，在第 4.2.2 节介绍控件的事件响应方法；接着，在第 4.2.3 节罗列 Android 系统常用控件；最后，在第 4.2.4~4.2.8 节依次介绍 Android 系统下的 5 种布局方式，即线性布局、相对布局、框架布局、表格布局和约束布局，其中，约束布局是最常用的布局方式，而 Android 应用程序常用的控件(视图)在介绍布局方法时借助于实例说明其用法。

4.2.1　布局软件 DroidDraw

　　DroidDraw 软件是 Android 应用程序界面设计与编辑器软件，可以运行在 Windows 或

Linux 环境下,该软件是用 Java 语言编写的,其最新版本为 r1b22,压缩后的文件大小约 845KB,登录网站 http://www.droiddraw.org/下载文件 droiddraw-r1b22.zip,解压后如图 4-1 所示,其中,droiddraw.exe 是 Windows 系统可执行文件,droiddraw.jar 是 Java 虚拟机可执行文件,双击图 4-1 中的 droiddraw.exe 文件进入 DroidDraw 软件主界面,如图 4-2 所示。

图 4-1　DroidDraw 软件包文件

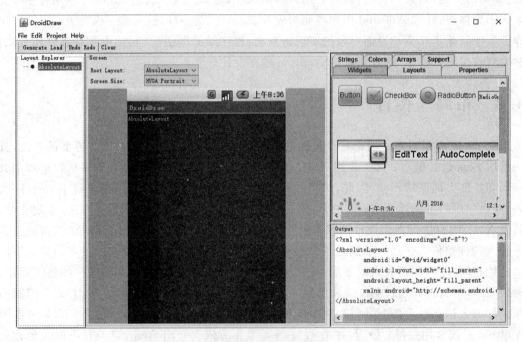

图 4-2　DroidDraw 主界面

在图 4-2 中,右边区域为多功能控件(Widgets)、布局(Layouts)、属性(Properties)等选项卡标签,这些选项卡选中后,可以向左边的黑色窗口中拖放布局和控件,并设置它们的属性。右边选项卡中的字符串(Strings)、颜色(Color)和数组(Arrays)用于创建字符串、颜色和数组资源。右边的支持(Support)选项卡选中后,将显示信息"DroidDraw 是免费的,依靠用户的捐助进行研发,并希望得到用户的支持"。在左边的屏幕(Screen)区中可以选择最底层的布局(Root Layout),它上面可以放置其他布局和控件;还可以选择屏幕的大小(Screen Size),选择 WVGA Portrait 对应窗口分辨率为 480×800(图 4-2 中选择了分辨率为 320×480 的 HVGA Portrait,因为选 WVGA Portrait 窗口太大,将使得缩小后的图 4-2 不清晰)。

在图 4-2 中,由右边的控件区随意拖几个控件到左边的黑色窗口中,然后单击左上角的 Generate,将在右下区域的 Output 中产生 XML 布局代码;任意选中某个控件,可以在属性页(Properties)设置它的属性,然后,单击 Apply 更新 XML 布局代码,如图 4-3 所示。将 XML 布局代码复制到 Android 应用程序工程中的布局文件中即可完成布局的设置。需要说明的是,Android Studio 集成的图形化布局功能比 DroidDraw 更强大,但是 DroidDraw 软件更适合于初学者学习 XML 语言语法。

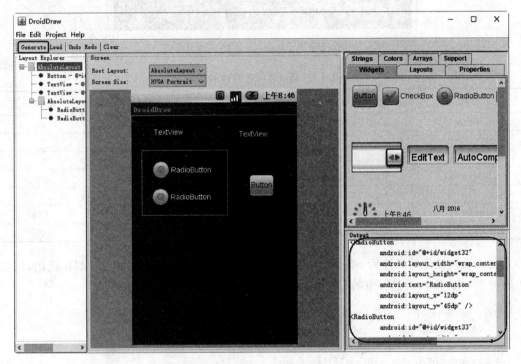

图 4-3　DroidDraw 生成 XML 布局代码

例 4-1　使用 DroidDraw 进行用户界面设计。

新建应用 MyDroidDrawGUIApp,包名为 cn. edu. jxufe. zhangyong. mydroiddrawguiapp,活动界面名为 MyDroidDrawGUIAct,保存目录为"D:\MyAndroidWorkSpace\MyDroidDrawGUIApp",SDK 版本号为 KitKat。

使用 DroidDraw 软件设计用户界面如图 4-4 所示,并设置各个控件的属性如表 4-1 所示。

图 4-4 中放置了两个 TextView 控件(静态文本框)、一个 Button 控件(命令按钮控件)、一个 RadioGroup 控件(单选钮组控件)和三个 RadioButton(单选钮),其中单选钮组控件将放置其内的单选钮作为一组,同一组中只能有一个单选钮处于选中状态。单选钮也称作收音机按钮,因为单选钮看上去像旧式的收音机调台旋钮。

图 4-4 DroidDraw 拖放的控件

表 4-1 图 4-4 中各控件的属性

控件名	文本(Text)	ID 号(@+id/)	宽(dp)	高(dp)
TextView	TextView1	tvSlt	140	20
TextView	TextView2	tvMsg	100	20
RadioButton	RadioButton1	rbPlane	130	40
RadioButton	RadioButton2	rbShip	130	40
RadioButton	RadioButton3	rbTrain	130	40
RadioGroup	—	rbgMeans	180	125
Button	Button1	btOK	100	40

完成设置控件的属性后，单击图 4-4 上的 Generate 可生成如下所示的 XML 代码，这些代码位于 DroidDraw 右下角的 Output 区域。

```
1   <?xml version = "1.0" encoding = "utf-8"?>
2   <AbsoluteLayout
3       android:id = "@+id/widget0"
4       android:layout_width = "fill_parent"
5       android:layout_height = "fill_parent"
6       xmlns:android = "http://schemas.android.com/apk/res/android">
7   <TextView
8       android:id = "@+id/tvSlt"
9       android:layout_width = "140dp"
10      android:layout_height = "20dp"
11      android:text = "TextView1"
12      android:layout_x = "10dp"
13      android:layout_y = "21dp" />
14  <TextView
15      android:id = "@+id/tvMsg"
16      android:layout_width = "100dp"
17      android:layout_height = "20dp"
18      android:text = "TextView2"
19      android:layout_x = "209dp"
```

```
20          android:layout_y = "106dp" />
21     < Button
22          android:id = "@ + id/btOK"
23          android:layout_width = "60dp"
24          android:layout_height = "40dp"
25          android:text = "Button1"
26          android:layout_x = "209dp"
27          android:layout_y = "140dp" />
28     < RadioGroup
29          android:id = "@ + id/rbgMeans"
30          android:layout_width = "180dp"
31          android:layout_height = "125dp"
32          android:layout_x = "9dp"
33          android:layout_y = "58dp">
34          < RadioButton
35               android:id = "@ + id/rbPlane"
36               android:layout_width = "130dp"
37               android:layout_height = "40dp"
38               android:text = "RadioButton1" />
39          < RadioButton
40               android:id = "@ + id/rbShip"
41               android:layout_width = "130dp"
42               android:layout_height = "40dp"
43               android:text = "RadioButton2" />
44          < RadioButton
45               android:id = "@ + id/rbTrain"
46               android:layout_width = "130dp"
47               android:layout_height = "40dp"
48               android:text = "RadioButton3" />
49     </RadioGroup >
50 </AbsoluteLayout >
```

为了增强可读性，上述代码做了缩进排版。第 2~6 行与第 50 行配对，描述绝对布局的属性，例如布局的宽（layout_width）和高（layout_height）为填满整个屏幕（fill_parent），而第 7~49 行的代码为布局上的控件。绝对布局是一种按控件的位置坐标确定控件在屏幕上位置的布局方法。因此，采用绝对布局时，放置在布局上的控件位置不变，是典型的所见即所得的构建界面方法。

第 7~13 行和第 14~20 行定义了两个静态文本框按钮，第 21~27 行定义了一个命令按钮控件，第 28~33 行和第 49 行配对，表示单选钮组的属性，例如，第 30、31 行分别定义了它的宽和高为 180dp 和 125dp，dp 是与设备显示屏无关的像素单位。而第 34~48 行的单选钮属于单选钮组。第 34~38 行定义文本为 RadioButton1 的单选钮，其宽和高分别为 130dp 和 40dp；类似地，第 39~43 行、第 44~48 行分别定义单选钮 RadioButton2 和 RadioButton3。

从上述这些控件的定义来看，定义一个控件的 XML 语言格式固定，即

```
<控件类型
控件属性
/>
</控件类型>
```

以命令按钮为例，其控件类型为 Button，控件的 ID 号为 btOK，其宽和高分别为 100dp 和 40dp，其显示的文本为 Button1，在布局平面上的位置坐标为（209dp，140dp），因此，其布局定义代码为

```
<Button              <!-- 控件类型 -->
    android:id = "@ + id/btOK"
    android:layout_width = "100dp"
    android:layout_height = "40dp"
    android:text = "Button1"
    android:layout_x = "209dp"
    android:layout_y = "140dp"       <!-- 控件属性书写不分先后顺序 -->
/>
</Button>           <!-- /控件类型 -->
```

控件的多个属性在代码中的位置不分先后顺序，每种属性均以"Android:"开头。在定义控件 ID 号时，必须在 ID 号前添加"@＋id/"。

图 4-5　应用 MyDroidDrawGUIApp 运行结果

将上述由 DroidDraw 产生的代码直接复制到应用 MyDroidDrawGUIApp 的布局文件 activity_my_droid_draw_gui.xml 中，覆盖其原来的代码，直接运行该工程，其结果如图 4-5 所示。

由于 Android 显示界面是由各种资源组成的，包括字符串资源，当 DroidDraw 显示界面使用的资源与 Android 应用程序不同时，在 DroidDraw 中显示良好的界面有可能在 Android 应用程序中显示不正常，仔细比较图 4-4 和图 4-5 可以发现两者的细微差别。通常由 DroidDraw 生成的布局代码应用于 Android 工程中时，其控件属性需要稍做调整，但是这些调整的工作量相对于从头至尾设计一个全新的布局文件而言，是微不足道的。

显示图 4-5 所示界面，并没有改动源程序文件 MyDroidDrawGUIAct.java，该文件的代码如下：

```
1  package cn.edu.jxufe.zhangyong.mydroiddrawguiapp;
2
3  import android.support.v7.app.AppCompatActivity;
4  import android.os.Bundle;
5
6  public class MyDroidDrawGUIAct extends AppCompatActivity {
7
8      @Override
9      protected void onCreate(Bundle savedInstanceState) {
10         super.onCreate(savedInstanceState);
11         setContentView(R.layout.activity_my_droid_draw_gui);
12     }
13 }
```

上述代码中的第 11 行将布局资源 activity_my_droid_draw_gui 显示在屏幕上，布局资源 activity_my_droid_draw_gui 对应于 activity_my_droid_draw_gui.xml 文件。事实上，图 4-5 所示界面中各控件显示的文本只表示了控件的类型，没有实际的意义。如果要求应用 MyDroidDrawGUIApp 显示如图 4-6 所示的界面，需要对应用 MyDroidDrawGUIApp 做进一步的完善工作，有两种方法修改用户界面控件的显示字符串，其一，是直接在 XML 布局文件中修改其 text 属性；其二，通过添加新的字符串资源文件实现，这种方法具有更好的通用性。

按照第 3.3.3 节介绍的方法新建字符串资源文件 strings_hz.xml，如图 4-7 所示添加 6 个字符串资源。每个字符串资源的格式固定，即

<string name = "字符串名">字符串的值</string>

其中字符串名由英文字母、下画线或数字组成，且首字符为英文字母或下画线。

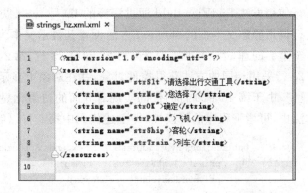

图 4-6 应用 MyDroidDrawGUIApp 界面　　　　图 4-7 字符串资源文件 strings_hz.xml

修改 MyDroidDrawGUIAct.java 的内容如下：

```
1   package cn.edu.jxufe.zhangyong.mydroiddrawguiapp;
2
3   import android.support.v7.app.AppCompatActivity;
4   import android.os.Bundle;
5   import android.widget.Button;
6   import android.widget.RadioButton;
7   import android.widget.TextView;
8
9   public class MyDroidDrawGUIAct extends AppCompatActivity {
10      private TextView tvSlt, tvMsg;
11      private RadioButton rbMeans;
12      private Button btOK;
13
14      @Override
15      protected void onCreate(Bundle savedInstanceState) {
16          super.onCreate(savedInstanceState);
17          setContentView(R.layout.activity_my_droid_draw_gui);
18          tvSlt = (TextView) findViewById(R.id.tvSlt);
19          tvSlt.setText(R.string.strSlt);
```

```
20        rbMeans = (RadioButton) findViewById(R.id.rbPlane);
21        rbMeans.setText(R.string.strPlane);
22        rbMeans = (RadioButton) findViewById(R.id.rbShip);
23        rbMeans.setText(R.string.strShip);
24        rbMeans = (RadioButton) findViewById(R.id.rbTrain);
25        rbMeans.setText(R.string.strTrain);
26        tvMsg = (TextView) findViewById(R.id.tvMsg);
27        tvMsg.setText(R.string.strMsg);
28        btOK = (Button) findViewById(R.id.btOK);
29        btOK.setText(R.string.strOK);
30    }
31 }
```

上述代码中,新添加的代码为第5~7行、第10~12行和第18~29行,其中第5~7行表示要引用的控件的类所在的包,通过相应控件的弹出菜单自动添加。第10~12行依次定义静态文本框控件对象tvSlt与tvMsg、单选钮对象rbMeans和命令按钮对象rbOK。第18行使用方法findViewById通过资源的ID号tvSlt获得静态文本框对象并赋给tvSlt;第19行调用静态文本框的setText方法设置其显示文本为字符串资源R.string.strSlt表示的字符串。同理,第20和第21行,第22和第23行,第24和第25行,第26和第27行和第28、29行采用相同的方法获得控件对象,并在对象上显示字符串。

由于第18~29行的代码实现对界面的初始化显示,因此,将它们单独放在一个私有方法中,可增强程序的可读性,即上述代码中第9行以后的代码改写如下:

```
9   public class MyDroidDrawGUIAct extends AppCompatActivity {
10      private TextView tvSlt, tvMsg;
11      private RadioButton rbMeans;
12      private Button btOK;
13
14      @Override
15      protected void onCreate(Bundle savedInstanceState) {
16          super.onCreate(savedInstanceState);
17          setContentView(R.layout.activity_my_droid_draw_gui);
18
19          myInitGUI();
20      }
21      private void myInitGUI(){
22          tvSlt = (TextView)findViewById(R.id.tvSlt);
23          tvSlt.setText(R.string.strSlt);
24          rbMeans = (RadioButton)findViewById(R.id.rbPlane);
25          rbMeans.setText(R.string.strPlane);
26          rbMeans = (RadioButton)findViewById(R.id.rbShip);
27          rbMeans.setText(R.string.strShip);
28          rbMeans = (RadioButton)findViewById(R.id.rbTrain);
29          rbMeans.setText(R.string.strTrain);
30          tvMsg = (TextView)findViewById(R.id.tvMsg);
31          tvMsg.setText(R.string.strMsg);
32          btOK = (Button)findViewById(R.id.btOK);
33          btOK.setText(R.string.strOK);
34      }
35  }
```

上述代码中,第 19 行调用类的私有方法 myInitGUI 对用户界面上的控件初始化,而第 21～34 行为 myInitGUI 方法的代码。此时执行应用 MyDroidDrawGUIApp,显示界面如图 4-6 所示。

本小节中使用了 DroidDraw 软件和绝对布局方法,由于绝对布局中,每个控制的位置是由其坐标值固定的,而 Android 智能设备的显示屏多种多样,分辨率各不相同,所以,绝对布局使得应用不能在多种 Android 设备上通用,现已经被 Android Studio 弃用,但是可以作为初学者学习 XML 语言的入门方法。

4.2.2 控件事件响应方法

在图 4-6 所示界面上,要求单击"确定"按钮,在"您选择了"静态文本框中显示选中的单选钮的信息,即如果"飞机"单选钮选中了,单击"确定"按钮,将显示"您选择了飞机"。为实现该功能,需要为"确定"按钮添加事件处理方法。在第 2.6 节介绍了 Java 语言中使用公有类、内部类和匿名内部类方法实现控件事件响应的方法,这些方法对于 Android 应用程序用户界面的控件来说,同样有效,而且这三种方法本质上是相同的,这里继续介绍一下使用匿名内部类响应控件事件的方法,同时,再介绍一种基于布局文件响应用户事件的方法。

例 4-2 使用匿名内部类响应控件事件。

在例 4-1 的基础上新建应用 MyClassEventApp,此时除了应用名为 MyClassEventApp 和活动界面名为 MyClassEventAct 外,其他与例 4-1 完全相同;然后修改源程序文件 MyClassEventAct.java。

新的 MyClassEventAct.java 内容如下:

```
1   package cn.edu.jxufe.zhangyong.myclasseventapp;
2
3   import cn.edu.jxufe.zhangyong.myclasseventapp.R.id;
4   import android.support.v7.app.AppCompatActivity;
5   import android.os.Bundle;
6   import android.widget.Button;
7   import android.widget.RadioButton;
8   import android.widget.RadioGroup;
9   import android.widget.TextView;
10  import android.view.View;
11  import android.view.View.OnClickListener;
12
13  public class MyClassEventAct extends AppCompatActivity {
14      private TextView tvSlt, tvMsg;
15      private RadioButton rbPlane, rbShip, rbTrain;
16      private RadioGroup rbgMeans;
17      private Button btOK;
18
19      @Override
20      protected void onCreate(Bundle savedInstanceState) {
21          super.onCreate(savedInstanceState);
22          setContentView(R.layout.activity_my_class_event);
23
24          myInitGUI();
```

```
25        }
26        private void myInitGUI(){
27            tvSlt = (TextView)findViewById(R.id.tvSlt);
28            tvSlt.setText(R.string.strSlt);
29            rbPlane = (RadioButton)findViewById(R.id.rbPlane);
30            rbPlane.setText(R.string.strPlane);
31            rbPlane.setChecked(true);
32            rbShip = (RadioButton)findViewById(R.id.rbShip);
33            rbShip.setText(R.string.strShip);
34            rbTrain = (RadioButton)findViewById(R.id.rbTrain);
35            rbTrain.setText(R.string.strTrain);
36            tvMsg = (TextView)findViewById(R.id.tvMsg);
37            tvMsg.setText(R.string.strMsg);
38            btOK = (Button)findViewById(R.id.btOK);
39            btOK.setText(R.string.strOK);
40            rbgMeans = (RadioGroup)findViewById(id.rbgMeans);
41            btOK.setOnClickListener(new OnClickListener(){
42                @Override
43                public void onClick(View v) {
44                    // TODO Auto-generated method stub
45                    tvMsg.setText(getResources().getString(R.string.strMsg) +
46                        ((RadioButton)findViewById(
47                            rbgMeans.getCheckedRadioButtonId())).getText());
48                }
49            });
50        }
51    }
```

上述代码中,第3~11行为声明引用的外部类,通过单击各个类名的弹出菜单自动添加。与例4-1中的文件MyDroidDrawGUIAct.java不同的地方有:第15行定义了三个单选按钮对象rbPlane、rbShip和rbTrain,第16行定义了一个单选按钮组对象rbgMeans;第29~35行使用了不同的单选按钮对象rbPlane、rbShip和rbTrain设置其各自的显示文本;第40行获取了单选按钮组对象rbgMeans。

第41~49行为使用匿名内部类方法实现btOK对象的事件响应方法,或者理解为对象btOK通过方法setOnClickListener注册了一个回调方法(或函数)onClickListener(其方法体为43~48行),当命令按钮被单击时,Android系统将调用该命令按钮的回调方法,即第43~47行的代码将得到执行。由于私有方法myInitGUI被onCreate调用,即完成一些界面的初始化工作,因此,按回调方法的理解方式更容易明白第41~49行的代码工作原理。所谓的回调方法(Callback)是指由Android操作系统调用执行的方法,用户应用程序只需要为对象的事件注册回调方法,当该对象事件发生后,Android系统自动去调用该对象注册的回调方法。需要补充说明的是,Android类中以"On"开头命名的方法,大都是由Android系统调用和管理的,例如,第41行的OnClickListener方法,因此,程序员在命名自己的方法时,尽可能不用"On"作为方法名的开头。

第41行的回调方法OnClickListener实际上是一个类,即第2.6.3节所阐述的匿名内部类,其中包括一个覆盖方法onClick,当对象btOK被单击时将执行onClick方法中的代码,即第45~47行,这是一句代码,即调用静态文本框对象tvMsg的setText方法将"getResources(). getString(R. string. strMsg)+((RadioButton)findViewById(rbgMeans. getCheckedRadioButtonId())). getText()"显示在静态文本框中,这里的语句"getResources(). getString(R. string. strMsg)"由字符串资源号获得字符串的值;而rbgMeans. getCheckedRadioButtonId()获得单选按钮组中被选中的单选按钮的ID号,再使用findViewById方法由单选按钮ID号获得其对象,最后调用其getText方法得到其文本值。

应用MyClassEventApp的运行结果如图4-8所示。

例4-3 基于布局文件属性响应用户控件事件。

在例4-1的基础上新建应用MyResourceEventApp,除了应用名为MyResourceEventApp和活动界面名为MyResourceEventAct外,其他与例4-1完全相同;然后,修改布局文件activity_my_resource_event.xml,即在Button属性(这里只有一个ID为btOK的命令按钮控件)中添加onClick属性,如图4-9所示,"android:onClick="myBtOKClick"",表示当命令按钮btOK被单击时,Android系统将调用myBtOKClick方法,该方法必须被定义为类MyResourceEventAct中的公有方法,且具有一个View类型参数。

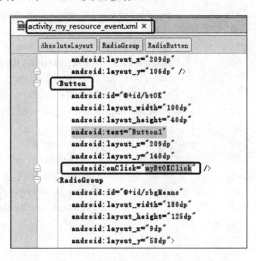

图4-8 应用MyClassEventApp执行结果　　图4-9 activity_my_resource_event.xml 文件中Button属性

修改后的MyResourceEventAct.java代码如下:

```
1  package cn.edu.jxufe.zhangyong.myresourceeventapp;
2
3  import android.support.v7.app.AppCompatActivity;
4  import android.os.Bundle;
5  import android.view.View;
6  import android.widget.Button;
7  import android.widget.RadioButton;
8  import android.widget.RadioGroup;
9  import android.widget.TextView;
10
11 public class MyResourceEventAct extends AppCompatActivity {
12     private TextView tvSlt, tvMsg;
13     private RadioButton rbPlane,rbShip,rbTrain;
```

```
14        private RadioGroup rgMeans;
15        private Button btOK;
16
17        @Override
18        protected void onCreate(Bundle savedInstanceState) {
19            super.onCreate(savedInstanceState);
20            setContentView(R.layout.activity_my_resource_event);
21
22            myInitGUI();
23        }
24        private void myInitGUI(){
25            tvSlt = (TextView)findViewById(R.id.tvSlt);
26            tvSlt.setText(R.string.strSlt);
27            rbPlane = (RadioButton)findViewById(R.id.rbPlane);
28            rbPlane.setText(R.string.strPlane);
29            rbPlane.setChecked(true);
30            rbShip = (RadioButton)findViewById(R.id.rbShip);
31            rbShip.setText(R.string.strShip);
32            rbTrain = (RadioButton)findViewById(R.id.rbTrain);
33            rbTrain.setText(R.string.strTrain);
34            tvMsg = (TextView)findViewById(R.id.tvMsg);
35            tvMsg.setText(R.string.strMsg);
36            btOK = (Button)findViewById(R.id.btOK);
37            btOK.setText(R.string.strOK);
38        }
39        public void myBtOKClick(View v){
40            rgMeans = (RadioGroup)findViewById(R.id.rbgMeans);
41            tvMsg.setText(getResources().getString(R.string.strMsg) +
42                    ((RadioButton)findViewById(
43                            rgMeans.getCheckedRadioButtonId())).getText());
44        }
45    }
```

图 4-10 应用 MyResourceEventApp 运行结果

对比例 4-2 中的 MyClassEventAct.java，上述代码没有使用命令控件的方法 setOnClickListener 注册回调方法，实际上私有方法 myInitGUI 几乎没有改变，而是添加了一个新的公有方法，即第 39 行的 myBtOKClick 方法，该方法有一个 View 类的实例 v 作为参数。第 40 行初始化单选钮组对象，第 41~43 行在静态文本框中显示所选择的字符串，方法同例 4-2。相对例 4-1，所做的修改包括添加第 4~9 行（借助类的弹出菜单自动添加）、修改了第 13 和第 14 行、添加了第 29 行（使"飞机"单选按钮启动时处于选中状态）、修改了第 27~33 行以及添加了公有方法 myBtOKClick。

应用 MyResourceEventApp 的运行结果如图 4-10 所示。

通过比较例 4-2 和例 4-3，可以看出例 4-3 中借助资源文件进行控件事件响应的方法更加简洁直观且易于理解，实际上这两种方法都非常流行，建议初学者使用后者。

例 4-4　单选按钮事件演示。

单选按钮的继承关系为：java. lang. Object→android. view. View→android. widget. TextView→android. widget. Button→android. widget. CompoundButton→android. widget. RadioButton。即单选按钮类间接继承了命令按钮类，因此单选按钮与命令按钮一样，也具有单击事件。实际上，几乎所有控件都具有单击事件。

例 4-3 的功能是选择了单选按钮之后，单击"确定"按钮，然后显示诸如"您选择了列车"这类提示信息。本例中将添加单选按钮的单击事件，当单击某个单选按钮时，将弹出一条诸如"您选择了列车"的信息，然后，这个信息框自动消失。Android 系统的 Toast 类可实现这种功能，该类具有一个静态方法 makeText 和一个动态方态 show，其原型如下：

```
1    public static Toast makeText(Context context, int resId, int duration);
2    public static Toast makeText(Context context, CharSequence text, int duration);
3    public void show();
```

这里，context 参数表示显示提示信息的对象；resId 或 text 分别表示字符串资源索引号或字符串；duration 表示显示提示信息的时间，只能输入 LENGTH_SHORT 或 LENGTH_LONG，这两个量为 Toast 类的静态常量，表示显示信息的时间比较短或比较长。使用 Toast 短暂显示一条信息的典型用法为：

```
Toast.makeText(this, "I am a toast message!",Toast. LENGTH_SHORT).show();
```

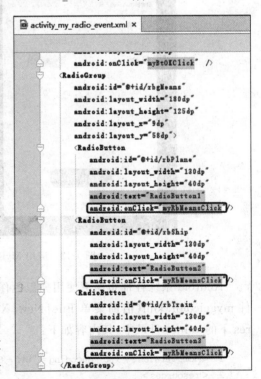

图 4-11　activity_my_radio_event. xml 文件中单选按钮添加 onClick 属性

这条语句首先使用 Toast. makeText 创建一个 Toast 类型对象（因为静态方法 makeText 返回一个 Toast 类型对象），然后再用这个对象调用其 show 方法显示字符串"I am a toast message!"。Toast 提示信息的位置一般位于屏幕的下方中央。

现在，在例 4-3 的基础上，新建应用 MyRadioEventApp，除了应用名为 MyRadioEventApp 和活动界面名为 MyRadioEventAct 外，其他与例 4-3 完全相同；然后，修改布局文件 activity_my_radio_event. xml，为每个单选按钮添加 onClick 属性，如图 4-11 所示。

从图 4-11 中可以看出，三个单选按钮的 onClick 属性均为 myRbMeansClick，表明在用户界面上无论单击哪个单选按钮，都将执行同一个公有方法，即 myRbMeansClick。然后，在 MyRadioEventAct. java 文件中添加方法 myRbMeansClick 即可（该文件其他代码没有变动），新添加的方法 myRbMeansClick 的代码如下：

```
1  public void myRbMeansClick(View v){
2      rgMeans = (RadioGroup)findViewById(R.id.rbgMeans);
3      Toast.makeText(this,getResources().getString(R.string.strMsg) +
4                  ((RadioButton)findViewById(
5                      rgMeans.getCheckedRadioButtonId())).getText(),
6          Toast.LENGTH_SHORT).show();
7  }
```

上述代码中，第3～6行为一条语句，即使用"Toast.makeText( ).show( )"语法结构在屏幕上短暂显示文本。由于使用了 Toast 类，在其弹出菜单中单击"Import 'Toast'（android.widget）"项自动在 MyRadioEventAct.java 文件头部添加"import android.widget.Toast;"，即引用 Toast 类。

应用 MyRadioEventApp 的执行结果如图4-12所示。当只单击单选按钮而不单击"确定"按钮，将弹出一个提示信息，稍后该信息自动消失；而"确定"按钮上方只显示"您选择了"。当单击"确定"按钮后，则显示诸如"您选择了列车"之类的信息。

图 4-12　应用 MyRadioEventApp 执行结果

Android 系统各种布局下使用亮黑色作为默认背景色，可以在工程中添加颜色资源文件 myguicolor.xml（借助菜单 File|New|XML|Values XML File 创建），该文件位于资源 res 下的 values 目录中，其内容如下：

```
1  <?xml version = "1.0" encoding = "utf-8"?>
2  <resources>
3      <color name = "mywincyan">#FF006666</color>
4  </resources>
```

定义颜色的方式主要是♯AARRGGBB 或♯RRGGBB,"♯"表示用十六进制数表示,"AA"表示透明程度的 alpha 值,"RRGGBB"依次表示红绿蓝的值。第 3 行定义的颜色为♯FF006666,即青色。定义颜色的另一个方式是用 drawable 关键字,第 3 行也可以写为:

< drawable name = "mywincyan">♯FF006666 </drawable >

使用 drawable 定义的颜色使用 R. drawable. mywincyan 访问。

然后,在 activity_my_radio_event. xml 文件中为 AbsoluteLayout 添加背景,即加入图 4-13 中圈住的一条语句,"android:background="@color/mywincyan""。

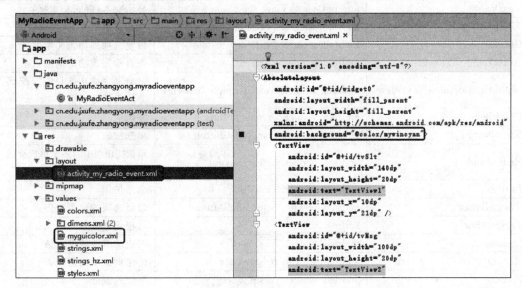

图 4-13　添加背景色

这时,运行应用 MyRadioEventApp,其结果如图 4-14 所示(图中以灰度显示青色)。

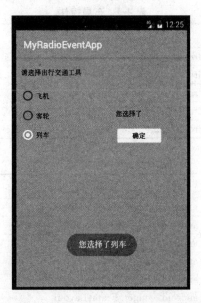

图 4-14　应用 MyRadioEventApp 采用青色背景(图中以灰度显示)

4.2.3 Android 常用控件

Android 系统与 Activity(活动界面)直接相关的控件(视图)可分为 7 大类,即窗体控件、布局控件、复合控件、图像与多媒体控件、时间与时历控件、切换控件和高级控件,分别列于表 4-2 至表 4-8 中。

表 4-2 窗体控件

类 名	直 接 父 类	功 能
TextView	android.view.View	显示文本(静态文本框)
EditText	android.widget.TextView	显示可编辑文本(编辑框)
CheckedTextView	android.widget.TextView	带复选项的静态文本框
AutoCompleteTextView	android.widget.EditText	带自动完成功能的编辑框
MultiAutoCompleteTextView	android.widget.AutoCompleteTextView	带自动完成功能且能匹配部分输入的编辑框
CheckBox	android.widget.CompoundButton	复选框
RadioButton	android.widget.CompoundButton	单选按钮
RadioGroup	android.widget.LinearaLayout	单选按钮组中的多个单选按钮中,只能有一个处于选中状态
Button	android.widget.TextView	命令按钮
ToggleButton	android.widget.CompoundButton	带复选项的命令按钮
ProgressBar	android.view.View	可视化进度条
SeekBar	android.widget.AbsSeekBar	可拖动滑块的进度条
RatingBar	android.widget.AbsSeekBar	带比例显示的进度条
Spinner	android.widget.AbsSpinner	下拉列表框
QuickContactBadge	android.widget.ImageView	显示联系人的电话、E-mail 和短信图标快捷方式

表 4-3 布局控件

类 名	直 接 父 类	功 能
LinearLayout	android.view.ViewGroup	沿行或列顺序摆放其中的控件
RelativeLayout	android.view.ViewGroup	其中的控件按相互间相对位置摆放
FrameLayout	android.view.ViewGroup	在屏幕中指定一块显示区域
TableLayout	android.widget.LinearLayout	将屏幕划分为表格区域
TableRow	android.widget.LinearLayout	TableLayout 表格中的一行

表 4-4 复合控件

类 名	直 接 父 类	功 能
ListView	android.widget.AbsListView	列表框
ExpandableListView	android.widget.ExpandableListView	垂直两级扩展列表框
TwoLineListItem	android.widget.RelativeLayout	列表项
GridView	android.widget.AbsListView	表格化显示数据
ScrollView	android.widget.FrameLayout	屏幕垂直滚动条
HorizontalScrollView	android.widget.FrameLayout	屏幕水平滚动条

续表

类 名	直 接 父 类	功 能
WebView	android.widget.AbsoluteLayout	网页显示控件
SlidingDrawer	android.widget.ViewGroup	滑动显示控件
TabHost	android.widget.FrameLayout	选项卡容器
TabWidget	android.widget.TabWidget	选项卡页面标签

表 4-5 图像与多媒体控件

类 名	直 接 父 类	功 能
ImageButton	android.widget.ImageView	具有图形界面的命令按钮
ImageView	android.view.View	图像显示控件
Gallery	android.widget.AbsSpinner	水平滚动显示图片(画廊)
MediaController	android.widget.FrameLayout	多媒体播放器控件容器
VideoView	android.widget.SurfaceView	播放视频文件

表 4-6 时间与时历控件

类 名	直 接 父 类	功 能
TimePicker	android.widget.FrameLayout	选择时间
DatePicker	android.widget.FrameLayout	选择日期
AnalogClock	android.view.View	模拟时钟
DigitalClock	android.widget.TextView	数字时钟
Chronometer	android.widget.TextView	计时器

表 4-7 切换控件

类 名	直 接 父 类	功 能
TextSwitcher	android.widget.ViewSwitcher	文本显示切换
ImageSwitcher	android.widget.ViewSwitcher	图像显示切换
ViewAnimator	android.widget.FrameLayout	动画切换 FrameLayout 视图
ViewFlipper	android.widget.ViewAnimator	动画显示
ViewSwitcher	android.widget.ViewAnimator	两个视图间动画切换

表 4-8 高级控件

类 名	直 接 父 类	功 能
View	java.lang.Object	用户界面元素容器,负责用户界面绘制和事件处理
ViewStub	android.view.View	不可见的 View,可在程序执行时控制其显示和隐藏
SurfaceView	android.view.View	专用绘图窗口视图
GestureOverlayView	android.widget.FrameLayout	手势输入透明窗口
ZoomButton	android.widget.ImageButton	缩放按钮(只能显示图像)
ZoomControls	android.widget.LinearLayout	缩小和放大控制按钮组
DialerFilter	android.widget.RelativeLayout	拨号窗口
AbsoluteLayout	android.widget.ViewGroup	根据控件坐标值布局控件位置

表 4-2～表 4-8 中的一部分控件将在本章第 4.2.4～4.2.8 节中通过实例的方式介绍其具体用法,其余控件将在第 5～8 章中介绍。Android 应用程序设计水平很大程度上取决于对本节中所列控件的熟练运用水平。

4.2.4 线性布局 LinearLayout

线性布局分为水平方向和垂直方向线性布局,当设置为水平方向布局时,控件水平方向上依次摆放,如果控件水平方向上超出屏幕显示范围,控件也不会换行摆放;当设置为垂直方向布局时,控件垂直方向依次摆放,即使有两个控件在同一行中能摆放下,它们也要按列摆放,如图 4-15 所示。

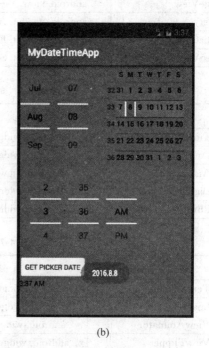

图 4-15 线性布局

例 4-5 实现图 4-15 的布局。

新建应用 MyDateTimeApp,应用名为 MyDateTimeApp,包名为 cn.edu.jxufe.zhangyong.mydatetimeapp,活动界面名为 MyDateTimeAct,SDK 版本为 KitKat。应用 MyDateTimeApp 中需要编辑的文件有 MyDateTimeAct.java、activity_my_date_time.xml、myguicolor.xml 和 strings.xml,其中,文件 myguicolor.xml 与例 4-4 中的同名文件内容相同,strings.xml 文件修改如下:

```
1  <?xml version = "1.0" encoding = "utf - 8"?>
2  < resources >
3      < string name = "str_gettime">Get Picker Date</string >
4      < string name = "app_name">MyDateTimeApp</string >
5  </resources >
```

第 3 行中设置名为 str_gettime 的字符串"Get Picker Date"。

文件 activity_my_date_time.xml 是图 4-15 的布局文件，其内容如下：

```
1    <?xml version = "1.0" encoding = "utf-8"?>
2    <LinearLayout
3        android:id = "@+id/widget32"
4        android:layout_width = "fill_parent"
5        android:layout_height = "fill_parent"
6        android:orientation = "vertical"
7        android:background = "@color/mywincyan"
8        xmlns:android = "http://schemas.android.com/apk/res/android"
9        android:weightSum = "1">
10
11       <DatePicker
12           android:id = "@+id/datePicker"
13           android:layout_width = "wrap_content"
14           android:layout_height = "99dp"
15           android:layout_weight = "0.82" />
16
17       <TimePicker
18           android:id = "@+id/timePicker"
19           android:layout_width = "269dp"
20           android:layout_height = "181dp" />
21
22       <Button
23           android:id = "@+id/btGetDate"
24           android:layout_width = "wrap_content"
25           android:layout_height = "wrap_content"
26           android:text = "@string/str_getdate"
27           android:onClick = "myDispPickerDate"/>
28
29       <TextClock
30           android:layout_width = "217dp"
31           android:layout_height = "50dp"
32           android:id = "@+id/textClock" />
33   </LinearLayout>
```

上述代码中，第 2～9 行与第 33 行配对，表示这是一个线性布局，第 6 行说明它是一个垂直线性布局，即整个屏幕视为一个多行一列的网格，每一行只能放置一个控件，多个控件按列依次摆放；第 7 行设置布局的背景色为青色。在垂直线性布局窗口中，首先放置一个日历选择控件，该控件用于设定某个日期值（第 11～15 行），其 ID 号为 datePicker；然后，其下放置一个时间选择控件，用于设定某个时间值（第 17～20 行），其 ID 号为 timePicker；接着，放置一个命令按钮（第 22～27 行），其事件方法为 myDispPickerDate；紧接下面放置了一个数字时钟（第 29～32 行），这个控件自动获取系统时间显示。

上述代码的布局显示结果如图 4-15 所示，其中，命令按钮 Get Picker Date 单击后短暂弹出当前日期的设定值（用 Toast 类实现）。

文件 MyDateTimeAct.java 的代码如下：

```
1    package cn.edu.jxufe.zhangyong.mydatetimeapp;
2
```

```
3      import android.support.v7.app.AppCompatActivity;
4      import android.os.Bundle;
5      import android.view.View;
6      import android.widget.DatePicker;
7      import android.widget.Toast;
8
9      public class MyDateTimeAct extends AppCompatActivity {
10         private DatePicker datePicker;
11         @Override
12         protected void onCreate(Bundle savedInstanceState) {
13             super.onCreate(savedInstanceState);
14             setContentView(R.layout.activity_my_date_time);
15         }
16         public void myDispPickerDate(View v){
17             datePicker = (DatePicker)findViewById(R.id.datePicker);
18             String str = String.valueOf(datePicker.getYear()) + "." +
19                     String.valueOf(datePicker.getMonth() + 1) + "." +
20                     String.valueOf(datePicker.getDayOfMonth());
21             Toast.makeText(this,str,Toast.LENGTH_LONG).show();
22         }
23     }
```

上述代码中,第 10 行定义了日历选择控件对象 datePicker;第 16 行为命令按钮 Get Picker Date 的事件响应方法 myDispPickerDate,该方法必须为公有方法;第 17 行使用 findViewById 方法从资源中获得控件对象 datePicker;第 18～20 行是一条语句,String.valueOf 方法将数值型变量转化为字符串,getYear、getMonth 和 getDayOfMonth 方法分别是获得日历选择控件设置的年、月和日;第 21 行使用 Toast 对象显示设置的日期,如图 4-15 所示。需要注意的是第 19 行,由于 getMonth 得到的月份整型变量比实际月份小 1,例如,2 月份对应整数 1,因此,在第 19 行中将得到的月份值加上 1。

日历选择控件和时间选择控件主要用于设置日期和时间,从而根据设定值修改系统时间,Android 应用程序修改系统时间需得到 Linux 内核访问的许可权限。事实上,由于 Android 系统设备都是联网的,Android 系统自动进行网络校时,因此,在 Android 应用中几乎不需要设置系统时间。

4.2.5 相对布局 RelativeLayout

相对布局中,第一个控件可以摆放在布局平面的 9 个位置,即左上、中上、右上、左中、中央、右中、左下、中下和右下方;第二个控件可以相对于第一个控件指定摆放位置,即它的上方、下方、左边和右边等;第三个控件则可以相对于前两个控件指定摆放的位置,依次类推。例如下面的布局文件 activity_my_chronoscope.xml,该文件中放置了两个控件 ImageButton 和 Chronometer。

```
1    <?xml version = "1.0" encoding = "utf - 8"?>
2    <RelativeLayout xmlns:app = "http://schemas.android.com/apk/res - auto"
3        android:id = "@ + id/widget39"
4        android:layout_width = "fill_parent"
5        android:layout_height = "fill_parent"
```

```
6          xmlns:android = "http://schemas.android.com/apk/res/android">
7          <ImageButton
8              android:id = "@ + id/imageBtn"
9              android:layout_width = "wrap_content"
10             android:layout_height = "wrap_content"
11             android:layout_centerVertical = "true"
12             android:layout_centerHorizontal = "true"
13             app:srcCompat = "@drawable/mychronostart"
14             android:onClick = "myChronoStart"/>
15
16         <Chronometer
17             android:layout_width = "wrap_content"
18             android:layout_height = "wrap_content"
19             android:layout_above = "@ + id/imageBtn"
20             android:layout_centerHorizontal = "true"
21             android:id = "@ + id/chronometer"
22             android:visibility = "visible" />
23 </RelativeLayout>
```

第 2～6 行与第 23 行配对,说明这是一个相对布局。第 7～14 行为一个图形按钮控件 ImageButton,其 ID 号为 imageBtn,摆放位置为中央(第 11、12 行),即水平和垂直都居中。它有一个事件方法,即 myChronoStart(第 14 行),其显示的图像为 mychronostart 资源图片。ImageButton 是相对布局的第一个控件。

第二个控件为 Chronometer(第 16～22 行),第 19 行指定它位于控件 imageBtn 的上方 (layout_above),且在水平方向上居中(layout_Horizontal 为真)(第 20 行)。即第二个控件相对于第一个控件布局。

这里的图形按钮控件可以显示图形,其功能与命令按钮控件相似。Chronometer 控件是一个定时器控件,当执行它的 start 方法时,开始定时;当执行 stop 方法时,停止定时。

例 4-6 相对布局演示。

新建应用 MyChronoscopeApp,使用上面的 activity_my_chronoscope.xml 布局文件,应用名为 MyChronoscopeApp,活动界面名为 MyChronoscopeAct。应用 MyChronoscopeApp 如图 4-16 所示,其中,为图形按钮控件添加了一个图片 mychronostart.png,文件 MyChronoscopeAct.java 的代码如图 4-16 右边所示。

在图 4-16 右边代码部分中,关于类 MyChronoscopeAct 的代码重新列在下面:

```
8  public class MyChronoscopeAct extends AppCompatActivity {
9      private Chronometer chronoMeter;
10     @Override
11     protected void onCreate(Bundle savedInstanceState) {
12         super.onCreate(savedInstanceState);
13         setContentView(R.layout.activity_my_chronoscope);
14     }
15     public void myChronoStart(View v){
16         chronoMeter = (Chronometer)findViewById(R.id.chronometer);
17         chronoMeter.setFormat("I have worked for: % s");
18         chronoMeter.start();
19     }
20 }
```

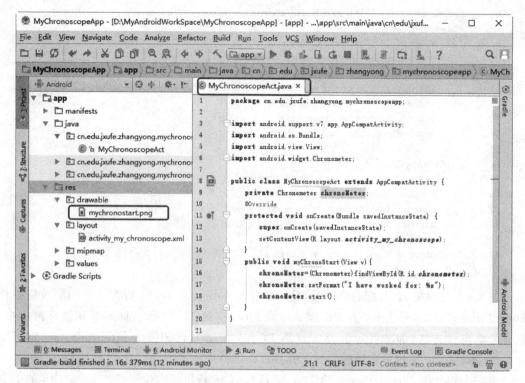

图 4-16　工程 ex04_06

上述代码的标号与图 4-16 中相同。第 9 行定义了私有的 Chronometer 类的对象 chronoMeter。第 11～14 行为覆盖父类的 onCreate 方法,用于界面初始化。第 15 行的方法 myChronoStart 为图形按钮控件的单击事件响应方法,必须为 public 公有方法,不能有返回值,且必须有一个 View 对象参数。第 16 行调用 findViewById 方法从资源中获得定时器控件对象;第 17 行调用方法 setFormat 设置定时器显示格式为"I have worked for：%s",其中的"%s"将被定时器的显示值"MM：SS"或"H：MM：SS"取代。在 setFormat 方法中,第一个出现的"%s"将被显示定时器取代,如果有第二个"%s"或者更多,则可以指定替换的字符串。第 18 行调用 start 方法启动定时器。

在图 4-16 中,添加了图片文件名为 mychronostart .png,位于资源 res 下的 drawable 目录下,分辨率为 300×224。

应用 MyChronoscopeApp 的执行结果如图 4-17 所示。单击"启动计时"按钮后,显示计时值,当显示时间超过 1 小时时,将以格式"小时：分：秒"的方式显示。

图 4-17　应用 MyChronoscopeApp 执行结果

4.2.6 框架布局 FrameLayout

前面介绍的线性布局和相对布局占有整个屏幕（也可占有部分屏幕），而框架布局 FrameLayout 则不同，它一般占有屏幕的一块区域，如图 4-18 所示。

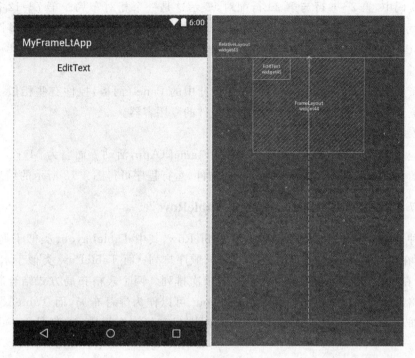

图 4-18 框架布局

在图 4-18 中，屏幕布局为相对布局（图中整个区域），在其上有一个框架布局（图中正方形区域），框架布局中放置了一个 EditText 文本编辑框（图中正方形区域的左上角区域），其布局文件 activity_my_frame_lt.xml 如下：

```
1   <?xml version = "1.0" encoding = "utf-8"?>
2   <RelativeLayout
3       android:id = "@ + id/widget43"
4       android:layout_width = "fill_parent"
5       android:layout_height = "fill_parent"
6       xmlns:android = "http://schemas.android.com/apk/res/android">
7       <FrameLayout
8           android:id = "@ + id/widget44"
9           android:layout_width = "221dp"
10          android:layout_height = "184dp"
11          android:layout_alignParentTop = "true"
12          android:layout_centerHorizontal = "true">
13          <EditText
14              android:id = "@ + id/widget45"
15              android:layout_width = "wrap_content"
16              android:layout_height = "wrap_content"
```

```
17                    android:text = "EditText"
18                    android:textSize = "18sp" />
19        </FrameLayout>
20  </RelativeLayout>
```

上述代码中,第 2~6 行与第 20 行配对,表示这是一个相对布局。第 7~12 行与第 19 行配对,表示这是一个框架布局,位于相对布局屏幕的顶端且水平居中,即属性 layout_alignParentTop 为真,layout_centerHorizontal 为真(第 11、12 行)。第 13~18 行为一个 EditText 控件,EditText 控件中可以输入文本。

框架布局的作用与 Windows 界面程序设计中的 Panel(面板)控件有些相似,相当于在用户界面上开辟一个新的用户界面,为某些控件的专用容器。

例 4-7 框架布局实例。

新建应用 MyFrameLtApp,应用名为 MyFrameLtApp,活动界面名为 MyFrameLtAct,添加上述的 activity_my_frame_lt.xml 布局文件,运行程序可得图 4-18 所示的结果。

4.2.7 表格布局 TableLayout 和 TableRow

有两种表格布局方式 TableLayout 和 TableRow,其中 TableLayout 类似于垂直方式的线性布局,在这种方式下,每个控件按垂直方向顺序排列;而 TableRow 类似于水平方式的线性布局,在这种方式下,各个控件水平方向依次排列。两个表格布局方式结合起来,实现类似于表格形式的布局。表格布局 TableLayout 可以作为屏幕布局,而 TableRow 只能放置在其他布局中,它可以放置在线性布局、相对布局和绝对布局中,通常配合 TableLayout 布局使用。

例 4-8 表格布局演示。

新建应用 MyTableLtApp,应用名为 MyTableLtApp,活动界面名为 MyTableLtAct。在工程文件中添加一个颜色资源文件 myguicolor.xml,其代码如下:

```
1   <?xml version = "1.0" encoding = "utf-8"?>
2   <resources>
3       <drawable name = "darkgray">#FF404040</drawable>
4       <drawable name = "black">#FF000000</drawable>
5       <drawable name = "red">#FFFF0000</drawable>
6       <drawable name = "green">#FF00FF00</drawable>
7       <drawable name = "lightgray">#FFC0C0C0</drawable>
8       <drawable name = "white">#FFFFFFFF</drawable>
9       <drawable name = "yellow">#FFFFFF00</drawable>
10      <drawable name = "blue">#FF0000FF</drawable>
11      <drawable name = "gray">#FF808080</drawable>
12      <drawable name = "magenta">#FFFF00FF</drawable>
13      <drawable name = "cyan">#FF00FFFF</drawable>
14  </resources>
```

上述代码中第 3~13 行依次表示暗灰、黑、红、绿、浅灰、白、黄、蓝、灰、品红和青色,每种颜色值的格式为"#AARRGGBB",其中,AA 表示透明程度,RR 表示红色分量的值,GG 表示绿色分量的值,BB 表示蓝色分量的值。

然后,修改布局文件 main.xml 如下:

```xml
1  <?xml version = "1.0" encoding = "utf-8"?>
2  <TableLayout
3      android:id = "@+id/tablelt"
4      android:layout_width = "fill_parent"
5      android:layout_height = "fill_parent"
6      android:orientation = "vertical"
7      xmlns:android = "http://schemas.android.com/apk/res/android"
8      android:background = "@drawable/darkgray"
9  >
10     <TableRow
11         android:id = "@+id/tablerow1"
12         android:layout_width = "fill_parent"
13         android:layout_height = "wrap_content"
14         android:orientation = "horizontal"
15         android:background = "@drawable/red"
16     >
17         <EditText
18             android:id = "@+id/edit1"
19             android:layout_width = "wrap_content"
20             android:layout_height = "wrap_content"
21             android:text = "EditText"
22             android:textSize = "18sp"
23         >
24         </EditText>
25     </TableRow>
26     <TableRow
27         android:id = "@+id/tablerow2"
28         android:layout_width = "fill_parent"
29         android:layout_height = "wrap_content"
30         android:orientation = "horizontal"
31         android:background = "@drawable/yellow"
32     >
33         <EditText
34             android:id = "@+id/edit2"
35             android:layout_width = "wrap_content"
36             android:layout_height = "wrap_content"
37             android:text = "EditText"
38             android:textSize = "18sp"
39         >
40         </EditText>
41         <EditText
42             android:id = "@+id/edit3"
43             android:layout_width = "wrap_content"
44             android:layout_height = "wrap_content"
45             android:text = "EditText"
46             android:textSize = "18sp"
47         >
48         </EditText>
49     </TableRow>
```

```
50        <TableRow
51           android:id = "@ + id/tablerow3"
52           android:layout_width = "fill_parent"
53           android:layout_height = "wrap_content"
54           android:orientation = "horizontal"
55           android:background = "@drawable/green"
56           >
57              <EditText
58                 android:id = "@ + id/edit4"
59                 android:layout_width = "wrap_content"
60                 android:layout_height = "wrap_content"
61                 android:text = "EditText"
62                 android:textSize = "18sp"
63                 >
64              </EditText>
65        </TableRow>
66        <TableRow
67           android:id = "@ + id/tablerow4"
68           android:layout_width = "fill_parent"
69           android:layout_height = "wrap_content"
70           android:orientation = "horizontal"
71           android:background = "@drawable/cyan"
72           >
73              <EditText
74                 android:id = "@ + id/edit5"
75                 android:layout_width = "wrap_content"
76                 android:layout_height = "wrap_content"
77                 android:text = "EditText"
78                 android:textSize = "18sp"
79                 >
80              </EditText>
81              <EditText
82                 android:id = "@ + id/edit6"
83                 android:layout_width = "wrap_content"
84                 android:layout_height = "wrap_content"
85                 android:text = "EditText"
86                 android:textSize = "18sp"
87                 >
88              </EditText>
89        </TableRow>
90        <TableRow
91           android:id = "@ + id/tablerow5"
92           android:layout_width = "fill_parent"
93           android:layout_height = "wrap_content"
94           android:orientation = "horizontal"
95           android:background = "@drawable/blue"
96           >
97              <EditText
98                 android:id = "@ + id/edit7"
99                 android:layout_width = "wrap_content"
100                android:layout_height = "wrap_content"
```

```
101             android:text = "EditText"
102             android:textSize = "18sp"
103             >
104         </EditText>
105     </TableRow>
106 </TableLayout>
```

上述代码中,第 2~9 行与第 106 行配对,说明屏幕总体布局为 TableLayout 布局,在其上放置了 5 个 TableRow 布局。第 10~16 行与第 25 行配对,说明这是一个 TableRow 布局,其中有一个 EditText 控件(第 17~24 行);第 26~32 行与第 49 行配对,说明这是一个 TableRow 布局,其中水平放置了两个 EditText 控件(第 33~48 行);第 50~56 行与第 65 行配对,说明这是一个 TableRow 布局,其中有一个 EditText 控件(第 57~64 行);第 66~72 行与第 89 行配对,该 TableRow 布局中有两个 EditText 控件(第 73~88 行);第 90~96 行与第 105 行配对,该 TableRow 布局中有一个 EditText 控件(第 97~104 行)。从第 8、15、31、55、71 和 95 行可以看出,TableLayout 布局和 5 个 TableRow 布局的背景色依次为暗灰、红、黄、绿、青和蓝色,如图 4-19 所示,在图中以灰度显示各种色彩。运行应用 MyTableLtApp,运行结果如图 4-19 所示。

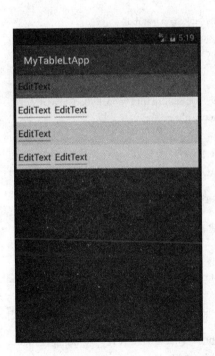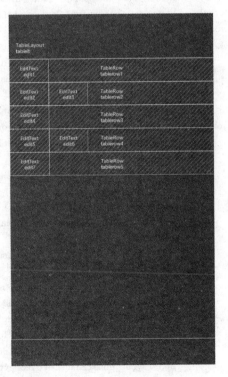

图 4-19 表格布局

4.2.8 约束布局 ConstraintLayout

约束布局是一种灵活的相对布局方式,屏幕上各个控件的位置由其相互间的约束关系决定,Android Studio 集成形成开发环境推荐底层使用这种布局方式。由于绝对布局下,屏幕上的控件位置由其横、纵坐标值决定,而 Android 程序设计针对各种屏幕大小的设备,采用绝对布局方式后,应用程序的通用性受到限制,因此,Android Studio 不建议使用绝对布局。例如,基于 480×800 分辨率的屏幕设计的应用程序运行在 240×320 屏幕上时会出现显示不完整等问题,而约束布局由于没有使用绝对坐标,则可以正常显示,约束布局更符合程序员的布局习惯。

例 4-9 约束布局演示。

新建应用 MyConstraintLyApp,应用名为 MyConstraintLyApp,活动界面名为 MyConstraintAct。该应用程序采用约束布局,其中放置了一个 ImageView 对象和一个静态文本框 TextView 对像,通过定时器周期设定 TextView 对象的计数值,并借助动画类 AnimationSet 实现 ImageView 对象的平移动画显示。此外,该工程也介绍了触屏事件的响应方法(在模拟器上触屏事件相当于单击鼠标,有按下、滑动和抬起触笔的事件)。

应用 MyConstraintLyApp 建立好后,添加两个图片文件,即 myschedule.png 和 mymaths.png,分别用作约束布局屏幕的背景图和 ImageView(图像显示控件)的背景图,然后,编辑布局文件 activity_my_constraint.xml 如下:

```
1   <?xml version = "1.0" encoding = "utf-8"?>
2   <android.support.constraint.ConstraintLayout xmlns:android = "http://schemas.android.com/apk/res/android"
3       xmlns:app = "http://schemas.android.com/apk/res-auto"
4       xmlns:tools = "http://schemas.android.com/tools"
5       android:id = "@+id/activity_my_constraint"
6       android:layout_width = "match_parent"
7       android:layout_height = "match_parent"
8       tools:context = "cn.edu.jxufe.zhangyong.myconstraintlyapp.MyConstraintAct"
9       android:background = "@drawable/myschedule">
10
11      <ImageView
12          android:layout_width = "65dp"
13          android:layout_height = "62dp"
14          app:srcCompat = "@drawable/mymaths"
15          android:id = "@+id/imageMath"
16          android:adjustViewBounds = "false"
17          android:cropToPadding = "false"
18          app:layout_constraintTop_toTopOf = "@+id/activity_my_constraint"
19          app:layout_constraintBottom_toBottomOf = "@+id/activity_my_constraint"
20          android:layout_marginEnd = "16dp"
21          app:layout_constraintRight_toRightOf = "@+id/activity_my_constraint"
22          android:layout_marginStart = "16dp"
23          app:layout_constraintLeft_toLeftOf = "@+id/activity_my_constraint"
24          app:layout_constraintHorizontal_bias = "0.18"
25          app:layout_constraintVertical_bias = "0.78" />
26
27      <TextView
28          android:text = "TextView"
29          android:layout_width = "185dp"
30          android:layout_height = "44dp"
31          android:id = "@+id/tvNumber"
```

32	app:layout_constraintBottom_toBottomOf = "@ + id/activity_my_constraint"
33	android:layout_marginStart = "16dp"
34	app:layout_constraintLeft_toLeftOf = "@ + id/activity_my_constraint"
35	android:layout_marginEnd = "16dp"
36	app:layout_constraintRight_toRightOf = "@ + id/activity_my_constraint"
37	app:layout_constraintTop_toBottomOf = "@ + id/imageMath"
38	app:layout_constraintHorizontal_bias = "0.28" />
39	</android.support.constraint.ConstraintLayout>

上述代码中，第 2～9 行与第 39 行配对，表示这是一个约束布局，第 9 行设置布局的背景为图像资源 myschedule。该布局中有一个图像显示控件（ImageView）（如第 11～25 行所示，其背景图为图像资源 mymaths）和一个静态文本框控制（TextView）（如第 27～38 行所示）。上述代码用如图 4-20 所示的图形化方法进行设置。

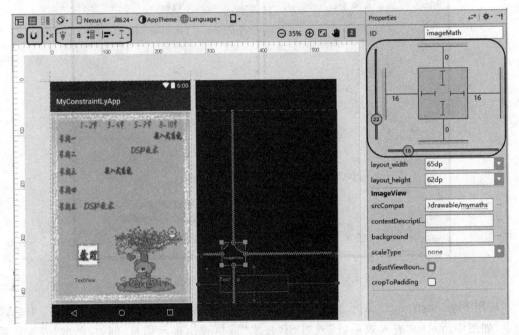

图 4-20　约束布局实例

在图 4-20 中，左上角的 U 形选中时，拖放到画布上的控件自动进行相互位置的调整，称为相互间的位置约束布局，称为自动约束布局；在上角"灯泡型"所在的框里，为推理约束布局，主要用于相邻控制位置的调整。中间的部分中有两个相同大小的画布，其中，左边为所见即所得的显示效果图，右边为画布中控件的相对位置关系图，即约束图。选中图中的 ImageView 控件，右上方出现它的位置约束关系，正方形外部的"0"和"16"表示画布离显示屏边缘的距离，正方形内部的 4 个线段表示控制离画布边缘或离与其有约束关系的控件的距离，左边的"22"和下边的"18"表示控制纵向或横向上的位置百分比，如果为"50"表示控制居中。距离的单位为 dp。

在图 4-20 中，先打开左上角的"U"，然后，放置 ImageView 控件，将会自动建立 ImageView 控制与画布边缘的约束，也可以按下控件 4 条边上的圆圈（称为锚点），拖动到画布边缘即完成约束，最后，在右上角的约束示意图中调整控件位置，约束示意图中的正方形内部 4 个小线段均可以单击而改变约束方式,直线表示固定比例,波浪线表示可变比例,箭头线表示控制大小以控件的内容为准。约束布局可以做到使任一控件放在画布的任一位置,而同时可满足任意分辨率的显示屏,即具有绝对布局的优点,同时消除了绝对布局的缺点。事实上,iPhone 应用程序的开发也是类似的布局方式。约束布局方式使得 Android 应用的界面设计进入了一个全新的时代,从而在 Android Studio 中建议尽可能使用约束布局,而不用其他布局方式,此外,约束布局可以在图 4-20 中的图形界面中完成,不用编写 XML 文件。但是,要熟练掌握约束布局,需要经过大量的应用。

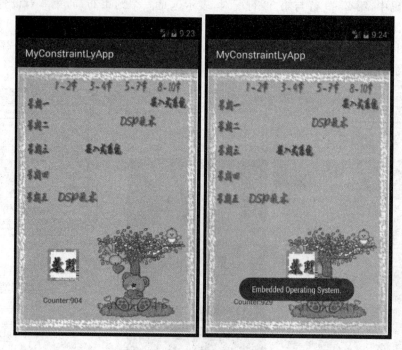

图 4-21 应用 MyConstraintLyApp 运行结果

应用 MyConstraintLyApp 的运行结果如图 4-21 所示。当单击课表(背景图片)中的课程名(如图中所示的"嵌入式系统"或"DSP 技术")时,弹出提示信息,例如,单击"嵌入式系统"弹出图 4-21 中所示的 Embedded Operating System。图中下方的图像显示控件以动画形式循环从左向右运动,文本框显示计数据器的值不断增长。源代码文件 MyConstraintAct.java 的代码如下:

```
1   package cn.edu.jxufe.zhangyong.myconstraintlyapp;
2
3   import java.util.Timer;
4   import java.util.TimerTask;
5   import cn.edu.jxufe.zhangyong.myconstraintlyapp.R.id;
6   import android.os.Message;
7   import android.support.v7.app.AppCompatActivity;
8   import android.os.Bundle;
9   import android.os.Handler;
10  import android.view.MotionEvent;
11  import android.view.animation.Animation;
```

```java
12  import android.view.animation.AnimationSet;
13  import android.view.animation.TranslateAnimation;
14  import android.widget.ImageView;
15  import android.widget.TextView;
16  import android.widget.Toast;
17
18  public class MyConstraintAct extends AppCompatActivity {
19      private ImageView imageMath;
20      private TextView tvNumber;
21      AnimationSet animationSet = new AnimationSet(true);
22      private int i = 0;
23      @Override
24      protected void onCreate(Bundle savedInstanceState) {
25          super.onCreate(savedInstanceState);
26          setContentView(R.layout.activity_my_constraint);
27
28          imageMath = (ImageView)findViewById(id.imageMath);
29          tvNumber = (TextView)findViewById(id.tvNumber);
30          tvNumber.setText("Counter:");
31          myInitGUI();
32      }
33      private final Timer timer = new Timer();
34      private void myInitGUI(){
35          timer.schedule(timerTask,2000,2000);
36          TranslateAnimation translateAnimation = new TranslateAnimation(
37                  Animation.RELATIVE_TO_SELF, 0f, Animation.RELATIVE_TO_SELF, 2.5f,
38                  Animation.RELATIVE_TO_SELF, 0f, Animation.RELATIVE_TO_SELF, 0f);
39          translateAnimation.setDuration(1000);
40          animationSet.addAnimation(translateAnimation);
41      }
42      private TimerTask timerTask = new TimerTask() {
43          @Override
44          public void run() {
45              Message msg = new Message();
46              msg.what = 1;
47              handler.sendMessage(msg);
48          }
49      };
50      private Handler handler = new Handler() {
51          @Override
52          public void handleMessage(Message msg){
53              int msgID = msg.what;
54              switch(msgID){
55                  case 1:
56                      imageMath.startAnimation(animationSet);
57                      tvNumber.setText("Counter:" + String.format("%d",i));
58                      i++;
59                      break;
60                  default:
61                      break;
62              }
```

```
63              super.handleMessage(msg);
64          }
65      };
66      public boolean onTouchEvent(MotionEvent event){
67          int nDsp = 0,nEos = 0;
68          int action = event.getAction();
69          int x = (int)event.getX();
70          int y = (int)event.getY();
71          if (((x>550) && (x<710)) && ((y>270) && (y<310))){
72              nEos = 1;
73          }
74          if (((x>280) && (x<440)) && ((y>450) && (y<490))){
75              nEos = 1;
76          }
77          if (((x>430) && (x<590)) && ((y>330) && (y<380))){
78              nDsp = 1;
79          }
80          if (((x>180) && (x<340)) && ((y>650) && (y<690))){
81              nDsp = 1;
82          }
83          switch(action){
84              case MotionEvent.ACTION_DOWN:
85                  String str1 = "Digital Signal Processer.";
86                  if (nDsp == 1){
87                      Toast.makeText(this, str1, Toast.LENGTH_SHORT).show();
88                  }
89                  break;
90              case MotionEvent.ACTION_UP:
91                  String str2 = "Embedded Operating System.";
92                  if (nEos == 1){
93                      Toast.makeText(this, str2, Toast.LENGTH_SHORT).show();
94                  }
95                  break;
96          }
97          return super.onTouchEvent(event);
98      }
99  }
```

上述代码中,第24～32行为onCreate方法,该方法中初始化图像按钮imageMath控制静态文本框tvNumber控件,同时调用myInitGUI方法(第34～41行)进行初始化定时器和动画类对象animationSet。第35行说明定时器timer的工作方法为第一次定时延时2000ms,以后每隔2000ms运行一次任务timerTask。建立定时器任务的步骤为:①定义定时器Timer对象timer,如第33行所示。②定义定时器任务TimerTask对象timerTask,并覆盖其虚方法run,在run方法中调用handler对象的sendMessage方法,如第42～48行所示。这里handler对象在第50行定义,Handler类的对象用于处理和发送消息。③定义一个Handler类对象handler,并覆盖其方法handleMessage,定时任务要实现的工作在handleMessage方法中。结合第46和第47行可知timerTask对象执行时将msg.what赋为1;第53和第54行根据收到的消息值做相应的处理,即第55～61行的代码。第56行启动动画；第 57 行设置静态文本框的计数值，第 58 行计数值累加 1。

第 66～98 行为屏幕触摸事件方法 onTouchEvent，该方法由 Android 系统调用，程序员只需要向该方法中添加代码即可。第 68 行得到触屏的动作 action；第 69、70 行得到触点坐标(x,y)。第 71～76 行表示当触点位于屏幕背景图"嵌入式系统"上时，令变量 nEos 等于 1；第 77～82 行表示当触点位于屏幕背景图"DSP 技术"上时，令变量 nDsp 等于 1。第 83～96 行为 switch 语句，当动作 action 为触笔按下时（第 84 行），如果 nDsp 为 1，则显示 "Digital Signal Processer."（第 85～88 行）；当动作 action 为触笔抬起时（第 90 行），则显示 "Embedded Operating System."（第 91～94 行）。这里为了介绍触笔事件的两种动作类型，故使用了 switch 语句，显然，这段程序应该被优化为一种动作类型。需要注意的是，第 71～82 行的判断语句中的常量值只能应用于 720×1280 的显示屏。

4.3 "计算器"工程

实现 Windows 系统附件中的科学计算器需要用到堆栈等数据结构，这里仅针对整数的加减乘除四则运算编写一个"计算器"工程，其目的在于说明 Android 单用户界面应用程序设计的思路，同时阐述 Android 应用程序的计算能力。

例 4-10 计算器工程。

新建应用 MyCalculatorApp，应用名为 MyCalculatorApp，活动界面名为 MyCalculatorAct。

设计一个计算器应用程序，首先需要进行界面设计，应用 MyCalculatorApp 使用约束布局方式，其布局后的界面如图 4-22 所示。

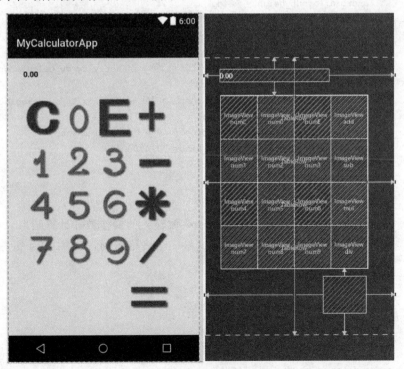

图 4-22 计算器工程布局

图 4-22 中的"0.00"处为一个静态文本框 TextView；图中的"C、0、E、+、1、2、3、-、4、5、6、*、7、8、9、/"是 16 个图像显示控件(ImageView)，采用表格布局方式，其中，每个控件都对应着两幅图像，一幅是没有被单击时的情况，另一幅是被单击时的情况。图中的"="是一个图像按钮控件(ImageButton)。图 4-22 左边为布局效果图，右边为约束关系图。布局文件 activity_my_calculator.xml 的代码如下：

```
1   <?xml version = "1.0" encoding = "utf-8"?>
2   <android.support.constraint.ConstraintLayout xmlns:android = "http://schemas.android.com/apk/res/android"
3       xmlns:app = "http://schemas.android.com/apk/res-auto"
4       xmlns:tools = "http://schemas.android.com/tools"
5       android:id = "@+id/activity_my_calculator"
6       android:layout_width = "match_parent"
7       android:layout_height = "match_parent"
8       tools:context = "cn.edu.jxufe.zhangyong.mycalculatorapp.MyCalculatorAct"
9       android:background = "@drawable/bkgrd"
10      android:clickable = "true">
11
```

第 2~10 行与第 218 行配对，表示这是一个约束布局，布局的背景为图像 bkgrd，是一个单色(黄色)背景。

```
12      <TextView
13          android:text = "0.00"
14          android:layout_width = "207dp"
15          android:layout_height = "24dp"
16          android:id = "@+id/tv_result"
17          app:layout_constraintTop_toTopOf = "@+id/activity_my_calculator"
18          android:layout_marginStart = "16dp"
19          app:layout_constraintLeft_toLeftOf = "@+id/activity_my_calculator"
20          android:layout_marginEnd = "16dp"
21          app:layout_constraintRight_toRightOf = "@+id/activity_my_calculator"
22          app:layout_constraintHorizontal_bias = "0.11"
23          app:layout_constraintBottom_toTopOf = "@+id/tableLayout"
24          app:layout_constraintVertical_bias = "0.47000003"
25          android:textColor = "@color/colorPrimary"
26          android:textStyle = "normal|bold" />
27
```

第 12~26 行为静态文本框 tv_result，用于存放输入的操作数和计算结果。

下面第 28~38 行与第 201 行配对，表示这是一个表格布局，表格中有 4 行。

```
28      <TableLayout
29          android:layout_width = "wrap_content"
30          android:layout_height = "wrap_content"
31          android:id = "@+id/tableLayout"
32          android:layout_marginStart = "16dp"
33          app:layout_constraintLeft_toLeftOf = "@+id/activity_my_calculator"
34          android:layout_marginEnd = "16dp"
35          app:layout_constraintRight_toRightOf = "@+id/activity_my_calculator"
```

```
36          app:layout_constraintBottom_toBottomOf = "@ + id/activity_my_calculator"
37          app:layout_constraintHorizontal_bias = "0.3"
38          app:layout_constraintTop_toTopOf = "@ + id/activity_my_calculator"
39          app:layout_constraintVertical_bias = "0.37">
40      < TableRow
41          android:layout_width = "match_parent"
42          android:layout_height = "match_parent"
43          android:orientation = "horizontal">
44          < ImageView
45              android:layout_width = "wrap_content"
46              android:layout_height = "wrap_content"
47              app:srcCompat = "@drawable/ivpressnumc"
48              android:id = "@ + id/numC"
49              android:onClick = "myNumClick"
50              android:clickable = "true" />
51          < ImageView
52              android:layout_width = "wrap_content"
53              android:layout_height = "wrap_content"
54              app:srcCompat = "@drawable/ivpressnum0"
55              android:id = "@ + id/num0"
56              android:onClick = "myNumClick"
57              android:clickable = "true" />
58          < ImageView
59              android:layout_width = "wrap_content"
60              android:layout_height = "wrap_content"
61              app:srcCompat = "@drawable/ivpressnume"
62              android:id = "@ + id/numE"
63              android:onClick = "myNumClick"
64              android:clickable = "true" />
65          < ImageView
66              android:layout_width = "wrap_content"
67              android:layout_height = "wrap_content"
68              app:srcCompat = "@drawable/ivpressadd"
69              android:id = "@ + id/add"
70              android:onClick = "myNumClick"
71              android:clickable = "true" />
72      </TableRow>
73
```

第 40~72 行为表格中的第一行,包括 4 个图像显示控件,依次为 "C"、"0"、"E" 和 "+",每个图像显示控制都有它的单击事件,均为 myNumClick。

```
74  < TableRow
75      android:layout_width = "match_parent"
76      android:layout_height = "match_parent"
77      android:orientation = "horizontal">
78      < ImageView
79          android:layout_width = "wrap_content"
80          android:layout_height = "wrap_content"
81          app:srcCompat = "@drawable/ivpressnum1"
82          tools:layout_editor_absoluteY = "162dp"
```

```
83              tools:layout_editor_absoluteX = "55dp"
84              android:id = "@ + id/num1"
85              android:onClick = "myNumClick"
86              android:clickable = "true"
87              android:contextClickable = "true" />
88          < ImageView
89              android:layout_width = "wrap_content"
90              android:layout_height = "wrap_content"
91              app:srcCompat = "@drawable/ivpressnum2"
92              tools:layout_editor_absoluteY = "199dp"
93              tools:layout_editor_absoluteX = "158dp"
94              android:id = "@ + id/num2"
95              android:onClick = "myNumClick"
96              android:contextClickable = "true" />
97          < ImageView
98              android:layout_width = "wrap_content"
99              android:layout_height = "wrap_content"
100             app:srcCompat = "@drawable/ivpressnum3"
101             tools:layout_editor_absoluteY = "240dp"
102             tools:layout_editor_absoluteX = "237dp"
103             android:id = "@ + id/num3"
104             android:onClick = "myNumClick"
105             android:contextClickable = "true" />
106         < ImageView
107             android:layout_width = "wrap_content"
108             android:layout_height = "wrap_content"
109             app:srcCompat = "@drawable/ivpresssub"
110             tools:layout_editor_absoluteY = "329dp"
111             tools:layout_editor_absoluteX = "137dp"
112             android:id = "@ + id/sub"
113             android:onClick = "myNumClick"
114             android:clickable = "true" />
115         </TableRow>
116
```

第 74～115 行为表格中的第二行，包括 4 个图像显示控件，依次为"1"、"2"、"3"和"一"，每个图像显示控件都有它的单击事件，均为 myNumClick。

```
117         < TableRow
118             android:layout_width = "match_parent"
119             android:layout_height = "match_parent"
120             android:orientation = "horizontal">
121             < ImageView
122                 android:layout_width = "wrap_content"
123                 android:layout_height = "wrap_content"
124                 app:srcCompat = "@drawable/ivpressnum4"
125                 tools:layout_editor_absoluteY = "290dp"
126                 tools:layout_editor_absoluteX = "47dp"
127                 android:id = "@ + id/num4"
128                 android:onClick = "myNumClick"
129                 android:clickable = "true" />
```

```
130        < ImageView
131            android:layout_width = "wrap_content"
132            android:layout_height = "wrap_content"
133            app:srcCompat = "@drawable/ivpressnum5"
134            tools:layout_editor_absoluteY = "307dp"
135            tools:layout_editor_absoluteX = "136dp"
136            android:id = "@ + id/num5"
137            android:onClick = "myNumClick"
138            android:clickable = "true" />
139        < ImageView
140            android:layout_width = "wrap_content"
141            android:layout_height = "wrap_content"
142            app:srcCompat = "@drawable/ivpressnum6"
143            tools:layout_editor_absoluteY = "363dp"
144            tools:layout_editor_absoluteX = "213dp"
145            android:id = "@ + id/num6"
146            android:onClick = "myNumClick"
147            android:clickable = "true" />
148        < ImageView
149            android:layout_width = "wrap_content"
150            android:layout_height = "wrap_content"
151            app:srcCompat = "@drawable/ivpressmul"
152            tools:layout_editor_absoluteY = "283dp"
153            tools:layout_editor_absoluteX = "228dp"
154            android:id = "@ + id/mul"
155            android:onClick = "myNumClick"
156            android:clickable = "true" />
157        </TableRow >
158
```

第 117～157 行为表格中的第三行,包括 4 个图像显示控件,依次为"4"、"5"、"6"和"*",每个图像显示控件都有它的单击事件,均为 myNumClick。

```
159    < TableRow
160        android:layout_width = "match_parent"
161        android:layout_height = "match_parent"
162        android:orientation = "horizontal">
163        < ImageView
164            android:layout_width = "wrap_content"
165            android:layout_height = "wrap_content"
166            app:srcCompat = "@drawable/ivpressnum7"
167            tools:layout_editor_absoluteY = "301dp"
168            tools:layout_editor_absoluteX = "31dp"
169            android:id = "@ + id/num7"
170            android:onClick = "myNumClick"
171            android:clickable = "true" />
172        < ImageView
173            android:layout_width = "wrap_content"
174            android:layout_height = "wrap_content"
175            app:srcCompat = "@drawable/ivpressnum8"
176            tools:layout_editor_absoluteY = "301dp"
```

```
177            tools:layout_editor_absoluteX = "101dp"
178            android:id = "@ + id/num8"
179            android:onClick = "myNumClick"
180            android:clickable = "true" />
181        < ImageView
182            android:layout_width = "wrap_content"
183            android:layout_height = "wrap_content"
184            app:srcCompat = "@drawable/ivpressnum9"
185            tools:layout_editor_absoluteY = "301dp"
186            tools:layout_editor_absoluteX = "171dp"
187            android:id = "@ + id/num9"
188            android:onClick = "myNumClick"
189            android:clickable = "true" />
190        < ImageView
191            android:layout_width = "wrap_content"
192            android:layout_height = "80dp"
193            app:srcCompat = "@drawable/ivpressdiv"
194            tools:layout_editor_absoluteY = "301dp"
195            tools:layout_editor_absoluteX = "241dp"
196            android:id = "@ + id/div"
197            android:onClick = "myNumClick"
198            android:clickable = "true"
199            android:baselineAlignBottom = "true" />
200        </TableRow>
201    </TableLayout>
202
```

第 159～200 行为表格中的第 4 行,包括 4 个图像显示控件,依次为"7"、"8"、"9"和"/",每个图像显示控件都有它的单击事件,均为 myNumClick。

```
203    < ImageButton
204        android:layout_width = "81dp"
205        android:layout_height = "68dp"
206        app:srcCompat = "@drawable/eql"
207        android:id = "@ + id/calc"
208        app:layout_constraintBottom_toBottomOf = "@ + id/activity_my_calculator"
209        android:layout_marginEnd = "16dp"
210        app:layout_constraintRight_toRightOf = "@ + id/activity_my_calculator"
211        android:layout_marginStart = "16dp"
212        app:layout_constraintLeft_toLeftOf = "@ + id/activity_my_calculator"
213        app:layout_constraintTop_toBottomOf = "@ + id/tableLayout"
214        app:layout_constraintHorizontal_bias = "0.85"
215        app:layout_constraintVertical_bias = "0.26"
216        android:onClick = "myCalcClick" />
217
218 </android.support.constraint.ConstraintLayout >
```

第 203～216 行为图像显示控制,显示图像"=",它的单击事件为 myCalcClick。

上述工程布局的设计需要制作 33 张图片,其中一张("=")用作图像按钮控件的显示图像;16 张用作数字"0～9"以及"C、E、+、-、*和/"的图像显示,另外 16 张用作这些数字和符号被单击时的图像显示。创建这些图像较快的方法是借助于免费的"可牛影像"软件。创建好的图像存放在资源 res 的 drawable 目录下。在 drawable 目录下,还要存放 ivpressnum0.xml、ivpressnum1.xml、ivpressnum2.xml、ivpressnum3.xml、ivpressnum4.xml、ivpressnum5.xml、ivpressnum6.xml、ivpressnum7.xml、ivpressnum8.xml、ivpressnum9.xml、ivpressadd.xml、ivpresssub.xml、ivpressmul.xml、ivpressdiv.mxl、ivpressnumc.xml 和 ivpressnume.xml(文件名均需要用小写字母、数字或下画线,不能使用大写字母),这些文件链接到布局文件 activity_my_calculator.xml 中,用于表示按下某个图像显示控件时图像将切换为按下状态的图像,例如,ivpressnum0.xml 链接到布局文件 activity_my_calculator.xml 的第 54 行,ivpressnum1.xml 链接到布局文件的第 81 行,等等。这些文件的格式比较固定,例如 ivpressnum0.xml:

```
1   <?xml version = "1.0" encoding = "utf-8"?>
2   <selector xmlns:android = "http://schemas.android.com/apk/res/android">
3       <item android:state_pressed = "false"
4           android:drawable = "@drawable/num0" />
5       <item android:state_pressed = "true"
6           android:drawable = "@drawable/num0p" />
7   </selector>
```

上述代码中,第 3 和第 4 行表示没有按下"数字 0"时的图标为 num0;第 5 和第 6 行表示按下"数字 0"时的图标为 nump0。drawable 目录下的其他文件,只需要修改第 4 行的 num0 和第 6 行的 nump0 为没有按下时和按下时的图标即可。

布局完成后,可开始 Java 语言源程序的设计,这里源文件 MyCalculatorAct.java 的代码较长,下面分段解释其工作原理:

```
1   package cn.edu.jxufe.zhangyong.mycalculatorapp;
2   
3   import android.support.v7.app.AppCompatActivity;
4   import android.os.Bundle;
5   import android.view.View;
6   import android.widget.ImageView;
7   import android.widget.TextView;
8   
9   public class MyCalculatorAct extends AppCompatActivity {
10      public ImageView num0,num1,num2,num3,num4;
11      public ImageView num5,num6,num7,num8,num9,numc,nume;
12      public ImageView myadd,mysub,mymul,mydiv;
13      public TextView tv_res;
14      public double op1,op2,res;
15      public int opmethod;
16      public int numpress;
17      
18      @Override
19      protected void onCreate(Bundle savedInstanceState) {
20          super.onCreate(savedInstanceState);
21          setContentView(R.layout.activity_my_calculator);
22
```

```
23        myInitGUI();
24    }
25
```

上述代码中第 10～16 行定义类 MyCalculatorAct 的公有数据成员(有些数据变量建议定义为私有成员),其中,第 10～13 行定义了各种控件对象,第 14～16 行定义的变量意义为:op1 和 op2 存储参与运算的两个操作数,res 存储运算结果,这三个变量均为 double 型;opmethod 的值为 1、2、3 或 4 时分别表示加、减、乘或除运算,当 opmethod 的值为 0 时,输入的值赋给操作数 op1,当 opmethod 的值为 1～4 时,输入的值赋给操作数 op2。变量 numpress 保存输入的数值(即单击计算器界面上的数字值),如果 numpress 等于 10,表示单击了计算器界面上的"C"或"E","C"表示取消本次计算;"E"表示撤销上一次的数字输入。

在第 23 行调用 myInitGUI 方法进行用户初始化工作,这是本书的程序风格,将这些初始化代码放在一个公有成员方法 myInitGUI 中,而不是放在 onCreate 方法中,这样程序的可读性较强。

```
26    public void myInitGUI(){
27        tv_res = (TextView)findViewById(R.id.tv_result);
28        tv_res.setText("");
29        op1 = 0;    op2 = 0;    res = 0;
30        opmethod = 0; //1-+,2--,3-*,4-/
31        numpress = 0;
32    }
33
```

上述的公有方法 myInitGUI 中,第 27 行获得静态文本框表示的控件对象 tv_result,第 28 行将静态文本框对象 tv_res 设为空白显示。第 29～31 行初始化操作数 op1 和 op2 为 0,运算结果 res 设为 0,numpress 变量设为 0。

```
34    public void myNumClick(View v){
35        switch(v.getId()){
36            case R.id.num0:
37                numpress = 0;
38                break;
```

如果按下的控件为"数字 0",则将数字 0 保存在按键值变量 numpress 中。下面的第 39～65 行依次判断按键是数字 1～9,并相应地将按键值保存在变量 numpress 中。

```
39            case R.id.num1:
40                numpress = 1;
41                break;
42            case R.id.num2:
43                numpress = 2;
44                break;
45            case R.id.num3:
46                numpress = 3;
47                break;
48            case R.id.num4:
```

```
49          numpress = 4;
50          break;
51     case R.id.num5:
52          numpress = 5;
53          break;
54     case R.id.num6:
55          numpress = 6;
56          break;
57     case R.id.num7:
58          numpress = 7;
59          break;
60     case R.id.num8:
61          numpress = 8;
62          break;
63     case R.id.num9:
64          numpress = 9;
65          break;
66     case R.id.add:
67          numpress = 10;
68          opmethod = 1;
69          break;
```

第66～69行表示如果按下的是"＋"号,则 opmethod 置为1,同时,numpress 置为10。下面的第70～81行表示如果按下的键是"－"、"＊"或"/",则相应地将 opmethod 置为2、3或4,同时 numpress 置为10。

```
70     case R.id.sub:
71          numpress = 10;
72          opmethod = 2;
73          break;
74     case R.id.mul:
75          numpress = 10;
76          opmethod = 3;
77          break;
78     case R.id.div:
79          numpress = 10;
80          opmethod = 4;
81          break;
82     case R.id.numC:
83          op1 = 0;
84          op2 = 0;
85          opmethod = 0;
86          numpress = 10;
87          tv_res.setText("0.00");
88          break;
```

第82～88行表示按下了"C"键(取消键),则清除两个运算数 op1 和 op2 的值,且运算方法 opmethod 清0,numpress 置为10,同时显示"0.00"。

```
89     case R.id.numE:
90          if(opmethod == 0){
```

```
 91            op1 = Math.floor(op1/10);
 92            tv_res.setText(String.format("%1$.2f",op1));
 93        }
 94        if(opmethod>=1 && opmethod<=4){
 95            op2 = Math.floor(op2/10);
 96            tv_res.setText(String.format("%1$.2f",op2));
 97        }
 98        numpress = 10;
 99        break;
100 }
```

第89～99行表示如果按下了"E"键（擦除键），则根据当前的情况，将运算数op1或op2的最低位清除，并将numpress置为10。

```
101 if(numpress<10){
102    switch(opmethod){
103        case 0: //op1
104            op1 = op1 * 10 + numpress;
105            tv_res.setText(String.format("%1$.2f",op1));
106            numpress = 10;
107            break;
108        case 1: //add - op2
109        case 2: //sub - op2
110        case 3: //mul - op2
111        case 4: //div - op2
112            op2 = op2 * 10 + numpress;
113            tv_res.setText(String.format("%1$.2f",op2));
114            numpress = 10;
115            break;
116        }
117    }
118 }
119
```

第101～117行表示，如果numpress小于10（即按下了有效的数字键），则第103～107行将numpress作为运算数op1的新的个位数；第108～115行将numpress作为运算数op2的新的个位数，并把它们显示出来。

```
120 public void myCalcClick(View v){
121    res = 0;
122    switch (opmethod){
123        case 1:
124            res = op1 + op2;
125            tv_res.setText(String.format("%1$.0f + %2$.0f = %3$.2f",op1,
                op2,res));
126            break;
127        case 2:
128            res = op1 - op2;
129            tv_res.setText(String.format("%1$.0f - %2$.0f = %3$.2f",op1,
                op2,res));
130            break;
```

```
131                case 3:
132                    res = op1 * op2;
133                    tv_res.setText(String.format("%1$.0f * %2$.0f = %3$.2f",op1,
                           op2,res));
134                    break;
135                case 4:
136                    res = op1 / op2;
137                    tv_res.setText(String.format("%1$.0f / %2$.0f = %3$.2f",op1,
                           op2,res));
138                    break;
139            }
140            op1 = 0;
141            op2 = 0;
142            opmethod = 0;
143            numpress = 10;
144        }
145    }
```

上述第 120~144 行为图像按钮"="的单击事件方法 myCalcClick，该方法首先将运算结果设置为 0（第 121 行），然后，根据 opmethod 的值选择相应的运算，如果 opmethod 变量值为 1，则第 124~126 行代码得到执行，即执行两个操作数相加的运算，将结果保存在 res 变量中，第 125 行显示运算过程及其结果，保留两位小数。

应用 MyCalculatorApp 的运行结果如图 4-23 所示，例如，计算 48/129 的值，其结果为 0.37。实验证明，应用 MyCalculatorApp 的计算器运行稳定，并且可以进一步扩充其功能。

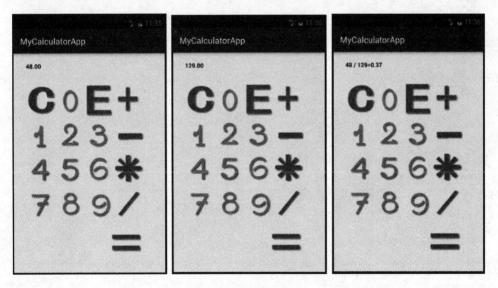

图 4-23　计算器功能演示

4.4　本章小结

Activity（活动界面）是 Android 应用程序与用户交互的界面，用于管理用户界面控件及其事件响应方法。Android 应用程序主要针对屏幕相对较小且分辨率相对较低的移动设备，一般借助 XML 格式布局文件进行快速界面设计。控件事件有两种响应方法，除了常用的匿名内部类监听方法外，还有一种借助布局文件指定 onClick 属性的方法。Activity 界面的布局方式主要有 5 种，即线性布局、相对布局、框架布局、表格布局和约束布局，此外还有几种常用的布局，如 ScrollView 滚动屏幕布局和 TabHost 选项卡布局等。但是 Android 4.4 以后，约束布局成为底层主要的布局方式。这些布局方式可以相互嵌套使用，例如线性布局中可以添加到约束布局等。Android 系统提供的用户控件是最基本的设计元素，几乎满足目前所有可能的应用需要，控件的应用设计包括两个方面，即设置它的属性和编写它的事件方法，多个控件协调工作组合成完整的用户界面。

第 5 章

多用户界面应用设计

在 Android 应用程序中,一个用户界面对应着一个 XML 布局文件,因此,对于多用户界面应用设计而言,需要设计多个 XML 布局文件。一般地,多用户界面应用的多个用户界面间有数据信息的交流,除了借助公有对象外,还广泛使用意图类(Intent)对象请求和管理这类数据。本章首先介绍 Intent 的概念,然后介绍对话框设计方法和菜单技术,最后讲述多界面应用设计的思想和方法。

5.1 Intent 概念

Intent(译为意图)提供了不同 Activity(活动界面)间数据交换的方法,被视为 Activity 之间的纽带,它所传递的信息主要是动作(Action)和数据(data),即要执行的动作和要操作的数据,动作使用 Android 系统预定义的常量表示,例如 ACTION_MAIN、ACTION_VIEW 和 ACTION_EDIT 等;数据使用 URI(统一资源标识符)表示。借助 Intent 对象,调用方法 startActivity 可启动一个新的界面;调用方法 startService 或 bindService 可与服务(没有用户界面的应用程序)通信;调用方法 sendBroadcast 与所有广播接收器通信。此外,Intent 还具有 category(分类)、type(类型)、component(组件)和 extras(附加信息)等属性,其中,category 为动作提供分类信息;type 用于显式指定 MIME(多用途网络邮件扩展)类型;component 显式指定 Intent 使用的组件类;extras 是一个 Bundle 对象,包括附加的数据信息。

有两种使用 Intent 的方法,即显式使用和隐式使用。显式 Intent 通过调用方法 setClass 或 setComponent 运行一个指定的类,这是应用程序装入新的活动界面(Activity) 常用的方法。隐式 Intent 不指定特定的组件或类,由 Android 系统寻找与该 Intent 描述的动作和数据匹配的组件执行,这一种过程称为 Intent 解析机制,Intent 解析器将 Intent 映射到一个匹配的 Activity、广播接收器(BroadcastReceiver)或服务(Service)。Intent 解析器根据应用程序的 AndroidManifest.xml 文件中 IntentFilter 包含的动作(action)、类型(type)或分类(category)等信息,决定 Intent 信息传递的对象。

下述内容摘译自 Android 开发者手册中关于 Intent 的描述。

为了介绍隐式 Intent 的使用方法,下面以"记事本"应用程序的 AndroidManifest.xml 为例说明,该文件内容如下:

```
1   <manifest xmlns:android = "http://schemas.android.com/apk/res/android"
2       package = "com.android.notepad">
3       <application android:icon = "@drawable/app_notes" android:label =
4           "@string/app_name">
5
6           <provider class = ".NotePadProvider"
7               android:authorities = "com.google.provider.NotePad" />
8
9           <activity class = ".NotesList" android:label = "@string/title_notes_list">
10              <intent - filter>
11                  <action android:name = "android.intent.action.MAIN" />
12                  <category android:name = "android.intent.category.LAUNCHER" />
13              </intent - filter>
14              <intent - filter>
15                  <action android:name = "android.intent.action.VIEW" />
16                  <action android:name = "android.intent.action.EDIT" />
17                  <action android:name = "android.intent.action.PICK" />
18                  <category android:name = "android.intent.category.DEFAULT" />
19                  <data android:mimeType = "vnd.android.cursor.dir/vnd.google.note" />
20              </intent - filter>
21              <intent - filter>
22                  <action android:name = "android.intent.action.GET_CONTENT" />
23                  <category android:name = "android.intent.category.DEFAULT" />
24                  <data android:mimeType = "vnd.android.cursor.item/vnd.google.note" />
25              </intent - filter>
26          </activity>
27
28          <activity class = ".NoteEditor" android:label = "@string/title_note">
29              <intent - filter android:label = "@string/resolve_edit">
30                  <action android:name = "android.intent.action.VIEW" />
31                  <action android:name = "android.intent.action.EDIT" />
32                  <category android:name = "android.intent.category.DEFAULT" />
33                  <data android:mimeType = "vnd.android.cursor.item/vnd.google.note" />
34              </intent - filter>
35
36              <intent - filter>
37                  <action android:name = "android.intent.action.INSERT" />
38                  <category android:name = "android.intent.category.DEFAULT" />
39                  <data android:mimeType = "vnd.android.cursor.dir/vnd.google.note" />
40              </intent - filter>
```

```
41
42                </activity>
43
44                <activity class = ".TitleEditor" android:label = "@string/title_edit_title"
                        android:theme = "@android:style/Theme.Dialog">
45                    <intent-filter android:label = "@string/resolve_title">
46                        <action android:name = "com.android.notepad.action.EDIT_TITLE" />
47                        <category android:name = "android.intent.category.DEFAULT" />
48                        <category android:name = "android.intent.category.ALTERNATIVE" />
49                        <category android:name = "android.intent.category.SELECTED_
                            ALTERNATIVE" />
50                        <data android:mimeType = "vnd.android.cursor.item/vnd.google.note" />
51                    </intent-filter>
52                </activity>
53
54        </application>
55 </manifest>
```

上述应用程序配置代码中，有三个 Activity，第 9~26 行为第一个 Activity，第 28~42 行为第二个 Activity，第 44~53 行为第三个 Activity。

在第一个 Activity 中有三个 IntentFilter，描述了三类动作和数据信息。第 10~13 行为第一个 IntentFilter，标准的 MAIN 动作表示这是应用程序的主程序入口点，android.intent.action.MAIN 是 ACTION_MAIN 常量在 AndroidManifest.xml 中的写法；LAUNCHER 分类说明该应用程序入口点应被放置在启动队列中（Android 操作系统是多任务多线程操作系统，可以同时执行多个应用程序，当前占用 CPU 的程序称为处于运行态，而处于请求 CPU 状态的应用程序称为就绪态，多个就绪态的应用程序组成队列等待 CPU）。

第 14~20 行为第二个 IntentFilter，第 19 行描述了数据类型，即 URI 地址 vnd.android.cursor.dir/vnd.google.note，表示光标选中的一个或多个数据将作为 Intent 传递的数据信息；第 15~17 行说明这些数据可以被查看、编辑或选取（并返回给它的调用者，由 PICK 动作发出）；第 18 行为 DEFAULT 分类，调用方法 startActivity 启动活动界面时需要该分类的支持。

第 21~25 行为第三个 IntentFilter，第 22 行的 GET_CONTENT 动作与第 17 行的 PICK 动作相似，都是选取数据并传递给它的调用者，只是 GET_CONTENT 动作选取的数据类型由调用者而不是用户指定。

上述三个 Intent 的动作在第一个 Activity 中都将被解析，即该 Activity 中可执行这些 Intent 描述的动作。该 Activity 中的三个 IntentFilter 将被解析成下述形式：

```
{ action = android.app.action.MAIN }
{ action = android.app.action.MAIN, category = android.app.category.LAUNCHER }
{ action = android.app.intent.action.VIEW data = content://com.google.provider.NotePad/notes }
{ action = android.app.action.PICK data = content://com.google.provider.NotePad/notes }
{ action = android.app.action.GET_CONTENT type = vnd.android.cursor.item/vnd.google.note }
```

第二个 Activity（第 28~42 行）中具有两个 IntentFilter，需要说明的是第 37 行的动作

INSERT,该动作允许用户插入新的内容。该 Activity 中的两个 IntentFilter 将被解析成如下形式：

{ action = android.intent.action.VIEW data = content://com.google.provider.NotePad/notes/{ID} }
{ action = android.app.action.EDIT data = content://com.google.provider.NotePad/notes/{ID} }
{ action = android.app.action.INSERT data = content://com.google.provider.NotePad/notes }

第三个 Activity（第 44～53 行）中只有一个 IntentFilter，第 47 行为自定义的 EDIT_TITLE 动作，表示可以修改记事本的标题，其数据类型必须为 vnd.android.cursor.item/vnd.google.note，第 49 和第 50 行的分类 ALTERNATIVE 和 SELECTED_ALTERNATIVE 允许用户借助方法 queryIntentActivityOptions 或 addIntentOptions 执行特殊的 Intent 动作。该 IntentFilter 将被解析成

{action = com.android.notepad.action.EDIT_TITLE data = content://com.google.provider.NotePad/notes/{ID} }

通过查阅 Android 开发者手册，可知 Intent 具有 ACTION_MAIN、ACTION_VIEW 和 ACTION_ATTACH_DATA 等 20 种标准 Activity 动作；具有 ACTION_TIME_TICK、ACTION_TIME_CHANGED 和 ACTION_BOOT_COMPLETED 等 14 种标准广播动作；具有 CATEGORY_DEFAULT、CATEGORY_BROWSABLE 和 CATEGORY_LAUNCHER 等 16 种标准分类（Category）；具有 EXTRA_TEXT、EXTRA_TITLE 和 EXTRA_UID 等 29 种标准附加（Extra）数据。在 Android 开发者手册（http://android.xsoftlab.net/reference/packages.html）中有关于这些内容的详细解释，在显示页面的 Search 栏中输入"Intent"，然后单击弹出的项 android.content.Intent，将显示 Intent 类及上述内容。此外，程序员还可以自定义 Intent 动作。

5.2 对话框

在 Windows 应用程序中，对话框属于窗口；而 Android 系统中，对话框类与视图类没有关系，对话框被视为活动界面的控件。与对话框有关的类的继承关系为：android.app.Dialog 类直接继承类 java.lang.Object，android.app.AlertDialog 类继承 Dialog 类，而 android.app.ProgressDialog 类继承 AlertDialog 类。对话框在 Android 应用程序中应用广泛，常用的创建对话框的方法有三种，即使用 AlertDialog.Builder 类创建 AlertDialog 对话框；使用自定义对话框布局创建具有复杂用户界面的对话框；借助 Dialog 类创建通用对话框。AlertDialog 常译为警示对话框，事实上，AlertDialog 可以创建各种类型的对话框。

5.2.1 AlertDialog 对话框

使用 AlertDialog.Builder 类创建 AlertDialog 对话框的步骤如下：
（1）创建 AlertDialog.Builder 对象。
（2）调用 AlertDialog.Builder 对象的方法 setTitle 和 setMessage 添加对话框的标题和提示信息。

(3) 调用 AlertDialog. Builder 对象的方法 setView 向对话框中添加控件(即设置控件)。

(4) 调用 AlertDialog. Builder 对象的方法 setPositiveButton、setNeutralButton 或 setNegativeButton 设置显示在对话框下方的左、中和右边的三个按钮,这三个方法都具有两个参数,其一为显示在按钮上的文字;其二为该按钮的监听事件方法。

(5) 在 setPositiveButton、setNeutralButton 或 setNegativeButton 按钮的监听事件方法中输入需要实现的操作代码。

(6) 调用 AlertDialog. Builder 对象的 create 方法创建一个 AlertDialog 对象。

(7) 调用 AlertDialog 对象的 show 方法显示对话框。

例 5-1 AlertDialog 对话框实例。

新建应用 MyAlertdialogApp,应用名为 MyAlertdialogApp,包名为 cn. edu. jxufe. zhangyong. myalertdialogapp,活动界面名为 MyAlertdialogAct。向工程中添加 myguicolor. xml 文件(与例 4-8 中的同名文件内容相同),添加 mystrings_hz. xml 文件,定义两个汉字字符串 strnameques 和 strnameinpt,其内容如下:

```
1  <?xml version = "1.0" encoding = "utf - 8"?>
2  < resources >
3      < string name = "strnameques">你的名字叫什么?</string >
4      < string name = "strnameinpt">请输入</string >
5  </resources >
```

应用 MyAlertdialogApp 的布局包括两个 TextView 控件和一个命令按钮控件 Button,Button 控件显示文字"请输入",TextView 控件 tvNameQuestion 显示提示文字"你的名字叫什么?",而另一个 TextView 控件显示输入的姓名。布局文件 activity_my_alertdialog. xml 的内容如下:

```
1   <?xml version = "1.0" encoding = "utf - 8"?>
2   < android. support. constraint. ConstraintLayout xmlns: android = "http://schemas. android.
    com/apk/res/android"
3       xmlns:app = "http://schemas. android. com/apk/res - auto"
4       xmlns:tools = "http://schemas. android. com/tools"
5       android:id = "@ + id/activity_my_alertdialog"
6       android:layout_width = "match_parent"
7       android:layout_height = "match_parent"
8       tools:context = "cn. edu. jxufe. zhangyong. myalertdialogapp. MyAlertdialogAct">
9
10      < Button
11          android:id = "@ + id/btInputName"
12          android:layout_width = "91dp"
13          android:layout_height = "wrap_content"
14          android:text = "@string/strnameinpt"
15          android:onClick = "myInputName"
16          app:layout_constraintTop_toBottomOf = "@ + id/tvNameQuestion"
17          android:layout_marginStart = "16dp"
18          app:layout_constraintLeft_toLeftOf = "@ + id/activity_my_alertdialog"
19          android:layout_marginEnd = "16dp"
```

```
20          app:layout_constraintRight_toRightOf = "@ + id/activity_my_alertdialog"
21          app:layout_constraintBottom_toBottomOf = "@ + id/activity_my_alertdialog"
22          app:layout_constraintHorizontal_bias = "0.07"
23          app:layout_constraintVertical_bias = "0.04"
24          style = "@android:style/Widget.Button">
25      </Button>
26
27      <TextView
28          android:id = "@ + id/tvNameQuestion"
29          android:layout_width = "165dp"
30          android:layout_height = "32dp"
31          android:text = "@string/strnameques"
32          app:layout_constraintTop_toTopOf = "@ + id/activity_my_alertdialog"
33          app:layout_constraintBottom_toBottomOf = "@ + id/activity_my_alertdialog"
34          android:layout_marginStart = "16dp"
35          app:layout_constraintLeft_toLeftOf = "@ + id/activity_my_alertdialog"
36          android:layout_marginEnd = "16dp"
37          app:layout_constraintRight_toRightOf = "@ + id/activity_my_alertdialog"
38          app:layout_constraintHorizontal_bias = "0.08"
39          app:layout_constraintVertical_bias = "0.07">
40      </TextView>
41
42      <TextView
43          android:id = "@ + id/tvNameAnswer"
44          android:layout_width = "123dp"
45          android:layout_height = "27dp"
46          android:text = " "
47          android:layout_marginStart = "8dp"
48          app:layout_constraintLeft_toRightOf = "@ + id/btInputName"
49          android:layout_marginEnd = "16dp"
50          app:layout_constraintRight_toRightOf = "@ + id/activity_my_alertdialog"
51          app:layout_constraintHorizontal_bias = "0.37"
52          app:layout_constraintVertical_bias = "0.060000002"
53          app:layout_constraintBaseline_toBaselineOf = "@ + id/btInputName">
54      </TextView>
55
56  </android.support.constraint.ConstraintLayout>
```

上述代码中,第10~25行为Button控件,其ID号为btInputName,显示的字符串来自资源"@string/strnameinpt",即"请输入",该控件的单击事件方法为myInputName。第27~40行为静态文本框TextView控件,其ID号为tvNameQuestion,显示资源"@string/strnameques"中的字符串"你的名字叫什么?"。第42~54行为另一个TextView控件,其ID号为tvNameAnswer,用于显示从对话框中得到的字符串。

程序文件MyAlertdialogAct.java的工作过程是这样的:程序运行后,系统将创建类MyAlertdialogAct的对象,通过这个对象启动onCreate方法,即程序启动后将自动执行onCreate方法;在onCreate方法中调用自定义的私有方法myInitGUI对程序进行初始化,这些初始化包括创建Builder类对象、设置该对象的显示信息、动作和界面,调用其create方法创建AlertDialog对话框对象;当用户单击命令按钮"请输入"时,弹出AlertDialog对话框界面,在对话框中的 EditText 控件中输入文字,单击 OK 按钮,关闭对话框的同时把输入的信息显示在活动界面上。应用 MyAlertdialogApp 的执行结果如图 5-1 所示。

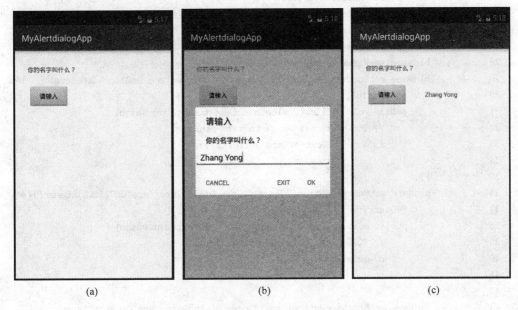

图 5-1　工程 ex05_01 运行结果

在图 5-1(a)中单击按钮"请输入"后弹出图 5-1(b)所示的对话框,在其中的编辑框中输入"Zhang Yong",然后单击 OK 按钮进入图 5-1(c)所示的界面,显示"你的名字叫什么? Zhang Yong"。

程序文件 MyAlertdialogAct.java 的内容如下:

```
1   package cn.edu.jxufe.zhangyong.myalertdialogapp;
2
3   import android.content.DialogInterface;
4   import android.support.v7.app.AlertDialog;
5   import android.support.v7.app.AppCompatActivity;
6   import android.os.Bundle;
7   import android.view.View;
8   import android.widget.EditText;
9   import android.widget.TextView;
10
11  public class MyAlertdialogAct extends AppCompatActivity {
12      private TextView tvName;
13      private AlertDialog.Builder bldName;
14      private AlertDialog dlgName;
15      private EditText etName;
16      @Override
17      protected void onCreate(Bundle savedInstanceState) {
18          super.onCreate(savedInstanceState);
19          setContentView(R.layout.activity_my_alertdialog);
20          myInitGUI();
21      }
```

```
22      private void myInitGUI(){
23          tvName = (TextView)findViewById(R.id.tvNameAnswer);
24          etName = new EditText(this);
25          bldName = new AlertDialog.Builder(MyAlertdialogAct.this);
26          bldName.setTitle(R.string.strnameinpt);
27          bldName.setMessage(R.string.strnameques);
28          bldName.setView(etName);
29          bldName.setPositiveButton("OK", new DialogInterface.OnClickListener() {
30              @Override
31              public void onClick(DialogInterface dialog, int which) {
32                  // TODO Auto-generated method stub
33                  tvName.setText(etName.getText());
34              }
35          });
36          bldName.setNeutralButton("Cancel", new DialogInterface.OnClickListener(){
37              @Override
38              public void onClick(DialogInterface dialog, int which) {
39                  // TODO Auto-generated method stub
40                  dlgName.dismiss();
41              }
42          });
43          bldName.setNegativeButton("Exit", new DialogInterface.OnClickListener() {
44              @Override
45              public void onClick(DialogInterface dialog, int which) {
46                  // TODO Auto-generated method stub
47                  MyAlertdialogAct.this.finish();
48              }
49          });
50          dlgName = bldName.create();
51      }
52      public void myInputName(View v){
53          dlgName.show();
54      }
55  }
```

上述代码中,第12~15行依次定义私有TextView型对象tvName、Builder类型对象bldName、AlertDialog类型对象dlgName和EditText类型对象etName。

第22~51行为myInitGUI方法。第23行得到tvName对象;第24行创建etName对象,该对象被放置在对话框中。第25行创建Builder对象bldName,第26行调用setTitle方法设置对话框的标题;第27行调用setMessage方法设置对话框显示的提示信息;第28行调用setView方法将etName设置为对话框中显示的编辑框;第29~35行调用setPositiveButton方法设置对话框下方左边显示的控件,其显示字符串为"OK",采用匿名内部类实现其单击的事件方法onClick,第33行表示如果该按钮被单击,则etName中的字符串将被显示在tvName静态文本框中;第36~42行调用setNeutralButton方法设置对话框下方靠右显示的控件,其显示字符串为"Cancel",其单击事件(第38~41行)为关闭对话框;第43~49行调用方法setNegativeButton设置对话框下方居中显示的按钮控件,其显示字符串为"Exit",其事件方法(第45~48行)为退出应用程序,即调用finish方法关闭整个应用程序。三个按钮的位置如图5-1(b)所示。第50行调用对象bldName的create方法创建AlertDialog对象dlgName。

第52~54行为命令按钮"请输入"的单击事件，即执行AlertDialog对象的show方法显示对话框。

使用AlertDialog.Builder类创建和显示对话框是最常用的一种对话框使用方法。第25~50行代码可以写成一条语句，如下所示：

```
1   dlgName = new AlertDialog.Builder(MyAlertdialogAct.this)
2           .setTitle(R.string.strnameinpt)
3           .setMessage(R.string.strnameques)
4           .setView(etName)
5           .setPositiveButton("OK", new DialogInterface.OnClickListener() {
6               @Override
7               public void onClick(DialogInterface dialog, int which) {
8                   // TODO Auto-generated method stub
9                   tvName.setText(etName.getText());
10              }
11          })
12          .setNeutralButton("Cancel", new DialogInterface.OnClickListener(){
13              @Override
14              public void onClick(DialogInterface dialog, int which) {
15                  // TODO Auto-generated method stub
16                  dlgName.dismiss();
17              }
18          })
19          .setNegativeButton("Exit", new DialogInterface.OnClickListener() {
20              @Override
21              public void onClick(DialogInterface dialog, int which) {
22                  // TODO Auto-generated method stub
23                  MyAlertdialogAct.this.finish();
24              }
25          }).create();
```

写成一条语句的优点在于可以不用定义Builder对象bldName，其缺点在于程序的可读性变差。

5.2.2 自定义对话框

对于具有复杂显示界面的对话框，可借助于布局文件实现，从而有效地节省创建对话框的代码。若要显示图5-2所示的对话框，可以使用下面的myfavdialog.xml布局文件。

```
1   <?xml version="1.0" encoding="utf-8"?>
2   <android.support.constraint.ConstraintLayout xmlns:android="http://schemas.android.com/apk/res/android"
3       xmlns:app="http://schemas.android.com/apk/res-auto"
4       xmlns:tools="http://schemas.android.com/tools"
5       android:orientation="vertical"
6       android:layout_width="match_parent"
```

```
7           android:layout_height = "match_parent"
8           android:id = "@ + id/myfavdlg">
9       <TextView
10          android:id = "@ + id/tvfav"
11          android:layout_width = "wrap_content"
12          android:layout_height = "wrap_content"
13          android:text = "@string/strfavsub"
14          app:layout_constraintLeft_toLeftOf = "@ + id/myfavdlg"
15          app:layout_constraintRight_toRightOf = "@ + id/myfavdlg"
16          app:layout_constraintTop_toTopOf = "@ + id/myfavdlg"
17          app:layout_constraintBottom_toBottomOf = "@ + id/myfavdlg"
18          app:layout_constraintHorizontal_bias = "0.11"
19          app:layout_constraintVertical_bias = "0.05">
20      </TextView>
21      <RadioGroup
22          android:id = "@ + id/rgfav"
23          android:layout_width = "131dp"
24          android:layout_height = "wrap_content"
25          app:layout_constraintTop_toBottomOf = "@ + id/tvfav"
26          app:layout_constraintLeft_toLeftOf = "@ + id/myfavdlg"
27          app:layout_constraintRight_toRightOf = "@ + id/myfavdlg"
28          app:layout_constraintBottom_toBottomOf = "@ + id/myfavdlg"
29          app:layout_constraintHorizontal_bias = "0.12"
30          app:layout_constraintVertical_bias = "0.06">
31          <RadioButton
32              android:id = "@ + id/rbchinese"
33              android:layout_width = "wrap_content"
34              android:layout_height = "wrap_content"
35              android:text = "@string/stryw"
36              android:checked = "true" >
37          </RadioButton>
38          <RadioButton
39              android:id = "@ + id/rbenglish"
40              android:layout_width = "wrap_content"
41              android:layout_height = "wrap_content"
42              android:text = "@string/stryy"
43              tools:layout_editor_absoluteY = "0dp"
44              tools:layout_editor_absoluteX = "0dp">
45          </RadioButton>
46          <RadioButton
47              android:id = "@ + id/rbmaths"
48              android:layout_width = "wrap_content"
49              android:layout_height = "wrap_content"
50              android:text = "@string/strsx" >
51          </RadioButton>
52      </RadioGroup>
53  </android.support.constraint.ConstraintLayout>
```

上述代码中,第9～20行为一个静态文本框,显示内容为资源"@string/strfavsub"中的字符串"我最喜欢的科目"。第 21~52 行为 RadioGroup 控件,其 ID 号为 rgfav,其中有三个单选钮控件,如第 31~37 行、38~45 行和第 46~51 行所示,三个单选钮显示的文本依次为"语文"、"英语"和"数学",如图 5-2 所示。而图 5-2 中的标题"请选择"和下方的两个按钮 OK 和 CANCEL 是在创建对话框时指定的。

创建自定义对话框的步骤为:

(1) 生成一个布局文件,例如上述的 myfavdialog.xml,该布局文件描述了对话框中所有的控件。

(2) 将该布局文件实例化,即调用 getLayoutInflater 方法得到一个 LayoutInflater 对象,然后,调用该对象的 inflate 方法将布局文件转化为 View 对象。

(3) 使用 AlertDialog.Builder 方法创建 AlertDialog 对话框,指定第(2)步生成的 View 对象为 setView 方法的参数,后续的步骤与创建普通的 AlertDialog 对话框相同。

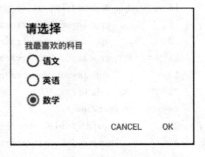

图 5-2 对话框

例 5-2 自定义对话框实例。

新建应用 MyDefinedialogApp,应用名为 MyDefinedialogApp,包名为 cn.edu.jxufe.zhangyong.mydefinedialogapp,活动界面名为 MyDefinedialogAct。向工程中添加上述的 myfavdialog.xml 布局文件,并添加汉字字符串资源文件 mystrings_hz.xml,该文件定义了 6 个字符串常量,其内容如下:

```
1  <?xml version = "1.0" encoding = "utf-8"?>
2  <resources>
3      <string name = "stryw">语文</string>
4      <string name = "strsx">数学</string>
5      <string name = "stryy">英语</string>
6      <string name = "strfavsub">我最喜欢的科目</string>
7      <string name = "strsubqus">你最喜欢的科目是什么?</string>
8      <string name = "strxz">请选择</string>
9  </resources>
```

活动主界面布局文件 activity_my_definedialog.xml 的内容如下:

```
1  <?xml version = "1.0" encoding = "utf-8"?>
2  <android.support.constraint.ConstraintLayout xmlns:android = "http://schemas.android.com/apk/res/android"
3      xmlns:app = "http://schemas.android.com/apk/res-auto"
4      xmlns:tools = "http://schemas.android.com/tools"
5      android:id = "@+id/activity_my_definedialog"
6      android:layout_width = "match_parent"
7      android:layout_height = "match_parent"
8      tools:context = "cn.edu.jxufe.zhangyong.mydefinedialogapp.MyDefinedialogAct">
9
10     <TextView
11         android:id = "@+id/tvHint"
12         android:layout_width = "257dp"
```

```
13          android:layout_height = "33dp"
14          android:text = "@string/strsubqus"
15          app:layout_constraintTop_toTopOf = "@ + id/activity_my_definedialog"
16          android:layout_marginStart = "16dp"
17          app:layout_constraintLeft_toLeftOf = "@ + id/activity_my_definedialog"
18          android:layout_marginEnd = "16dp"
19          app:layout_constraintRight_toRightOf = "@ + id/activity_my_definedialog"
20          app:layout_constraintHorizontal_bias = "0.04"
21          app:layout_constraintBottom_toBottomOf = "@ + id/activity_my_definedialog"
22          app:layout_constraintVertical_bias = "0.060000002">
23      </TextView>
24      <TextView
25          android:id = "@ + id/tvRes"
26          android:layout_width = "130dp"
27          android:layout_height = "33dp"
28          android:layout_marginStart = "8dp"
29          app:layout_constraintLeft_toRightOf = "@ + id/btSelect"
30          android:layout_marginEnd = "16dp"
31          app:layout_constraintRight_toRightOf = "@ + id/activity_my_definedialog"
32          app:layout_constraintBaseline_toBaselineOf = "@ + id/btSelect"
33          app:layout_constraintHorizontal_bias = "0.33">
34      </TextView>
35      <Button
36          android:id = "@ + id/btSelect"
37          android:layout_width = "88dp"
38          android:layout_height = "wrap_content"
39          android:text = "@string/strxz"
40          android:onClick = "mySelectMTD"
41          app:layout_constraintTop_toBottomOf = "@ + id/tvHint"
42          android:layout_marginStart = "16dp"
43          app:layout_constraintLeft_toLeftOf = "@ + id/activity_my_definedialog"
44          android:layout_marginEnd = "16dp"
45          app:layout_constraintRight_toRightOf = "@ + id/activity_my_definedialog"
46          app:layout_constraintHorizontal_bias = "0.02"
47          app:layout_constraintBottom_toBottomOf = "@ + id/activity_my_definedialog"
48          app:layout_constraintVertical_bias = "0.050000012">
49      </Button>
50
51  </android.support.constraint.ConstraintLayout>
```

上述代码中定义了两个 TextView 控件和一个 Button 按钮,其中 ID 号为 tvHint 的 TextView 控件用来显示提示字符串"你最喜欢的科目是什么？"(第 10～23 行);ID 号为 tvRes 的 TextView 控件显示从对话框返回的选择结果(第 24～34 行);Button 按钮显示的文本为"请选择",其事件方法为 mySelectMTD(第 35～49 行)。第 14 和 39 行的字符串资源"@string/strsubqus"和"@string/strxz"定义在文件 mystrings_hz.xml 中。向应用 MyDefinedialogApp 中添加颜色资源文件 myguicolor.xml,该文件与例 5-1 中的同名文件内容相同(该文件中的颜色用于设定视图背景色,但是由于插图采用灰度印刷,所以该文件在本章的工程里没有使用)。

应用 MyDefinedialogApp 的执行结果如图 5-3 所示。在图 5-3(a)中单击"请选择"按钮,弹出图 5-3(b)所示的对话框,选择"数学"后单击 OK 按钮,进入图 5-3(c)所示的界面,即显示"你最喜欢的科目是什么？　数学"。

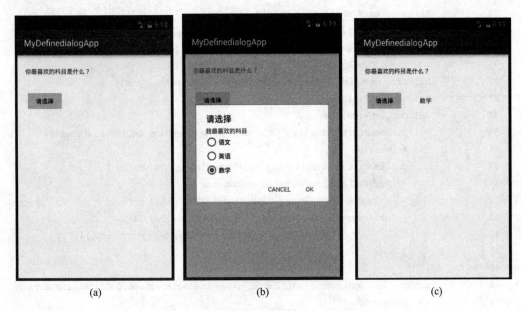

图 5-3　应用 MyDefinedialogApp 执行结果

源程序文件 MyDefinedialogAct.java 的内容如下:

```
1    package cn.edu.jxufe.zhangyong.mydefinedialogapp;
2
3    import android.content.DialogInterface;
4    import android.support.v7.app.AlertDialog;
5    import android.support.v7.app.AppCompatActivity;
6    import android.os.Bundle;
7    import android.view.LayoutInflater;
8    import android.view.View;
9    import android.view.ViewGroup;
10   import android.widget.RadioButton;
11   import android.widget.RadioGroup;
12   import android.widget.TextView;
13
14   public class MyDefinedialogAct extends AppCompatActivity {
15       private LayoutInflater fact;
16       private View view;
17       private AlertDialog.Builder builder;
18       private AlertDialog dlg;
19       private TextView tvres;
20       private RadioGroup rgsel;
21       @Override
22       protected void onCreate(Bundle savedInstanceState) {
```

```
23          super.onCreate(savedInstanceState);
24          setContentView(R.layout.activity_my_definedialog);
25          myInitGUI();
26      }
27      private void myInitGUI(){
28          tvres = (TextView)findViewById(R.id.tvRes);
29          fact = getLayoutInflater();
30          view = fact.inflate(R.layout.myfavdialog,
31                  (ViewGroup) findViewById(R.id.myfavdlg));
32          builder = new AlertDialog.Builder(MyDefinedialogAct.this);
33          builder.setTitle(R.string.strxz);
34          builder.setView(view);
35          builder.setPositiveButton("OK", new DialogInterface.OnClickListener(){
36              @Override
37              public void onClick(DialogInterface dialog, int which) {
38                  //TODO Auto-generated method stub
39                  rgsel = (RadioGroup) view.findViewById(R.id.rgfav);
40                  tvres.setText(((RadioButton)view.findViewById(
41                          rgsel.getCheckedRadioButtonId())).getText());
42              }
43          });
44          builder.setNegativeButton("Cancel", new DialogInterface.OnClickListener() {
45              @Override
46              public void onClick(DialogInterface dialog, int which) {
47                  // TODO Auto-generated method stub
48                  dialog.cancel();
49              }
50          });
51          dlg = builder.create();
52      }
53      public void mySelectMTD(View v){
54          dlg.show();
55      }
56  }
```

上述代码中,第 25 行表示在 onCreate 方法中调用自定义的私有方法 myInitGUI,第 27~52 行为 myInitGUI 方法。第 29 行调用方法 getLayoutInflater 得到一个 LayoutInflater 对象 fact,第 30 和第 31 行调用 fact 对象的 inflate 方法将布局资源 myfavdialog 转化为 View 对象,称为布局的视图实例化,inflate 方法有两个参数,第一个参数传递布局资源的 ID 号;第二个参数是传递实例化后的根视图。第 32 行创建一个 builder 对象,第 33 行设定对话框的标题,第 34 行设置对话框的视图(即界面)为 view 对象,即自定义的布局资源实例化的视图。第 35~43 行设置对话框的 OK 按钮,并添加其 onClick 事件,这里第 39 行取得对话框中的单选钮组对象 rgsel,第 40 行将单选钮组中选中的单选钮的文本传递给 tvres 静态文本框;第 44~50 行设置对话框的 Cancel 按钮,并添加其 onClick 事件,语句 dialog.cancel()将关闭对话框。第 51 行调用 builder 对象的 create 方法创建 dlg 对象,此时对话框创建好了。

第 53~55 行为图 5-3(a)中按钮"请选择"的单击事件方法 mySelectMTD,第 56 行调用

show 方法显示对话框 dlg。对话框显示后,单击对话框中的按钮,Android 系统将为用户管理单击事件,并自动跳转到相应的事件方法去执行。

5.2.3 Dialog 类

使用 Android 的 Dialog 类创建对话框的方法与 Java 语言标准对话框创建方法类似。对于 Android 应用程序而言,首先设计对话框的布局,例如,要创建图 5-4 所示的对话框,其布局文件 mycdialog.xml 如下:

图 5-4 对话框

```
1   <?xml version = "1.0" encoding = "utf-8"?>
2   <android.support.constraint.ConstraintLayout xmlns:android = "http://schemas.android.com/apk/res/android"
3       xmlns:app = "http://schemas.android.com/apk/res-auto"
4       android:orientation = "vertical"
5       android:layout_width = "match_parent"
6       android:layout_height = "match_parent">
7       <CheckBox
8           android:id = "@+id/cbwq"
9           android:layout_width = "100dp"
10          android:layout_height = "32dp"
11          android:text = "@string/strwq"
12          app:layout_constraintTop_toTopOf = "@+id/constraintLayout"
13          app:layout_constraintBottom_toBottomOf = "@+id/constraintLayout"
14          android:layout_marginEnd = "16dp"
15          app:layout_constraintRight_toRightOf = "@+id/constraintLayout"
16          android:layout_marginStart = "16dp"
17          app:layout_constraintLeft_toLeftOf = "@+id/constraintLayout"
18          app:layout_constraintHorizontal_bias = "0.09"
19          app:layout_constraintVertical_bias = "0.07">
20      </CheckBox>
21      <CheckBox
22          android:id = "@+id/cbgq"
23          android:layout_width = "100dp"
24          android:layout_height = "32dp"
25          android:text = "@string/strgq"
26          app:layout_constraintTop_toBottomOf = "@+id/cbwq"
27          app:layout_constraintBottom_toBottomOf = "@+id/constraintLayout"
28          android:layout_marginEnd = "16dp"
29          app:layout_constraintRight_toRightOf = "@+id/constraintLayout"
30          app:layout_constraintLeft_toLeftOf = "@+id/cbwq"
31          app:layout_constraintHorizontal_bias = "0.0"
32          app:layout_constraintVertical_bias = "0.060000002">
33      </CheckBox>
34      <CheckBox
35          android:id = "@+id/cbsf"
36          android:layout_width = "100dp"
37          android:layout_height = "32dp"
```

```
38              android:text = "@string/strsf"
39              app:layout_constraintTop_toBottomOf = "@ + id/cbgq"
40              app:layout_constraintLeft_toLeftOf = "@ + id/cbgq"
41              android:layout_marginEnd = "16dp"
42              app:layout_constraintRight_toRightOf = "@ + id/constraintLayout"
43              app:layout_constraintBottom_toBottomOf = "@ + id/constraintLayout"
44              app:layout_constraintHorizontal_bias = "0.0"
45              app:layout_constraintVertical_bias = "0.1">
46         </CheckBox>
47         <CheckBox
48              android:id = "@ + id/cbms"
49              android:layout_width = "100dp"
50              android:layout_height = "32dp"
51              android:text = "@string/strms"
52              app:layout_constraintLeft_toLeftOf = "@ + id/cbsf"
53              android:layout_marginEnd = "16dp"
54              app:layout_constraintRight_toRightOf = "@ + id/constraintLayout"
55              app:layout_constraintTop_toBottomOf = "@ + id/cbsf"
56              app:layout_constraintBottom_toBottomOf = "@ + id/constraintLayout"
57              app:layout_constraintHorizontal_bias = "0.0"
58              app:layout_constraintVertical_bias = "0.12">
59         </CheckBox>
60         <Button
61              android:id = "@ + id/btOK"
62              android:layout_width = "wrap_content"
63              android:layout_height = "48dp"
64              android:text = "@string/strok"
65              android:layout_marginStart = "16dp"
66              app:layout_constraintLeft_toLeftOf = "@ + id/constraintLayout"
67              android:layout_marginEnd = "16dp"
68              app:layout_constraintRight_toRightOf = "@ + id/constraintLayout"
69              app:layout_constraintTop_toBottomOf = "@ + id/cbms"
70              app:layout_constraintBottom_toBottomOf = "@ + id/constraintLayout"
71              app:layout_constraintHorizontal_bias = "0.12"
72              app:layout_constraintVertical_bias = "0.22">
73         </Button>
74         <Button
75              android:id = "@ + id/btCancel"
76              android:layout_width = "88dp"
77              android:layout_height = "48dp"
78              android:text = "@string/strcancel"
79              android:layout_marginStart = "8dp"
80              app:layout_constraintLeft_toRightOf = "@ + id/btOK"
81              android:layout_marginEnd = "16dp"
82              app:layout_constraintRight_toRightOf = "@ + id/constraintLayout"
83              app:layout_constraintBaseline_toBaselineOf = "@ + id/btOK">
84         </Button>
85  </android.support.constraint.ConstraintLayout>
```

结合图 5-4,可见上述代码创建了 4 个复选框(第 7~59 行)和两个命令按钮(第 60~84 行),每个控件都指定了 ID 号、宽度、高度、显示文本和约束位置。图 5-4 的标题"请选择"是在程序中设置的。

然后，创建一个继承 Dialog 类的对话框类，例如 MyCDialog，在该类中调用 setContentView 方法将上述布局文件设为其视图，并编写图 5-4 中命令按钮"确定"和"取消"的单击事件方法。

最后，把 MyCDialog 类作为 Activity 活动界面的内部类使用。在活动界面中创建 MyCDialog 类的对象 mydlg，调用该对象的方法 show 显示图 5-4 所示的对话框。

采用 Dialog 类创建对话框的最大好处在于：Dialog 类创建的对话框类是活动界面的内部类，可以直接调用活动界面类的私有数据成员，因此，对话框和 Activity 活动界面间可直接进行数据交换。

例 5-3 Dialog 类对话框实例。

新建应用 MyCDialogApp，应用名为 MyCDialogApp，包名为 cn.edu.jxufe.zhangyong.mycdialogapp，活动界面名为 MyCDialogAct。应用 MyCDialogApp 包括程序文件 MyCDialogAct.java、对话框布局文件 mycdialog.xml、主窗口布局文件 activity_my_cdialog.xml、汉字字符串资源文件 mystrings_hz.xml 和颜色资源文件 myguicolor.xml 等，其中，myguicolor.xml 文件与例 5-1 中的同名文件相同，mystrings_hz.xml 定义了 9 个汉字字符串，如下所示：

```
1   <?xml version = "1.0" encoding = "utf-8"?>
2   <resources>
3       <string name = "strwq">围棋</string>
4       <string name = "strgq">钢琴</string>
5       <string name = "strsf">书法</string>
6       <string name = "strms">美术</string>
7       <string name = "strmyent">我的业余爱好</string>
8       <string name = "stryrent">你的业余爱好是什么?</string>
9       <string name = "strqxz">请选择</string>
10      <string name = "strok">确定</string>
11      <string name = "strcancel">取消</string>
12  </resources>
```

主窗口布局文件 activity_my_cdialog.xml 包含两个 TextView 控件和一个命令按钮控件，其内容如下：

```
1   <?xml version = "1.0" encoding = "utf-8"?>
2   <android.support.constraint.ConstraintLayout xmlns:android = "http://schemas.android.com/apk/res/android"
3       xmlns:app = "http://schemas.android.com/apk/res-auto"
4       xmlns:tools = "http://schemas.android.com/tools"
5       android:id = "@+id/activity_my_cdialog"
6       android:layout_width = "match_parent"
7       android:layout_height = "match_parent"
8       tools:context = "cn.edu.jxufe.zhangyong.mycdialogapp.MyCDialogAct">
9       <TextView
10          android:id = "@+id/tvHint"
11          android:text = "@string/stryrent"
12          android:layout_height = "40dp"
```

```
13          android:layout_width = "160dp"
14          android:layout_marginEnd = "16dp"
15          app:layout_constraintRight_toRightOf = "@ + id/activity_my_cdialog"
16          android:layout_marginStart = "16dp"
17          app:layout_constraintLeft_toLeftOf = "@ + id/activity_my_cdialog"
18          app:layout_constraintTop_toTopOf = "@ + id/activity_my_cdialog"
19          app:layout_constraintBottom_toBottomOf = "@ + id/activity_my_cdialog"
20          app:layout_constraintHorizontal_bias = "0.11"
21          app:layout_constraintVertical_bias = "0.08">
22      </TextView>
23      <TextView
24          android:id = "@ + id/tvRes"
25          android:text = " "
26          android:layout_height = "40dp"
27          android:layout_width = "140dp"
28          android:textAppearance = "@style/TextAppearance.AppCompat"
29          android:layout_marginStart = "8dp"
30          app:layout_constraintLeft_toRightOf = "@ + id/btSel"
31          android:layout_marginEnd = "16dp"
32          app:layout_constraintRight_toRightOf = "@ + id/activity_my_cdialog"
33          app:layout_constraintBaseline_toBaselineOf = "@ + id/btSel">
34      </TextView>
35      <Button
36          android:id = "@ + id/btSel"
37          android:text = "@string/strqxz"
38          android:onClick = "myOpenDlgMTD"
39          android:layout_height = "48dp"
40          android:layout_width = "88dp"
41          android:layout_marginStart = "16dp"
42          app:layout_constraintLeft_toLeftOf = "@ + id/activity_my_cdialog"
43          android:layout_marginEnd = "16dp"
44          app:layout_constraintRight_toRightOf = "@ + id/activity_my_cdialog"
45          app:layout_constraintTop_toBottomOf = "@ + id/tvHint"
46          app:layout_constraintBottom_toBottomOf = "@ + id/activity_my_cdialog"
47          app:layout_constraintHorizontal_bias = "0.1"
48          app:layout_constraintVertical_bias = "0.13">
49      </Button>
50  </android.support.constraint.ConstraintLayout>
```

上述代码中第 2～8 行与第 50 行配对,说明这是一个约束布局,其中,第 9～34 行为两个 TextView 控件,第 35～49 行为一个 Button 控件,布局文件中指明了各个控件的 ID 号、宽度、高度、显示文本和约束位置等信息,其中,Button 控件的单击事件方法为 myOpenDlgMTD。

应用 MyCDialogApp 的执行结果如图 5-5 所示。

图 5-5(a)为 activity_my_cdialog.xml 布局文件的显示界面,其中 ID 号为 tvRes 的 TextView 控件显示空格(即不显示,见 activity_my_cdialog.xml.xml 第 25 行代码),所以图 5-5(a)中仅显示了一个 TextView 和一个 Button 按钮。在图 5-5(a)中单击"请选择"命令按钮,弹出图 5-5(b)所示的对话框,在其中选择几个复选框,然后单击"确定"按钮,则进

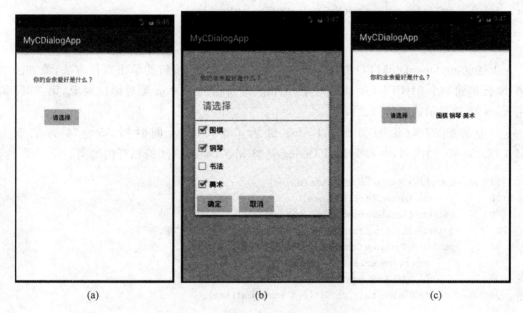

图 5-5 应用 MyCDialogApp 执行结果

入到图 5-5(c)所示的界面,显示"你的业余爱好是什么? 围棋 钢琴 美术"。

程序文件 MyCDialogAct.java 的代码如下:

```
1   package cn.edu.jxufe.zhangyong.mycdialogapp;
2
3   import android.app.Dialog;
4   import android.content.Context;
5   import android.support.v7.app.AppCompatActivity;
6   import android.os.Bundle;
7   import android.view.View;
8   import android.widget.Button;
9   import android.widget.CheckBox;
10  import android.widget.TextView;
11
12  public class MyCDialogAct extends AppCompatActivity {
13      private TextView tvres;
14      @Override
15      protected void onCreate(Bundle savedInstanceState) {
16          super.onCreate(savedInstanceState);
17          setContentView(R.layout.activity_my_cdialog);
18          myInitGUI();
19      }
```

程序启动后首先调用第 15 行的 onCreate 方法,执行第 17 行设置显示界面如图 5-5(a)所示,然后调用 myInitGUI 方法,即第 20~22 行代码得到静态文本框对象 tvres。

```
20      private void myInitGUI(){
21          tvres = (TextView)findViewById(R.id.tvRes);
22      }
23      public void myOpenDlgMTD(View v){
```

```
24      MyCDialog mydlg = new MyCDialog(MyCDialogAct.this);
25      mydlg.show();
26  }
```

上述的 myOpenDlgMTD 方法为图 5-5(a) 中 "请选择" 按钮的单击事件方法,该方法中第 24 行创建一个 MyCDialog 类的对象 mydlg,即所谓的 Dialog 类对话框对象,第 25 行调用 show 方法显示该对话框。

下面的第 27 行说明类 MyCDialog 继承了类 Dialog,同时第 28～74 行位于类 MyCDialogAct 的内部,表示类 MyCDialog 是类 MyCDialogAct 的私有内部类。

```
27  private class MyCDialog extends Dialog{
28      private Button btok,btcancel;
29      private CheckBox cbwq,cbgq,cbsf,cbms;
30      private String[] strah = { "围棋","钢琴","书法","美术"};
31      public MyCDialog(Context context) {
32          super(context);
33          // TODO Auto-generated constructor stub
34          MyCDialog.this.setTitle(R.string.strqxz);
35      }
```

第 34 行代码为指定对话框的标题为资源字符串 R.string.strqxz,即 "请选择"。

```
36      @Override
37      protected void onCreate(Bundle savedInstanceState){
38          super.onCreate(savedInstanceState);
39          setContentView(R.layout.mycdialog);
40          myInitDlg();
41      }
```

对话框启动时将首先调用 onCreate 方法,该方法第 39 行调用 setContentView 设置对话框界面如图 5-4 所示,第 40 行调用方法 myInitDlg 进行对话框初始化。

```
42      private void myInitDlg(){
43          btok = (Button)findViewById(R.id.btOK);
44          btcancel = (Button)findViewById(R.id.btCancel);
45          cbwq = (CheckBox)findViewById(R.id.cbwq);
46          cbgq = (CheckBox)findViewById(R.id.cbgq);
47          cbsf = (CheckBox)findViewById(R.id.cbsf);
48          cbms = (CheckBox)findViewById(R.id.cbms);
```

上述第 43～48 行调用方法 findViewById 分别获得对话框中 2 个 Button 控件和 4 个 CheckBox 控件的对象。下面第 49～72 行设置两个 Button 控件的事件监听方法。

```
49      btok.setOnClickListener(new View.OnClickListener(){
50          @Override
51          public void onClick(View v) {
52              // TODO Auto-generated method stub
53              String strv = "";
54              if(cbwq.isChecked())
55                  strv = strv + strah[0];
56              if(cbgq.isChecked())
```

```
57                strv = strv + strah[1];
58            if(cbsf.isChecked())
59                strv = strv + strah[2];
60            if(cbms.isChecked())
61                strv = strv + strah[3];
62            tvres.setText(strv);
63            MyCDialog.this.dismiss();
64        }
65    });
```

当单击图 5-5(b)中的"确定"按钮时将执行第 49~65 行的代码,第 54~61 行依次判断各个 CheckBox 控件是否选中,如果选中,则将其对应的文本添加到字符串 strv 中,第 62 行将字符串 strv 显示在活动界面的 TextView 控件中,第 63 行关闭对话框。

```
66    btcancel.setOnClickListener(new View.OnClickListener() {
67        @Override
68        public void onClick(View v) {
69            // TODO Auto-generated method stub
70            MyCDialog.this.cancel();
71        }
72    });
73    }
74  }
75 }
```

当单击图 5-5(b)中的"取消"按钮时,执行第 66~72 行代码,即关闭对话框(第 70 行)。

使用例 5-3 所示的方法创建对话框和创建普通的活动界面的过程相似,因此,与其他两种创建对话框的方法相比,这种创建方法最直观。

5.2.4 ProgressDialog 对话框

除了 AlertDialog 类外,Android 系统还提供了 DatePickerDialog(选择日期对话框)、TimePickerDialog(选择时间对话框)和 ProgressDialog(进度条对话框)三个类,以简化程序员在相关方面的应用程序设计。使用这些对话框类创建相应的对话框,主要的工作在于设置其对话框对象的属性,下面仅介绍 ProgressDialog 对话框的使用方法。

ProgressDialog 对话框主要用于提示程序事件处理的进度,例如网络文件的下载进度提示等,该类对话框中集成了一个进度条,进度条可以显示为圆形或长条形。对于长条形进度条对话框而言,需要设置其标题、提示信息、进度条样式、进度条长度和按钮等,接着,调用进度条对话框对象的 show 方法显示对话框,然后,在事件的处理过程中不断调用 setProgress 方法设置进度条的当前位置,指示事件的进度。

例 5-4 进度条对话框实例。

新建应用 MyProgressDlgApp,其执行结果如图 5-6 所示,应用名为 MyProgressDlgApp,包名为 cn.edu.jxufe.zhangyong.myprogressdlgapp,活动界面名为 MyProgressDlgAct。应用 MyProgressDlgApp 包括程序文件 MyProgressDlgAct.java、汉字字符串资源文件 mystrings_hz.xml、主界面布局文件 activity_my_progress_dlg.xml 和颜色资源文件 myguicolor.xml 等。

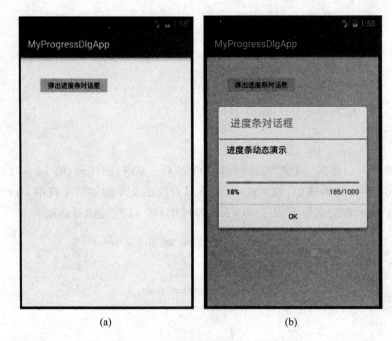

图 5-6 进度条对话框实例显示结果

由图 5-6(a)可知,主界面布局文件 activity_my_progress_dlg.xml 中只有一个命令按钮,显示文本为"弹出进度条对话框",其事件处理方法为 myOpenPDlgMTD。资源文件 myguicolor.xml 与例 5-3 中的同名文件相同,mystrings_hz.xml 文件中定义了三个汉字字符串常量,如下所示:

```
1  <?xml version = "1.0" encoding = "utf - 8"?>
2  < resources >
3      < string name = "stropen">弹出进度条对话框</string >
4      < string name = "strcap">进度条对话框</string >
5      < string name = "strprg">进度条动态演示</string >
6  </resources >
```

在图 5-6(a)中单击"弹出进度条对话框"按钮,弹出图 5-6(b)所示的进度条对话框,这里使用一个线程和延时方法控制进度条的进度,程序文件 MyProgressDlgAct.java 的代码如下:

```
1  package cn.edu.jxufe.zhangyong.myprogressdlgapp;
2
3  import android.app.ProgressDialog;
4  import android.content.DialogInterface;
5  import android.os.Handler;
6  import android.os.Message;
7  import android.support.v7.app.AppCompatActivity;
8  import android.os.Bundle;
9  import android.util.Log;
10 import android.view.View;
11
```

```
12  public class MyProgressDlgAct extends AppCompatActivity {
13      private ProgressDialog prgdlg;
14      private MyProgressThread myprgthd;
15      private final boolean RUNNING = true;
16      private final boolean STOPPING = false;
```

第 13 和第 14 行定义了进度条对话框对象 prgdlg 和自定义类 MyProgressThread 的对象 myprgthd；第 15 和第 16 行定义两个布尔常量 RUNNING 和 STOPPING，其值分别为真和假。自定义类 MyProgressThread 继承线程类 Thread，是类 MyProgressDlgAct 的内部类，如第 56 行所示。

```
17      @Override
18      protected void onCreate(Bundle savedInstanceState) {
19          super.onCreate(savedInstanceState);
20          setContentView(R.layout.activity_my_progress_dlg);
21          myInitGUI();
22      }
23      private void myInitGUI(){
24          prgdlg = new ProgressDialog(MyProgressDlgAct.this);
25          prgdlg.setTitle(R.string.strcap);
26          prgdlg.setMessage(getString(R.string.strprg));
27          prgdlg.setProgressStyle(ProgressDialog.STYLE_HORIZONTAL);
28          prgdlg.setMax(1000);
29          prgdlg.setIndeterminate(false);
30          prgdlg.setButton(DialogInterface.BUTTON_POSITIVE,"OK", new DialogInterface.
    OnClickListener() {
31              @Override
32              public void onClick(DialogInterface dialog, int which) {
33                  // TODO Auto-generated method stub
34                  myprgthd.setState(STOPPING);
35                  dialog.cancel();
36              }
37          });
38      }
```

第 18～22 行为 onCreate 方法，其中调用了自定义私有方法 myInitGUI（第 23～38 行），第 24 行创建进度条对话框对象 prgdlg；第 25 行设置对话框标题为资源字符串 R.string.strcap，即"进度条对话框"；第 26 行设置对话框提示文字，即"进度条动态演示"，这里使用 getString 方法从资源字符串 ID 中读取其字符串值；第 27 行设置进度条为长条形；第 28 行设置进度条长度为 1000；第 29 行设置进度条显示进度值；第 30 行为进度条对话框设置一个按钮 OK，其单击事件为"停止线程"（第 34 行）和退出对话框（第 35 行）。

在 Android 中一个线程启动后，程序员或用户无法停止或删除它，这里所谓的"停止线程"只是让线程不去执行特定的代码而永远处于休眠态，Android 系统会自动清除无用的线程。

```
39  public void myOpenPDlgMTD(View v){
40      myprgthd = new MyProgressThread(handler);
41      myprgthd.setState(RUNNING);
```

```
42         myprgthd.start(); // It cannot be stopped
43         prgdlg.show();
44     }
```

第 39～44 行为图 5-6(a)中按钮"弹出进度条对话框"的单击事件,第 40 行为开辟一个新的线程,第 41 行设置该线程对象的私有数据 state(第 59 行)为 RUNNING,即 true,允许第 66～82 行无限循环,直到 state 变量被设置为 STOPPING,即 flase。然后,第 42 行调用 start 方法启动该线程对象,第 43 行显示进度条对话框 prgdlg。注意,每单击图 5-6(a)中的按钮一次,第 40～43 行就会执行一次,则就会创建一个新的线程,因此,该应用程序在执行时会创建多个线程,Android 系统会帮助程序员或用户撤销无用的线程。

```
45     final Handler handler = new Handler(){
46         @Override
47         public void handleMessage(Message msg){
48             int curv = msg.getData().getInt("value");
49             prgdlg.setProgress(curv);
50             if(curv >= 1000){
51                 myprgthd.setState(STOPPING);
52                 prgdlg.dismiss();
53             }
54         }
55     };
```

第 45～55 行创建一个 Handler 类对象 handler,该对象用于管理和接收 Message 类型的消息,第 48 行接收来自线程发送的消息(第 80 行),getData 方法可获得 Bundle 类型的任意数据,每个数据有一个键值对应,这里 getInt 方法获得键值为 value 的整型数值,与第 77～79 行对应。第 49 行调用 setProgress 方法设置进度条的当前值为 curv;第 50～53 行判断 curv 的值是否大于 1000(进度条的长度),如果大于 1000,则"停止线程",然后关闭对话框。

```
56     private class MyProgressThread extends Thread{
57         private Handler h;
58         private Message msg;
59         private boolean state;
60         private int val;
61         MyProgressThread(Handler h){
62             this.h = h;
63             val = 0;
64         }
65         @Override
66         public void run(){
67             Log.i("MyDebug","I am running.");
68             while(state){
69                 val++;
70                 try{
71                     Thread.sleep(50);
72                 }
73                 catch(InterruptedException e){
74                     Log.e("MyDebug", "Thread was interrupted.");
```

```
75              }
76              msg = h.obtainMessage();
77              Bundle bd = new Bundle();
78              bd.putInt("value",val);
79              msg.setData(bd);
80              h.sendMessage(msg);
81          }
82      }
83      public void setState(boolean state){
84          this.state = state;
85      }
86   }
87 }
```

第 56～86 行的类 MyProgressThread 继承了类 Thread,是一个内部类,该类中覆盖了方法 run,线程每次启动后将执行 run 中的代码。当 state 为真时,第 68～81 行是一个死循环,每隔 50ms 循环执行一次(第 71 行),每执行一次,val 变量的值加 1,第 76～80 行向 handler 对象发送消息 msg,其中包含了 val 的值。第 83～85 行为公有方法 setState,即通常意义的 set 方法,用于设置类私有成员的值。

当 state 为真时,第 66～82 行的 run 方法与第 45～55 行的 handleMessage 方法每隔 50ms 就通信一次,当循环的值 val 为 1000 时,即 run 方法循环了 1000 次时,第 50 行条件表达式为真,第 51 行调用 setState 方法将 state 设为假,于是 run 方法不再执行,而此时进度条也步进到最大值 1000,第 52 行调用 dismiss 方法关闭进度条对话框。

细心的读者会觉得此处有些画蛇添足,即去掉第 45～55 行的 handler 对象方法,将 49～53 行的代码改写为如下形式:

```
1  prgdlg.setProgress(val);
2  if(val>=1000){
3      myprgthd.setState(STOPPING);
4      prgdlg.dismiss();
5  }
```

放在第 69 行后,并删除消息处理的代码(第 76～80 行),也可以实现同样的功能。的确如此,但是线程是程序的基本执行单元,直接占用 CPU 使用权,往往希望线程中的操作和处理越简单越好,这样线程的执行时间最短。修改后的程序中 setProgress 方法的执行时间远远超过原程序中第 76～80 行消息处理代码的执行时间,所以尽管修改后的程序可以执行,但不是一种好的方法。从这段程序可以体会到,线程中应该尽可能放置一些消息设定和发送方法,需要进行的处理放在接收消息的方法中。

5.3 菜单

Android 应用程序菜单与 Windows CE 的菜单差别很大,Windows CE 沿用了桌面 Windows 系统的菜单风格,分为下拉式菜单和上下文弹出菜单;Android 应用程序菜单采用底部浮动式和中央弹出式,分别称为选项菜单和上下文菜单,这种菜单风格主要是适应移动设备屏幕小和分辨率低的特点。调查研究发现,大多数 Windows 用户不喜欢这种菜单,但是这种菜单受到了年轻用户甚至小学生的喜爱。此外,Android 菜单只支持两级菜单结构,即顶级菜单和一级子菜单结构。

Android 应用程序中创建菜单的方法有两种,其一是借助 XML 布局文件生成菜单,其二为借助菜单对象的 add 方法等动态添加菜单项。当单击菜单项时,Android 系统会自动调用 onOptionsItemSelected 方法响应菜单单击事件,程序员需要覆盖该方法。下面依次介绍 XML 布局静态菜单、动态菜单和上下文菜单的创建与使用方法。

5.3.1 XML 布局菜单

借助 XML 布局文件创建菜单的步骤如下:

(1) 创建菜单布局文件,在布局文件中指定菜单项和菜单标题。

(2) 将菜单布局实例化。定义 MenuInflater 对象,并调用该对象的 inflate 方法将菜单布局转化为菜单对象。

(3) 编写菜单项的单击事件方法。

例 5-5 XML 布局菜单实例。

新建应用 MyXMLMenuApp,应用名为 MyXMLMenuApp,包名为 cn.edu.jxufe.zhangyong.myxmlmenuapp,活动界面名为 MyXMLMenuAct。向工程中添加菜单布局文件的方法为:在 res 资源中新建 menu 子目录,然后创建文件名为 mymenu.xml 的资源文件(不需要输入扩展名.xml),如图 5-7 所示。

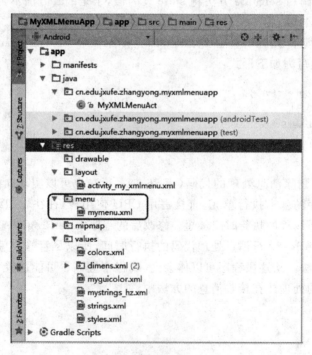

图 5-7 新建菜单资源文件

创建好的菜单布局文件 mymenu.xml 内容如下:

```
1    <?xml version = "1.0" encoding = "utf - 8"?>
2    < menu
```

```
3       xmlns:android = "http://schemas.android.com/apk/res/android">
4       < item android:id = "@ + id/m_fruit" android:title = "@string/myfruit">
5         < menu >
6           < item android:id = "@ + id/m_apple" android:title = "@string/myapple"></item>
7           < item android:id = "@ + id/m_straw" android:title = "@string/mystrawberry"></item>
8         </menu>
9       </item>
10      < item android:id = "@ + id/m_vege" android:title = "@string/myvegetable"></item>
11      < item android:id = "@ + id/m_food" android:title = "@string/myfood">
12        < menu >
13          < item android:id = "@ + id/m_steak" android:title = "@string/mysteak"></item>
14          < item android:id = "@ + id/m_pig" android:title = "@string/mypig"></item>
15        </menu>
16      </item>
17      < item android:id = "@ + id/m_sweet" android:title = "@string/mysweet"></item>
18    </menu>
```

Android 应用程序只能创建两级菜单，上述代码创建了一个两级菜单，菜单使用关键字 <menu> 和 </menu> 包括，其中的菜单项用关键字 <item> 和 </item> 包括。第 2～18 行说明顶级菜单，其中有 4 个菜单项，即 m_fruit(第 4～9 行)、m_vege(第 10 行)、m_food(第 11～16 行)和 m_sweet(第 17 行)。其中，菜单 m_fruit 包括一级子菜单，即第 6 行的 m_apple 和第 7 行的 m_straw；菜单 m_food 包括一级子菜单，即第 13 行的 m_steak 和第 14 行的 m_pig。结合下述的资源文件 mystrings_hz.xml 的代码

```
1    <?xml version = "1.0" encoding = "utf - 8"?>
2    < resources >
3        < string name = "myfruit">水果</string>
4        < string name = "myapple">苹果</string>
5        < string name = "mystrawberry">草莓</string>
6        < string name = "myvegetable">蔬菜</string>
7        < string name = "myfood">主食</string>
8        < string name = "mysteak">牛排</string>
9        < string name = "mypig">猪排</string>
10       < string name = "mysweet">甜点</string>
11       < string name = "mychoice">我点了</string>
12   </resources>
```

可知，顶级菜单显示样式如图 5-8(b) 所示，菜单是单击右上角的菜单项浮动出来的，而其中的"水果"菜单的子菜单如图 5-8(c) 所示，子菜单显示在屏幕右上角，子菜单窗口名为其上一级菜单名(这里是"水果")。

应用 MyXMLMenuApp 包括源程序文件 MyXMLMenuAct.java、活动界面布局文件 acitivity_my_xmlmenu.xml、菜单布局文件 mymenu.xml、汉字字符串资源文件 mystrings_hz.xml 和颜色资源文件 myguicolor.xml 等。其中，文件 mymenu.xml 和 mystrings_hz.xml 文件内容已经讨论了，文件 myguicolor.xml 与例 5-4 中的同名文件内容相同。布局文件 acitivity_my_xmlmenu.xml 中只有一个 TextView 控件，用于显示单击的菜单项内容，其 ID 号为 tvdisp，显示内容为"我点了"。

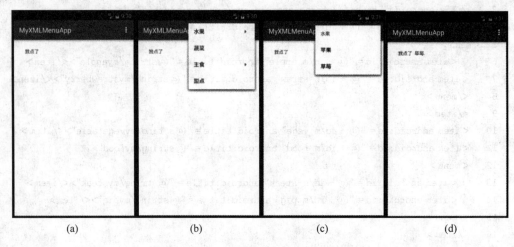

(a)   (b)   (c)   (d)

图 5-8　工程 ex05_05 运行结果

源程序文件 MyXMLMenuAct.java 的内容如下：

```
1   package cn.edu.jxufe.zhangyong.myxmlmenuapp;
2
3   import android.support.v7.app.AppCompatActivity;
4   import android.os.Bundle;
5   import android.view.Menu;
6   import android.view.MenuInflater;
7   import android.view.MenuItem;
8   import android.widget.TextView;
9
10  public class MyXMLMenuAct extends AppCompatActivity {
11      private TextView tvorder;
12
13      @Override
14      protected void onCreate(Bundle savedInstanceState) {
15          super.onCreate(savedInstanceState);
16          setContentView(R.layout.activity_my_xmlmenu);
17
18          myInitGUI();
19      }
20      private void myInitGUI(){
21          tvorder = (TextView)findViewById(R.id.tvdisp);
22      }
```

第 21 行表示取得静态文件框对象 tvorder。

```
23      @Override
24      public boolean onCreateOptionsMenu(Menu menu){
25          MenuInflater inflater = getMenuInflater();
26          inflater.inflate(R.menu.mymenu, menu);
27          return true;
28      }
```

第 24～28 行将菜单布局文件（即菜单资源 R.menu.mymenu）实例化，第 25 行创建

MenuInflater 对象 inflater,第 26 行调用 inflater 的 inflate 方法得到菜单对象 menu(menu 对象传递给 Android 系统,由 Android 系统维护)。

```
29      @Override
30      public boolean onOptionsItemSelected(MenuItem item){
31          switch(item.getItemId()){
32              case R.id.m_apple:
33                  tvorder.setText(getString(R.string.mychoice) + " "
34                      + getString(R.string.myapple) + ".");
35                  break;
36              case R.id.m_straw:
37                  tvorder.setText(getString(R.string.mychoice) + " "
38                      + getString(R.string.mystrawberry) + ".");
39                  break;
40              case R.id.m_vege:
41                  tvorder.setText(getString(R.string.mychoice) + " "
42                      + getString(R.string.myvegetable) + ".");
43                  break;
44              case R.id.m_steak:
45                  tvorder.setText(getString(R.string.mychoice) + " "
46                      + getString(R.string.mysteak) + ".");
47                  break;
48              case R.id.m_pig:
49                  tvorder.setText(getString(R.string.mychoice) + " "
50                      + getString(R.string.mypig) + ".");
51                  break;
52              case R.id.m_sweet:
53                  tvorder.setText(getString(R.string.mychoice) + " "
54                      + getString(R.string.mysweet) + ".");
55                  break;
56          }
57          return true;
58      }
59  }
```

第 30～58 行为单击菜单项时 Android 系统将调用的方法 onOptionsItemSelected,如果某个菜单项被单击,Android 系统将自动执行该方法,第 31 行根据单击的菜单项 ID 号,跳转到相应的分支去执行,如果单击了子菜单项"草莓"(如图 5-8(c)所示),则第 31 行 item. getItemId 方法将得到 R.id.m_straw,然后,程序跳转到第 37 行执行,静态文本框 tvorder 将显示字符串"我点了草莓。"(如图 5-8(d)所示)。

借助 XML 布局文件创建菜单直观方便,是最常用的菜单创建方式。应用 MyXMLMenuApp 的执行结果如图 5-8 所示,单击模拟器右上角的菜单项(如图 5-8(a)所示)即可弹出图 5-8(b)所示的菜单;在图 5-8(b)中单击"水果"或"主食"将弹出子菜单,图 5-8(c)为单击"水果"后弹出的子菜单;单击"草莓"子菜单项后显示图 5-8(d)所示的结果。

5.3.2 动态菜单

Android 应用程序执行过程可以动态创建菜单(也可以动态清除菜单),这种方法不需要创建菜单布局文件,菜单项(或子菜单项)均由程序代码创建。

例 5-6 动态菜单实例。

例 5-6 与例 5-5 实现的功能完全相同。新建应用 MyLiveMenuApp,应用名为 MyLiveMenuApp,活动界面名为 MyLiveMenuAct。应用 MyLiveMenuApp 中包括文件 MyLiveMenuAct.java、activity_my_live_menu.xml、myguicolor.xml 和 mystrings_hz.xml 等,其中后两个文件与例 5-5 中的同名文件完全相同,而布局文件 activity_my_live_menu.xml 中只有一个静态文本框 TextView 控件,ID 号为 tvdisp。与例 5-5 相比,例 5-6 中没有菜单布局文件。应用 MyLiveMenuApp 的执行结果与例 5-5 相同,如图 5-8 所示。

源程序文件 MyLiveMenuAct.java 的代码如下:

```
1   package cn.edu.jxufe.zhangyong.mylivemenuapp;
2
3   import android.support.v7.app.AppCompatActivity;
4   import android.os.Bundle;
5   import android.view.Menu;
6   import android.view.MenuItem;
7   import android.view.SubMenu;
8   import android.widget.TextView;
9
10  public class MyLiveMenuAct extends AppCompatActivity {
11      private TextView tvorder;
12      @Override
13      protected void onCreate(Bundle savedInstanceState) {
14          super.onCreate(savedInstanceState);
15          setContentView(R.layout.activity_my_live_menu);
16
17          myInitGUI();
18      }
19      private void myInitGUI(){
20          tvorder = (TextView)findViewById(R.id.tvdisp);
21      }
22      @Override
23      public boolean onCreateOptionsMenu(Menu menu){
24          SubMenu subMenu1 = menu.addSubMenu(Menu.NONE,
25                  Menu.FIRST + 20,1,R.string.myfruit);
26          subMenu1.add(1,Menu.FIRST + 1,2,R.string.myapple);
27          subMenu1.add(1,Menu.FIRST + 2,2,R.string.mystrawberry);
28          menu.add(Menu.NONE,Menu.FIRST + 3,2,
29                  R.string.myvegetable);
30          SubMenu subMenu2 = menu.addSubMenu(Menu.NONE,
31                  Menu.FIRST + 21,3,R.string.myfood);
32          subMenu2.add(2,Menu.FIRST + 4,1,R.string.mysteak);
33          subMenu2.add(2,Menu.FIRST + 5,2,R.string.mypig);
34          menu.add(Menu.NONE,Menu.FIRST + 6,4,R.string.mysweet);
35          return true;
36      }
37      @Override
38      public boolean onOptionsItemSelected(MenuItem item){
```

```
39          switch(item.getItemId()){
40              case Menu.FIRST + 1:
41                  tvorder.setText(getString(R.string.mychoice) + " "
42                          + getString(R.string.myapple) + ".");
43                  break;
44              case Menu.FIRST + 2:
45                  tvorder.setText(getString(R.string.mychoice) + " "
46                          + getString(R.string.mystrawberry) + ".");
47                  break;
48              case Menu.FIRST + 3:
49                  tvorder.setText(getString(R.string.mychoice) + " "
50                          + getString(R.string.myvegetable) + ".");
51                  break;
52              case Menu.FIRST + 4:
53                  tvorder.setText(getString(R.string.mychoice) + " "
54                          + getString(R.string.mysteak) + ".");
55                  break;
56              case Menu.FIRST + 5:
57                  tvorder.setText(getString(R.string.mychoice) + " "
58                          + getString(R.string.mypig) + ".");
59                  break;
60              case Menu.FIRST + 6:
61                  tvorder.setText(getString(R.string.mychoice) + " "
62                          + getString(R.string.mysweet) + ".");
63                  break;
64          }
65          return true;
66      }
67  }
```

对比工程 ex05_05 中 MyXMLMenuAct.java 程序的代码,上述代码多了第 22~36 行,这 15 行代码组成的方法 onCreateOptionsMenu 为 Android 系统自动调用的创建菜单方法。创建不带子菜单的菜单的方法和带子菜单的菜单的方法不同,对于创建不带子菜单的菜单,直接调用 add 方法,如第 28 行所示,其中带有四个参数,第一个参数表示菜单所在的组号;第二个参数表示菜单的 ID 号,必须唯一(其中,Menu.FIRST 为常数 1);第三个参数为菜单出现的位置号;第四个参数为菜单显示的标题。如果创建带有子菜单项的菜单,需要先创建带有子菜单项的菜单,即需要先定义 SubMenu 对象,如第 24 行所示,调用 addSubMenu 方法添加顶级菜单(第 24 和第 25 行),然后,调用 SubMenu 对象的 add 方法添加子菜单,如第 26 和第 27 行所示。这样,第 24~34 行的菜单与工程 ex05_05 中的 mymenu.xml 文件定义的菜单完全相同。

5.3.3 上下文菜单

Android 应用程序有两种菜单,除了上述介绍的单击浮动菜单外,Android 也具有上下文关联菜单,即当长时间按下触摸屏时(相当于桌面 Windows 系统的鼠标右键单击)弹出的提示菜单。

创建上下文菜单的步骤为:

(1) 在方法 onCreateContextMenu 中创建上下文菜单,该方法由 Android 系统自动调用。

(2) 在方法 onContextItemSelected 中创建菜单单击事件的响应代码,当上下文菜单被单击时,该方法由 Android 系统调用。

(3) 在活动界面的 onCreate 方法中调用方法 setOnCreateContextMenuListener 监听上下文菜单的单击事件。

例 5-7 上下文菜单实例。

本例中使用 ListActivity 类活动界面,此时的布局文件不再是默认情况下的约束布局,而是如下所示的布局文件(文件名为 activity_my_context_menu.xml):

```
1  <?xml version = "1.0" encoding = "utf-8"?>
2  <TextView xmlns:android = "http://schemas.android.com/apk/res/android"
3      android:layout_width = "fill_parent"
4      android:layout_height = "fill_parent"
5      android:padding = "10dp"
6      android:textSize = "16sp" >
7  </TextView>
```

即 ListView 活动界面的布局文件中只需包括一个 TextView。ListActivity 类的继承关系为 java.lang.Object→Android.content.Context→android.content.ContextWrapper→android.view.ContextThemeWrapper→android.app.Activity→android.app.ListActivity,ListActivity 类管理了一个 ListView 对象,可用于在屏幕上显示垂直滚动列表,显示列表内容一般来自 ArrayAdapter 类型数组对象。ArrayAdapter 类是一个泛型数组类,可以存放各种数据类型的对象数据。

新建应用 MyContextMenuApp,应用名为 MyContextMenuApp,活动界面名为 MyContextMenuAct。应用 MyContextMenuApp 包括程序文件 MyContextMenuAct.java、布局文件 activity_my_context_menu.xml、汉字字符串资源文件 mystrings_hz.xml 和颜色资源文件 myguicolor.xml 等,其中 myguicolor.xml 文件与例 5-5 中的同名文件相同。汉字字符串和字符串数组资源文件 mystrings_hz.xml 的内容如下:

```
1   <?xml version = "1.0" encoding = "utf-8"?>
2   <resources>
3       <string name = "myusual">普通-----60元</string>
4       <string name = "myrefine">精品-----80元</string>
5       <string name = "myspecial">特品-----120元</string>
6       <string name = "myorganic">有机-----200元</string>
7       <string name = "mychoice">我点了</string>
8       <string name = "myselect">请选择</string>
9       <string-array name = "dinner">
10          <item>水果套餐</item>
11          <item>蔬菜套餐</item>
12          <item>牛排套餐</item>
13          <item>甜点套餐</item>
```

```
14      </string-array>
15 </resources>
```

上述代码中第 3~8 行定义了 6 个汉字字符串常量,资源名依次为 myusual、myrefine、myspecial、myorganic、mychoice 和 myselect。第 9~14 行为一个字符串数组,使用关键字<string-array>和</string-array>包括,其中的数组项使用关键字<item>和</item>包括,因此,第 9~14 行定义了一个包含 4 个字符串的字符串数组,数组资源名为 dinner。

应用 MyContextMenuApp 的执行结果如图 5-9 所示。图 5-9(a)为应用 MyContextMenuApp 的主界面,该界面为 ListView 活动界面,长按任一项时,将弹出图 5-9(b)所示的上下文菜单,单击菜单项,例如单击"特品—120元",将显示点菜情况,如图 5-9(c)所示。从图 5-9(b)中可以看出,上下文菜单显示在屏幕中央位置。

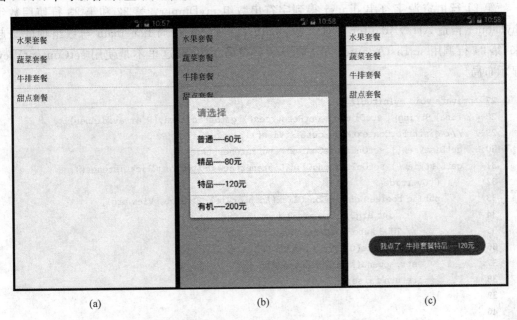

图 5-9 应用 MyContextMenuApp 执行结果

程序文件 MyContextMenuAct.java 的代码如下:

```
1  package cn.edu.jxufe.zhangyong.mycontextmenuapp;
2
3  import android.app.ListActivity;
4  import android.os.Bundle;
5  import android.view.ContextMenu;
6  import android.view.Menu;
7  import android.view.ContextMenu.ContextMenuInfo;
8  import android.view.MenuItem;
9  import android.view.View;
10 import android.widget.AdapterView;
11 import android.widget.AdapterView.OnItemLongClickListener;
12 import android.widget.ArrayAdapter;
13 import android.widget.Toast;
14
```

```
15  public class MyContextMenuAct extends ListActivity {
16    StringBuilder str = new StringBuilder(" ");
17    /** Called when the activity is first created. */
18    @Override
19    public void onCreate(Bundle savedInstanceState) {
20      super.onCreate(savedInstanceState);
21      String[] myDinner = getResources().getStringArray(R.array.dinner);
22      ArrayAdapter<String> lvadapter = new ArrayAdapter<String>(this,
23                                        R.layout.list_main,myDinner);
24      setListAdapter(lvadapter);
25      myInitGUI();
26    }
```

第 21 行由资源字符串 dinner 得到字符串数组 myDinner；第 22 和第 23 行使用数组 myDinner 构造 ArrayAdapter 型对象 lvadapter，该对象将布局 list_main 与数组项联系起来，第 24 行调用 setListAdapter 方法将显示该布局。注意，这里不是使用 setContentView 显示布局。

```
27  private void myInitGUI(){
28    final String[] strDinner = getResources().getStringArray(R.array.dinner);
29    //registerForContextMenu(getListView());
30    getListView().setOnCreateContextMenuListener(this);
31    getListView().setOnItemLongClickListener(new OnItemLongClickListener(){
32      @Override
33      public boolean onItemLongClick(AdapterView<?> arg0, View arg1,
34            int arg2, long arg3) {
35        // TODO Auto-generated method stub
36        str.delete(0, str.length());
37        str.append(strDinner[arg2]);
38        return false;
39      }
40    });
41  }
```

在第 30 行调用 setOnCreateContextMenuListener 方法设置上下文菜单监听事件，注释掉的第 29 行代码与第 30 行代码功能相同，如果使用第 29 行的方法，则将第 30 行注释掉。第 31 行设置图 5-9(a)中的列表项长按事件(相当于右键菜单)监听方法，第 33 行的方法 onItemLongClick 中第 3 个参数 arg2 为单击项的位置，从 0 开始索引，即如果单击图 5-9(a) 中的"牛排套餐"，则 arg2 的值为 2。第 36 行的 srt 为 StringBuilder 对象(第 16 行)，因为 String 类型对象不能修改，故这里使用 StringBuilder 对象，第 36 行删除该对象中的所有字符串，第 37 行在其末尾添加字符串 strDinner[arg2]，对于图 5-9(a)中的"牛排套餐"项而言，就是 strDinner[2]，参考第 28 行和 mystrings_hz.xml 文件可知，strDinner[2]刚好是 "牛排套餐"。需要特别注意的是，第 38 行必须返回 false，绝不能用 true！

```
42  @Override
43  public void onCreateContextMenu(ContextMenu menu,
44        View v, ContextMenuInfo menuInfo){
45    menu.setHeaderTitle((CharSequence)getString(R.string.myselect));
```

```
46        menu.add(1,Menu.FIRST + 1,1,R.string.myusual);
47        menu.add(1,Menu.FIRST + 2,2,R.string.myrefine);
48        menu.add(1,Menu.FIRST + 3,3,R.string.myspecial);
49        menu.add(1,Menu.FIRST + 4,4,R.string.myorganic);
50    }
```

第 43～50 行用于创建上下文菜单,第 45 行设置上下文菜单的标题,如图 5-9(b)所示,第 46～49 行依次添加 4 个菜单项。

```
51    @Override
52    public boolean onContextItemSelected(MenuItem item){
53        String stritem = getString(R.string.mychoice) + ": "
54                + str.toString() + item.getTitle().toString();
55        switch(item.getItemId()){
56        case Menu.FIRST + 1:
57            Toast.makeText(this, stritem, Toast.LENGTH_SHORT).show();
58            break;
59        case Menu.FIRST + 2:
60            Toast.makeText(this, stritem, Toast.LENGTH_SHORT).show();
61            break;
62        case Menu.FIRST + 3:
63            Toast.makeText(this, stritem, Toast.LENGTH_SHORT).show();
64            break;
65        case Menu.FIRST + 4:
66            Toast.makeText(this, stritem, Toast.LENGTH_SHORT).show();
67            break;
68        }
69        return true;
70    }
71 }
```

上述代码为当上下文菜单被单击时的事件方法 onContextSelected,第 53 行将字符串"我点了"(字符串资源 R.string.mychoice)、":"、列表项的值(str.toString())和菜单项的值(item.getTitle().toString())合成为一个字符串 stritem。第 55～68 行根据选择的菜单项跳转到相应的分支去执行,如果单击了图 5-9(a)中的"牛排套餐"和图 5-9(b)中的"特品—120 元",则 Toast 将短暂显示"我点了:牛排套餐特品—120 元",如图 5-9(c)所示。

5.4 多用户界面设计

在 Android 应用程序中,一个 Activity(活动界面)就是一个用户界面,一个用户界面对应着一个用户界面布局文件。本节介绍一个应用程序工程中同时有多个 Activity 的情况,并阐述它们的显示和数据通信方法。

5.4.1 简单多用户界面显示

Android 应用程序启动后,Android 系统将显示应用程序配置文件 AndroidManifest.xml 中 Intent-Filter 为 ACTION_MAIN 和 ACTION_LAUCHER 指定的 Activity,该 Activity 被称为主程序界面。主程序界面显示后,通过 Intent 对象显示其他的 Activity,其

方法如下:

```
1  Intent it = new Intent();
2  it.setClass(this, MyCourse.class);
3  startActivity(it);
4  this.finish();
```

第1行定义一个Intent对象it;第2行中的MyCourse为要显示的Activity类,第2行调用it对象的setClass方法将MyCourse类链接到Intent对象中;第3行调用方法startActivity显示MyCourse界面;第4行调用finish方法关闭当前正在显示的Activity。

例5-8 多用户界面显示实例。

新建应用MyMGUIDispApp,应用名为MyMGUIDispApp,包名为cn.edu.jxufe.zhangyong.mymguidispapp,活动界面名为MyMGUIDispAct。应用MyMGUIDispApp的执行结果如图5-10所示。应用MyMGUIDispApp启动后进入图5-10(a)所示的界面,输入密码"123456"(在图5-10(b)中显示为"……")后,选中"查看课表"单选按钮,然后单击"登录"按钮,进入图5-10(c)所示的界面;如果在图5-10(b)中选择"查看日程"单选按钮,单击"登录"按钮,进入图5-10(d)所示的界面;在图5-10(c)和图5-10(d)中单击"返回"按钮,回到图5-10(a)所示的界面。图5-10(a)、图5-10(c)和图5-10(d)间没有数据的传递,只是简单地切换显示界面。

应用MyMGUIDispApp包括程序文件MyCourse.java、MyMGUIDispAct.java、MySchedule.java、布局文件activity_my_mguidisp.xml、mycourse.xml、myschedule.xml、汉字字符串资源文件mystrings_hz.xml、颜色资源文件myguicolor.xml以及工程配置文件AndroidManifest.xml等,其中,文件myguicolor.xml与例5-7中的同名文件相同。

应用MyMGUIDispApp中有三个Activity(活动界面),这三个Activity都需要在配置文件AndroidManifest.xml中声明,配置文件AndroidManifest.xml的代码如下:

```
1  <?xml version = "1.0" encoding = "utf-8"?>
2  <manifest xmlns:android = "http://schemas.android.com/apk/res/android"
3      package = "cn.edu.jxufe.zhangyong.mymguidispapp">
4      <application
5          android:allowBackup = "true"
6          android:icon = "@mipmap/ic_launcher"
7          android:label = "@string/app_name"
8          android:supportsRtl = "true"
9          android:theme = "@style/AppTheme">
10         <activity android:name = ".MyMGUIDispAct">
11             <intent-filter>
12                 <action android:name = "android.intent.action.MAIN" />
13                 <category android:name = "android.intent.category.LAUNCHER" />
14             </intent-filter>
15         </activity>
16         <activity android:name = "MyCourse" android:label = "@string/mycourse">
17         </activity>
18         <activity android:name = "MySchedule" android:label = "@string/myschedule">
19         </activity>
20     </application>
21 </manifest>
```

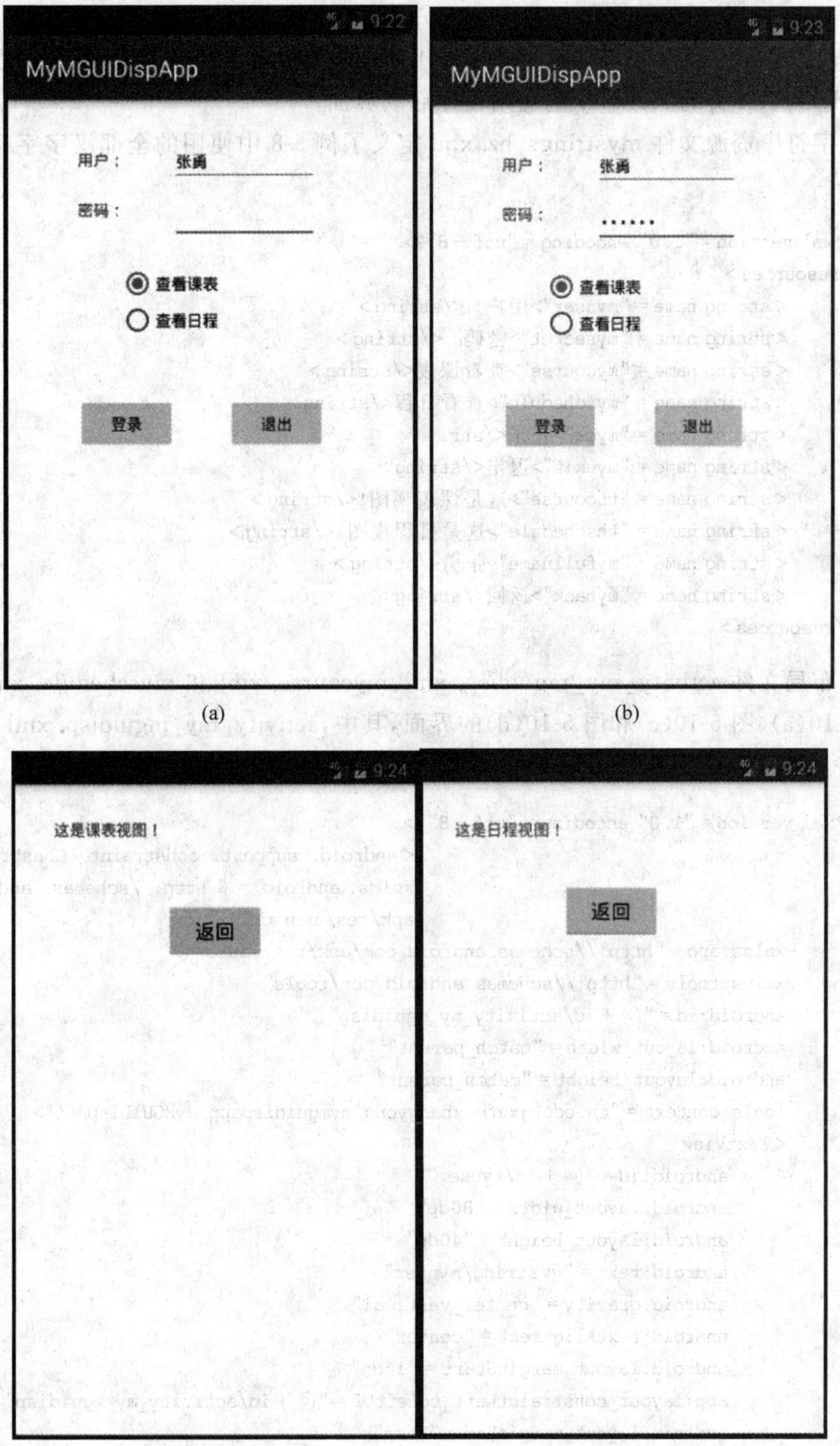

图 5-10 应用 MyMGUIDispApp 执行结果

上述代码第 10~15 行为 MyMGUIDispAct 类对应的 Activity，即图 5-10(a)的界面；第 16、17 行定义 MyCourse 类对应的 Activity，即图 5-10(c)的界面；第 18、19 行定义 MySchedule 类对应的 Activity，即图 5-10(d) 对应的界面。第 11~14 行说明 MyMGUIDispAct 类的 Activity 是应用程序启动界面。

汉字字符串资源文件 mystrings_hz.xml 定义了例 5-8 中使用的全部汉字字符串,其代码如下：

```
1   <?xml version = "1.0" encoding = "utf-8"?>
2   <resources>
3       <string name = "myuser">用户：</string>
4       <string name = "mysecret">密码：</string>
5       <string name = "mycourse">查看课表</string>
6       <string name = "myschedule">查看日程</string>
7       <string name = "myok">登录</stri
8       <string name = "myexit">退出</string>
9       <string name = "thcourse">这是课表视图!</string>
10      <string name = "thschedule">这是日程视图!</string>
11      <string name = "myfullname">张勇</string>
12      <string name = "myback">返回</string>
13  </resources>
```

三个布局文件 activity_my_mguidisp.xml、mycourse.xml 和 myschedule.xml 分别对应着图 5-10(a)、图 5-10(c)和图 5-10(d)的界面,其中,activity_my_mguidisp.xml 文件的代码如下：

```
1   <?xml version = "1.0" encoding = "utf-8"?>
2                                   <android.support.constraint.ConstraintLayout
                                    xmlns:android = "http://schemas.android.com/
                                    apk/res/android"
3       xmlns:app = "http://schemas.android.com/apk/res-auto"
4       xmlns:tools = "http://schemas.android.com/tools"
5       android:id = "@+id/activity_my_mguidisp"
6       android:layout_width = "match_parent"
7       android:layout_height = "match_parent"
8       tools:context = "cn.edu.jxufe.zhangyong.mymguidispapp.MyMGUIDispAct">
9       <TextView
10          android:id = "@+id/tvuser"
11          android:layout_width = "80dp"
12          android:layout_height = "40dp"
13          android:text = "@string/myuser"
14          android:gravity = "center_vertical"
15          android:textAlignment = "center"
16          android:layout_marginStart = "16dp"
17          app:layout_constraintLeft_toLeftOf = "@+id/activity_my_mguidisp"
18          android:layout_marginEnd = "16dp"
19          app:layout_constraintRight_toRightOf = "@+id/activity_my_mguidisp"
20          app:layout_constraintHorizontal_bias = "0.11"
21          app:layout_constraintTop_toTopOf = "@+id/activity_my_mguidisp"
22          app:layout_constraintBottom_toBottomOf = "@+id/activity_my_mguidisp"
```

```
23              app:layout_constraintVertical_bias = "0.07">
24      </TextView>
25      <EditText
26              android:id = "@ + id/etuser"
27              android:layout_width = "130dp"
28              android:layout_height = "40dp"
29              android:text = "@string/myfullname"
30              android:textSize = "14sp"
31              android:gravity = "center_vertical"
32              android:layout_marginStart = "8dp"
33              app:layout_constraintLeft_toRightOf = "@ + id/tvuser"
34              android:layout_marginEnd = "16dp"
35              app:layout_constraintRight_toRightOf = "@ + id/activity_my_mguidisp"
36              app:layout_constraintTop_toTopOf = "@ + id/activity_my_mguidisp"
37              app:layout_constraintBottom_toTopOf = "@ + id/rgdirc"
38              app:layout_constraintHorizontal_bias = "0.18"
39              app:layout_constraintVertical_bias = "0.32999998">
40      </EditText>
41      <TextView
42              android:id = "@ + id/tvsecret"
43              android:layout_width = "80dp"
44              android:layout_height = "40dp"
45              android:text = "@string/mysecret"
46              android:layout_marginStart = "16dp"
47              app:layout_constraintLeft_toLeftOf = "@ + id/activity_my_mguidisp"
48              android:layout_marginEnd = "16dp"
49              app:layout_constraintRight_toRightOf = "@ + id/activity_my_mguidisp"
50              app:layout_constraintHorizontal_bias = "0.11"
51              app:layout_constraintTop_toBottomOf = "@ + id/tvuser"
52              app:layout_constraintBottom_toTopOf = "@ + id/rgdirc"
53              android:textAlignment = "center"
54              app:layout_constraintVertical_bias = "0.43">
55      </TextView>
56      <EditText
57              android:id = "@ + id/etsecret"
58              android:layout_width = "130dp"
59              android:layout_height = "40dp"
60              android:text = ""
61              android:textSize = "14sp"
62              android:inputType = "textPassword"
63              android:layout_marginStart = "8dp"
64              app:layout_constraintLeft_toRightOf = "@ + id/tvsecret"
65              android:layout_marginEnd = "16dp"
66              app:layout_constraintRight_toRightOf = "@ + id/activity_my_mguidisp"
67              app:layout_constraintHorizontal_bias = "0.18"
68              app:layout_constraintTop_toBottomOf = "@ + id/etuser"
69              app:layout_constraintBottom_toTopOf = "@ + id/rgdirc"
70              app:layout_constraintVertical_bias = "0.32999998">
71      </EditText>
72      <Button
73              android:id = "@ + id/btOK"
```

```
74          android:layout_width = "88dp"
75          android:layout_height = "48dp"
76          android:text = "@string/myok"
77          android:onClick = "myLoginMD"
78          android:layout_marginStart = "16dp"
79          app:layout_constraintLeft_toLeftOf = "@ + id/activity_my_mguidisp"
80          android:layout_marginEnd = "16dp"
81          app:layout_constraintRight_toRightOf = "@ + id/activity_my_mguidisp"
82          app:layout_constraintTop_toBottomOf = "@ + id/rgdirc"
83          app:layout_constraintBottom_toBottomOf = "@ + id/activity_my_mguidisp"
84          app:layout_constraintHorizontal_bias = "0.19"
85          app:layout_constraintVertical_bias = "0.17">
86      </Button>
87      <Button
88          android:id = "@ + id/btExit"
89          android:layout_width = "88dp"
90          android:layout_height = "48dp"
91          android:text = "@string/myexit"
92          android:onClick = "myExitMD"
93          app:layout_constraintBaseline_toBaselineOf = "@ + id/btOK"
94          android:layout_marginStart = "8dp"
95          app:layout_constraintLeft_toRightOf = "@ + id/btOK"
96          android:layout_marginEnd = "16dp"
97          app:layout_constraintRight_toRightOf = "@ + id/activity_my_mguidisp"
98          app:layout_constraintHorizontal_bias = "0.39">
99      </Button>
100     <RadioGroup
101         android:id = "@ + id/rgdirc"
102         android:layout_width = "146dp"
103         android:layout_height = "73dp"
104         app:layout_constraintTop_toTopOf = "@ + id/activity_my_mguidisp"
105         app:layout_constraintBottom_toBottomOf = "@ + id/activity_my_mguidisp"
106         android:layout_marginStart = "16dp"
107         app:layout_constraintLeft_toLeftOf = "@ + id/activity_my_mguidisp"
108         android:layout_marginEnd = "16dp"
109         app:layout_constraintRight_toRightOf = "@ + id/activity_my_mguidisp"
110         app:layout_constraintHorizontal_bias = "0.46"
111         app:layout_constraintVertical_bias = "0.33">
112         <RadioButton
113             android:id = "@ + id/rbcourse"
114             android:layout_width = "wrap_content"
115             android:layout_height = "wrap_content"
116             android:text = "@string/mycourse"
117             android:checked = "true" >
118         </RadioButton>
119         <RadioButton
120         android:id = "@ + id/rbschedule"
121         android:layout_width = "wrap_content"
122         android:layout_height = "wrap_content"
123         android:text = "@string/myschedule"
124         tools:layout_editor_absoluteY = "190dp"
```

```
125             tools:layout_editor_absoluteX = "132dp">
126         </RadioButton>
127     </RadioGroup>
128 </android.support.constraint.ConstraintLayout>
```

结合图 5-10(a),上述代码中第 9~24 行定义了显示"用户"的静态文本框,第 25~40 行定义了显示"张勇"的编辑框,第 41~55 行定义了显示"密码"的静态文本框,第 56~71 行定义了显示空白的编辑框。第 72~88 行定义了显示文字为"登录"的按钮,其事件方法为 myLoginMD(第 77 行),第 87~99 行定义了显示文字为"退出"的按钮,其事件方法为 myExitMD(第 92 行)。第 100~127 行定义了具有"查看课表"和"查看日程"两个单选钮的单选钮组。

布局文件 mycourse.xml 对应于图 5-10(c),其代码如下:

```
1  <?xml version = "1.0" encoding = "utf-8"?>
2  <android.support.constraint.ConstraintLayout xmlns:app = "http://schemas.android.com/apk/res-auto"
3      xmlns:tools = "http://schemas.android.com/tools"
4      android:id = "@+id/widget0"
5      android:layout_width = "fill_parent"
6      android:layout_height = "fill_parent"
7      xmlns:android = "http://schemas.android.com/apk/res/android"
8      android:background = "@drawable/white">
9      <TextView
10         android:id = "@+id/tvhint"
11         android:layout_width = "100dp"
12         android:layout_height = "25dp"
13         android:text = "@string/thcourse"
14         tools:layout_constraintTop_creator = "1"
15         tools:layout_constraintBottom_creator = "1"
16         app:layout_constraintBottom_toTopOf = "@+id/btback"
17         android:layout_marginStart = "24dp"
18         tools:layout_constraintLeft_creator = "1"
19         app:layout_constraintLeft_toLeftOf = "@+id/widget0"
20         app:layout_constraintTop_toTopOf = "@+id/widget0"
21         android:layout_marginEnd = "16dp"
22         app:layout_constraintRight_toRightOf = "@+id/widget0"
23         app:layout_constraintHorizontal_bias = "0.05"
24         app:layout_constraintVertical_bias = "0.38">
25     </TextView>
26     <Button
27         android:id = "@+id/btback"
28         android:layout_width = "89dp"
29         android:layout_height = "48dp"
30         android:text = "@string/myback"
31         android:onClick = "myBackMD"
32         tools:layout_constraintTop_creator = "1"
33         app:layout_constraintTop_toTopOf = "@+id/widget0"
34         app:layout_constraintBottom_toBottomOf = "@+id/widget0"
35         app:layout_constraintVertical_bias = "0.2"
```

```
36              app:layout_constraintRight_toRightOf = "@ + id/widget0"
37              app:layout_constraintLeft_toLeftOf = "@ + id/widget29"
38              app:layout_constraintHorizontal_bias = "0.43">
39       </Button>
40  </android.support.constraint.ConstraintLayout>
```

结合图 5-10(c),上述代码第 9~25 行为显示"这是课表视图"的静态文本框,第 26~39 行为显示文本"返回"的按钮,其单击事件方法为 myBackMD(第 31 行)。

布局文件 myschedule.xml 对应于图 5-10(d),其代码如下:

```
1   <?xml version = "1.0" encoding = "utf-8"?>
2   <android.support.constraint.ConstraintLayout xmlns:app = "http://schemas.android.com/apk/res-auto"
3       xmlns:tools = "http://schemas.android.com/tools"
4       android:id = "@ + id/widget0"
5       android:layout_width = "fill_parent"
6       android:layout_height = "fill_parent"
7       xmlns:android = "http://schemas.android.com/apk/res/android"
8       android:background = "@drawable/white">
9       <TextView
10          android:id = "@ + id/tvdispsch"
11          android:layout_width = "100dp"
12          android:layout_height = "25dp"
13          android:text = "@string/thschedule"
14          tools:layout_constraintTop_creator = "1"
15          tools:layout_constraintBottom_creator = "1"
16          app:layout_constraintBottom_toTopOf = "@ + id/btback"
17          android:layout_marginStart = "10dp"
18          tools:layout_constraintLeft_creator = "1"
19          app:layout_constraintLeft_toLeftOf = "@ + id/widget0"
20          app:layout_constraintTop_toTopOf = "@ + id/widget0"
21          android:layout_marginEnd = "16dp"
22          app:layout_constraintRight_toRightOf = "@ + id/widget0"
23          app:layout_constraintHorizontal_bias = "0.08">
24      </TextView>
25      <Button
26          android:id = "@ + id/btback"
27          android:layout_width = "88dp"
28          android:layout_height = "48dp"
29          android:text = "@string/myback"
30          android:onClick = "myBackMD"
31          tools:layout_constraintTop_creator = "1"
32          tools:layout_constraintLeft_creator = "1"
33          app:layout_constraintLeft_toLeftOf = "@ + id/widget29"
34          app:layout_constraintTop_toTopOf = "@ + id/widget0"
35          android:layout_marginEnd = "16dp"
36          app:layout_constraintRight_toRightOf = "@ + id/widget0"
37          app:layout_constraintBottom_toBottomOf = "@ + id/widget0"
38          app:layout_constraintVertical_bias = "0.17"
39          app:layout_constraintHorizontal_bias = "0.42">
40      </Button>
```

```
41  </android.support.constraint.ConstraintLayout>
```

结合图 5-10(d)，上述代码第 9～24 行定义了显示"这是日程视图"的静态文本框，第 25～40 行定义了显示"返回"的按钮控件，其单击事件方法为 myBackMD(第 30 行)。

应用 MyMGUIDispApp 启动后首先执行文件 MyMGUIDispAct.java，其代码如下：

```
1   package cn.edu.jxufe.zhangyong.mymguidispapp;
2
3   import android.content.Intent;
4   import android.support.v7.app.AppCompatActivity;
5   import android.os.Bundle;
6   import android.view.Gravity;
7   import android.view.View;
8   import android.widget.EditText;
9   import android.widget.RadioButton;
10  import android.widget.Toast;
11
12  public class MyMGUIDispAct extends AppCompatActivity {
13      private RadioButton rbcourse,rbschedule;
14      private EditText etuser,etsecret;
15      @Override
16      protected void onCreate(Bundle savedInstanceState) {
17        super.onCreate(savedInstanceState);
18        setContentView(R.layout.activity_my_mguidisp);
19
20        myInitGUI();
21      }
22      private void myInitGUI(){
23          rbcourse = (RadioButton)findViewById(R.id.rbcourse);
24          rbschedule = (RadioButton)findViewById(R.id.rbschedule);
25          etuser = (EditText)findViewById(R.id.etuser);
26          etsecret = (EditText)findViewById(R.id.etsecret);
27      }
28      public void myLoginMD(View v){
29          if(getResources().getString(R.string.myfullname)
30                  .equals(etuser.getText().toString())
31                  && "123456".equals(etsecret.getText().toString())){
32              if(rbcourse.isChecked()){
33                  Intent it = new Intent();
34                  it.setClass(this, MyCourse.class);
35                  startActivity(it);
36              }
37              if(rbschedule.isChecked()){
38                  Intent it = new Intent();
39                  it.setClass(this, MySchedule.class);
40                  startActivity(it);
41              }
42              this.finish();
43          }
44          else{
45              Toast toa = Toast.makeText(this, "Input right code!", Toast.LENGTH_LONG);
```

```
46              toa.setGravity(Gravity.TOP, 0, 440);
47              toa.show();
48          }
49      }
50      public void myExitMD(View v){
51          this.finish();
52      }
53  }
```

第13、14行定义了两个单选按钮对象 rbcourse 和 rbschedule 以及两个编辑框对象 etuser 和 etsecret，第22~27行的 myInitGUI 方法为这些对象赋值。当单击图5-10(a)中的"登录"按钮时，Android 系统将执行第28~49行的代码：第29~31行判断输入的用户名是否为"张勇"(R.string.myfullname 对应的字符串)，且输入的密码是否为"123456"，如果成立则执行第32~42行的代码；否则提示输入错误"Input right code!"(第45~47行)。当输入用户名和密码都正确时，第32行判断"查看课表"单选按钮是否选中，如果选中则执行第33~35行代码，即借助 Intent 对象显示"查看课表"的 Activity，如图5-10(c)所示；如果第32行为假，而第37行为真，即"查看日程"单选按钮选中时，第38~40行的代码得到执行，借助 Intent 对象显示"查看日程"的 Activity，如图5-10(d)所示。第42行调用 finish 方法关闭当前的 Activity。

程序文件 MyCourse.java 的代码如下：

```
1   package cn.edu.jxufe.zhangyong.mymguidispapp;
2
3   import android.app.Activity;
4   import android.content.Intent;
5   import android.os.Bundle;
6   import android.view.View;
7
8   public class MyCourse extends Activity{
9       @Override
10      public void onCreate(Bundle savedInstanceState) {
11          super.onCreate(savedInstanceState);
12          setContentView(R.layout.mycourse);
13          myInitGUI();
14      }
15      private void myInitGUI(){
16
17      }
18      public void myBackMD(View v){
19        Intent it = new Intent();
20        it.setClass(this, MyMGUIDispAct.class);
21        startActivity(it);
22        this.finish();
23      }
24  }
```

这里第13行的 myInitGUI 方法为空方法，如第15~17行所示，这是全书的程序设计风格，即在 onCreate 方法中调用 myInitGUI 方法进行用户初始化设计，而不是把这些初始化代码直接放在 onCreate 方法中。第 18～23 行为图 5-10(c)中"返回"按钮的单击事件方法,其中,第 19～21 行借助 Intent 对象显示 MyMGUIDispAct 类对应的界面(即图 5-10(a)),第 22 行关闭 MyCourse 类对应的界面(即图 5-10(c))。

程序文件 MySchedule.java 的代码如下:

```
1   package cn.edu.jxufe.zhangyong.mymguidispapp;
2
3   import android.app.Activity;
4   import android.content.Intent;
5   import android.os.Bundle;
6   import android.view.View;
7
8   public class MySchedule extends Activity{
9       @Override
10      public void onCreate(Bundle savedInstanceState) {
11          super.onCreate(savedInstanceState);
12          setContentView(R.layout.myschedule);
13          myInitGUI();
14      }
15      private void myInitGUI(){
16
17      }
18      public void myBackMD(View v){
19          Intent it = new Intent();
20          it.setClass(this, MyMGUIDispAct.class);
21          startActivity(it);
22          this.finish();
23      }
24  }
```

比较上述代码与 MyCourse.java 文件的代码,除了第 8 行的 MyCourse 改为 MySchedule 外,其余代码完全相同,其工作原理也相同,这里不再赘述。

5.4.2 多用户界面数据传递

例 5-8 演示了借助 Intent 对象显示多个 Activity 的方法,本小节介绍借助于 Intent 对象由正在显示的 Activity 向将要显示的 Activity 传递数据的方法。一般地,有两种传递数据的方法,其一是调用 Intent 对象的 putExtra 方法;其二是调用 Intent 对象的 putExtras 方法,此时,需要建立一个 Bundle 类的对象。例如,把字符串"张勇"传递到另一个 Activity 中,其步骤如下:

(1) 建立 Bundle 对象,如"Bundle bd = new Bundle();"。

(2) 为 Bundle 类对象 bd 赋值,如"bd.putString("USER","张勇");",这里的"USER"是键值,Bundle 类维护了一个数据表,键值是这个数据表的索引值,是一个自定义常量。

(3) 创建 Intent 对象,并连接要显示的类,如"Intent it = new Intent(); it.setClass(this, MyCourse.class);"。

(4) 将 Bundle 对象添加到 Intent 对象的数据空间中,如"it.putExtras(bd);"。

(5) 调用方法 startActivity 显示新的 Activity,如"startActivity(it);"。

(6) 在新显示的 Activity 的 onCreate 方法中，从 Intent 对象中得到 Bundle 对象数据，如"Bundle bd＝getIntent().getExtras();"。

(7) 通过 Bundle 对象取出传递的数据，如"String struser＝bd.getString("USER");"，这里必须指定键值为"USER"，即与第(2)步中的键值对应。

而直接使用 Intent 对象的 putExtra 方法更简洁，即省略第(1)步和第(2)步，第(4)步改为"it.putExtra("USER","张勇")"；第(6)步和第(7)步改为"String struser = data.getExtras().getString("USER");"。这里的"USER"仍是键值常量。

例 5-9 多用户界面数据传递实例。

新建应用 MyMGUISCommApp，应用名为 MyMGUISCommApp，活动界面名为 MyMGUISCommAct。应用 MyMGUISCommApp 的执行结果如图 5-11 所示。在图 5-11(a)中输入用户"张勇"和密码"123456"，输入的密码显示为"……"，如图 5-11(b)所示。选中图 5-11(b)中的"查看课表"单选按钮，单击"登录"按钮，进入图 5-11(c)所示的界面，显示了"张勇，这是课表视图！"，其中的"张勇"是由图 5-11(b)的界面传递过来的。在图 5-11(b)中选中"查看日程"单选按钮，则显示图 5-11(d)所示的界面，同样，这里的"张勇"也是由图 5-11(b)所示的界面传递过来的。

应用 MyMGUISCommApp 包括源文件 MyMGUISCommAct.java、MySchedule.java、MyCourse.java、布局文件 activity_my_mguiscomm.xml、mycourse.xml、myschedule.xml、汉字字符串资源文件 mystrings_hz.xml、颜色资源文件 myguicolor.xml 和工程配置文件 AndroidManifest.xml 等，除了包名改为 cn.eud.jxufe.zhangyong.mymguiscommapp 和活动视图改为 MyMGUISCommAct 之外，文件 activity_my_mguiscomm.xml、mycourse.xml、myschedule.xml、mystrings_hz.xml、myguicolor.xml 和 AndroidManifest.xml 与例 5-8 中的布局文件或同名文件内容相同，内容有所改动的文件是 MyMGUISCommAct.java、MySchedule.java 和 MyCourse.java。相比于例 5-8 中的源文件 MyMGUIDispAct.java，这里的 MyMGUISCommAct.java 文件中包名、活动视图名和方法 myLoginMD 的内容有所改动，其代码如下：

```
1    package cn.edu.jxufe.zhangyong.mymguiscommapp;
2
3    import android.content.Intent;
4    import android.support.v7.app.AppCompatActivity;
5    import android.os.Bundle;
6    import android.view.Gravity;
7    import android.view.View;
8    import android.widget.EditText;
9    import android.widget.RadioButton;
10   import android.widget.Toast;
11
12   public class MyMGUISCommAct extends AppCompatActivity {
13       private RadioButton rbcourse,rbschedule;
14       private EditText etuser,etsecret;
15       @Override
16       protected void onCreate(Bundle savedInstanceState) {
17           super.onCreate(savedInstanceState);
```

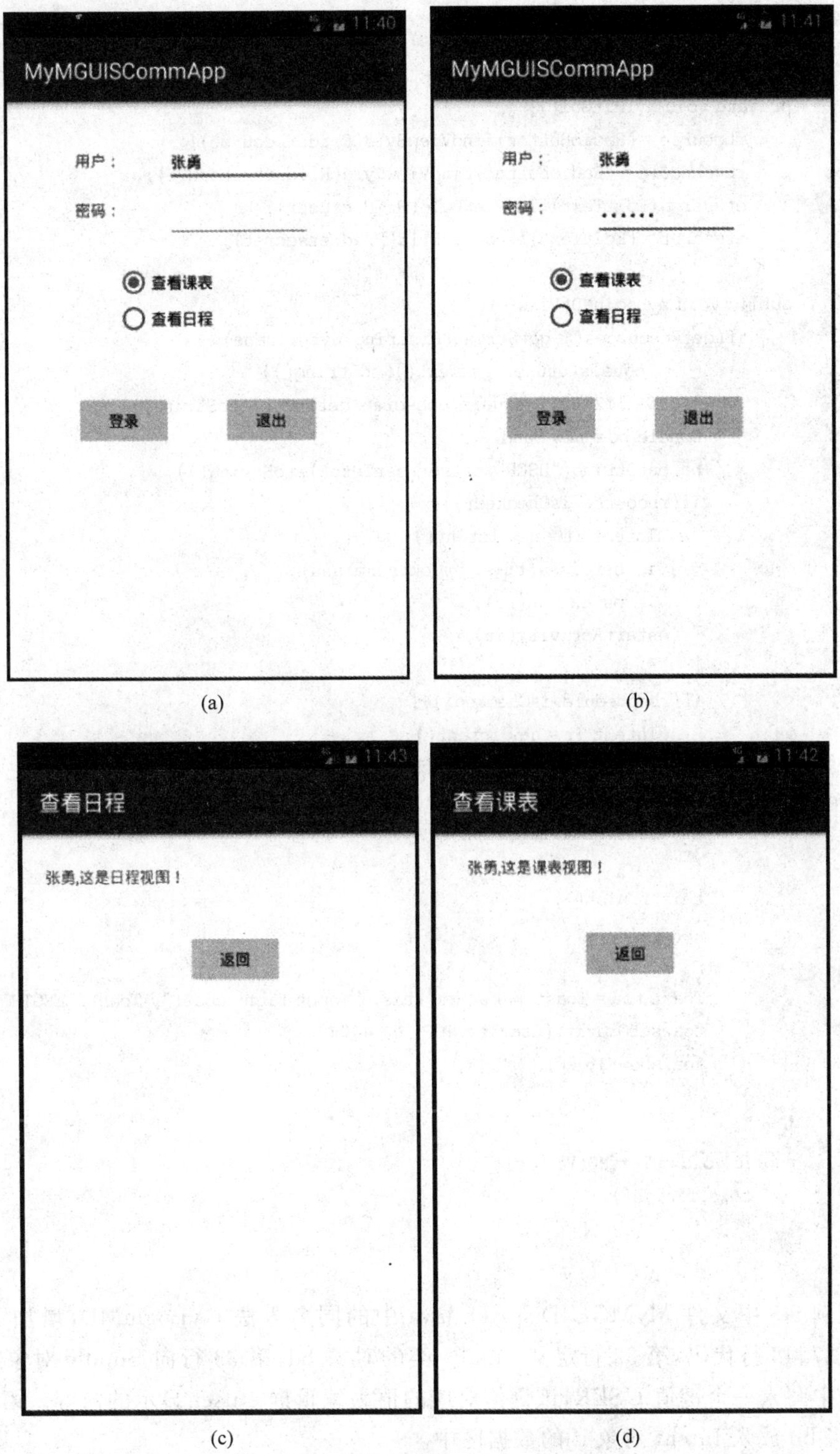

图 5-11 应用 MyMGUISCommApp 运行结果

```
18              setContentView(R.layout.activity_my_mguiscomm);
19
20              myInitGUI();
21          }
22      private void myInitGUI(){
23              rbcourse = (RadioButton)findViewById(R.id.rbcourse);
24              rbschedule = (RadioButton)findViewById(R.id.rbschedule);
25              etuser = (EditText)findViewById(R.id.etuser);
26              etsecret = (EditText)findViewById(R.id.etsecret);
27          }
28      public void myLoginMD(View v){
29              if(getResources().getString(R.string.myfullname)
30                      .equals(etuser.getText().toString())
31                      && "123456".equals(etsecret.getText().toString())){
32                  Bundle bd = new Bundle();
33                  bd.putString("USER",etuser.getText().toString());
34                  if(rbcourse.isChecked()){
35                      Intent it = new Intent();
36                      it.setClass(this, MyCourse.class);
37                      it.putExtras(bd);
38                      startActivity(it);
39                  }
40                  if(rbschedule.isChecked()){
41                      Intent it = new Intent();
42                      it.setClass(this, MySchedule.class);
43                      it.putExtras(bd);
44                      startActivity(it);
45                  }
46                  this.finish();
47              }
48              else{
49                  Toast toa = Toast.makeText(this, "Input right code!", Toast.LENGTH_LONG);
50                  toa.setGravity(Gravity.TOP, 0, 440);
51                  toa.show();
52              }
53          }
54      public void myExitMD(View v){
55              this.finish();
56          }
57  }
```

对比例 5-8 中文件 MyMGUIDispAct.java 中的同名方法 myLoginMD，增加了第 32、33 行、第 37、43 行代码，第 32 行定义 Bundle 类的对象 bd，第 33 行向 Bundle 对象 bd 维护的数据表中写入一个键值 USER，该键值对应的值为编辑框 etuser 显示的内容。第 37 行或第 43 行将 bd 放入 Intent 对象 it 的数据区中。

文件 MyCourse.java 的内容如下：

```
1   package cn.edu.jxufe.zhangyong.mymguiscommapp;
```

```
2
3    import android.content.Intent;
4    import android.os.Bundle;
5    import android.support.v7.app.AppCompatActivity;
6    import android.view.View;
7    import android.widget.TextView;
8
9    public class MyCourse extends AppCompatActivity {
10       private TextView tv;
11       @Override
12       public void onCreate(Bundle savedInstanceState) {
13           super.onCreate(savedInstanceState);
14           setContentView(R.layout.mycourse);
15           myInitGUI();
16       }
17       private void myInitGUI(){
18           tv = (TextView)findViewById(R.id.tvhint);
19           Bundle bd = getIntent().getExtras();
20           String struser = bd.getString("USER");
21           tv.setText(struser + "," + getResources().getString(R.string.thcourse));
22       }
23       public void myBackMD(View v){
24           Intent it = new Intent();
25           it.setClass(this, MyMGUISCommAct.class);
26           startActivity(it);
27           this.finish();
28       }
29   }
```

对比于例 5-8 中的同名文件,增加了第 10 行、第 18~21 行代码,其中第 10 行定义静态文本框对象 tv;第 19、20 行从 Intent 对象中数据区得到 Bundle 对象,进而得到键值为 "USER" 的字符串值,即图 5-11(c)中的 "张勇",然后,第 21 行将 "张勇" 显示出来,如图 5-11(c)所示。

文件 MySchedule.java 与文件 MyCourse.java 的代码不同的地方在于:① 文件 MySchedule.java 对应的类为 MySchedule,即上述代码第 9 行的 MyCourse 改为 MySchedule;② 第 18 行的 tvhint 改为 tvdispsch;③ 第 21 行的 thcourse 改为 thschedule。这两个文件进行的处理方式完全相同。

5.4.3 活动界面间双向数据通信

第 5.4.2 节介绍了由一个活动界面向另一个活动界面传递数据的方法,即借助于 Intent 对象进行数据的传递,借助于 Intent 对象可以实现活动界面间的双向数据传递,但是第 5.4.2 节的方法却无法实现活动界面间的双向数据传递。原因是这样的:假设程序启动后,界面 A 向界面 B 发送了一个数据,即在界面 A 中建立了一个 Intent 对象 O,调用 Intent 对象 O 的 putExtras 方法向 Intent 对象写入数据,然后执行 startActivity 进入到界面 B,界面 B 在其 onCreate 方法中取得 O 对象,读出由界面 A 传递的数据;接着,由于某个事件(例如单击界面 B 的按钮事件），界面 B 创建一个 Intent 对象 P，调用对象 P 的 putExtras 方法向 P 写入数据，然后调用 startActivity 方法显示界面 A，在界面 A 的 onCreate 方法中取得对象 P，读出由界面 B 传来的数据。读者多读几遍就会发现其中的问题。问题在于无论是界面 A 还是界面 B，都是在 onCreate 方法中读取 Intent 中的数据。由于界面 A 是先创建的，所以，如果仅考虑由界面 A 向界面 B 传递数据，当界面 B 启动时执行其 onCreate 方法时，由界面 A 向界面 B 发送数据的 Intent 对象已经存在了，这就是第 5.4.2 节的方法；然而，正是由于界面 A 是先创建的，所以，当程序第一次执行时，由界面 B 向界面 A 发送数据的 Intent 对象并不存在，而为了实现双向数据传递，界面 A 和界面 B 的 onCreate 方法中都含有提取 Intent 对象数据的方法，因此，程序启动时，界面 A 无法读取界面 B 传递来的数据（事实上，此时根本没有界面 B，更不会有界面 B 发送的 Intent 对象和数据）。

  由上所述，解决界面间双向数据传递的矛盾，只需要让先创建的界面 A 接收来自后创建的界面 B 的数据这一动作，不是在 onCreate 方法中接收即可。也就是说，先创建的界面 A 向界面 B 传递数据依然是第 5.4.2 节的方法，后创建的界面 B 向界面 A 传递数据的方法需要通过一个专门的方法通知界面 A，而不是使用 startActivity 方法让界面 A 在 onCreate 方法中接收数据。

  因此，Android 应用程序实现两个界面间传递数据的方法为（假定界面 A 先创建，界面 B 后创建）：

  （1）界面 A 中创建 Intent 对象并写入数据，然后调用 startActivityForResult 方法而不是 startActivity 方法显示界面 B，并且不能调用 finish 方法关闭界面 A。

  （2）界面 B 通过 Intent 对象取得界面 A 传递来的数据；界面 B 创建新的 Intent 对象并写入数据，然后执行 setResult 方法而不是 startActivity 方法显示界面 A，这时界面 A 将自动调用 onActivityResult 方法而不是 onCreate 方法，界面 A 在 onActivityResult 方法中接收界面 B 通过 Intent 对象发送的数据。

  方法 startActivityForResult 的原型如下：

```
void startActivityForResult(Intent intent, int requestCode);
```

其中，将启动的 Activity 界面对应的 Intent 对象传递给 intent 参数，而 requestCode 为一个请求码，一个界面可以从多个界面取回数据，至于从哪个界面取回数据，由 requestCode 决定。

  例 5-10 用户界面间数据双向传递实例。

  新建应用 MyMGUIDCommApp，应用名为 MyMGUIDCommApp，活动界面名为 MyMGUIDCommAct。

  应用 MyMGUIDCommApp 如图 5-12 所示，包含源文件 MyMGUIDCommAct.java、MyCourse.java、MySchedule.java、布局文件 activity_my_mguidcomm.xml、mycourse.xml、myschedule.xml、汉字字符串资源文件 mystrings_hz.xml、颜色文件 myguicolor.xml 以及项目配置文件 AndroidManifest.xml 等，其中，myguicolor.xml 文件与例 5-9 中的同名文件内容相同；除了包名、应用名和活动界面名不同外，AndroidManifest.xml 文件与例 5-9 中的同名文件内容相同。

  汉字字符串资源文件 mystrings_hz.xml 定义了工程中使用的字符串资源，其内容

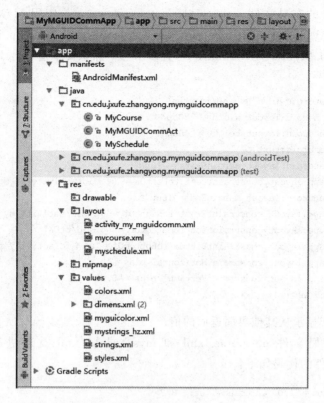

图 5-12　应用 MyMGUIDCommApp

如下：

```
1   <?xml version = "1.0" encoding = "utf - 8"?>
2   <resources>
3       <string name = "myuser">用户：</string>
4       <string name = "mysecret">密码：</string>
5       <string name = "mycourse">查看课表</string>
6       <string name = "myschedule">查看日程</string>
7       <string name = "myok">登录</string>
8       <string name = "myexit">退出</string>
9       <string name = "thcourse">这是课表视图！</string>
10      <string name = "thschedule">这是日程视图！</string>
11      <string name = "myfullname">张勇</string>
12      <string name = "myback">返回</string>
13      <string name = "mydsp">DSP 技术</string>
14      <string name = "myeos">嵌入式操作系统</string>
15      <string name = "myeda">EDA 技术</string>
16      <string name = "myyk">今天有课：</string>
17      <string name = "mymk">今天没有课！</string>
18      <string name = "myckkb">您刚查看了课表，</string>
19      <string name = "myckrc">您刚查看了日程表，</string>
20      <string name = "myyh">今天有会！</string>
21      <string name = "mywh">今天没有会！</string>
22  </resources>
```

相对于例 5-9 的布局文件 activity_my_mguiscomm.xml，应用 MyMGUIDCommApp 的布局文件 activity_my_mguidcomm.xml 中 tools:context 为"cn.edu.jxufe.zhangyong. mymguidcommapp.MyMGUIDCommAct"，且多了一个静态文本框，即

```
1   <TextView
2          android:id="@+id/tvatten"
3          android:layout_width="280dp"
4          android:layout_height="40dp"
5          android:text=" "
6          android:layout_marginStart="16dp"
7          app:layout_constraintLeft_toLeftOf="@+id/activity_my_mguidisp"
8          android:layout_marginEnd="16dp"
9          app:layout_constraintRight_toRightOf="@+id/activity_my_mguidisp"
10         app:layout_constraintTop_toBottomOf="@+id/btOK"
11         app:layout_constraintBottom_toBottomOf="@+id/activity_my_mguidisp"
12         app:layout_constraintHorizontal_bias="0.42"
13         app:layout_constraintVertical_bias="0.25">
14  </TextView>
```

该静态文本框用于显示从其他界面返回的值。

例 5-10 的布局文件 mycourse.xml 和 myschedule.xml 重新进行了设计，其中，mycourse.xml 文件的代码如下：

```
1   <?xml version="1.0" encoding="utf-8"?>
2   <android.support.constraint.ConstraintLayout xmlns:app="http://schemas.android.com/apk/res-auto"
3       xmlns:tools="http://schemas.android.com/tools"
4       android:id="@+id/widget0"
5       android:layout_width="fill_parent"
6       android:layout_height="fill_parent"
7       xmlns:android="http://schemas.android.com/apk/res/android"
8       android:background="@drawable/darkgray">
9       <TextView
10          android:id="@+id/tvhint"
11          android:layout_width="133dp"
12          android:layout_height="28dp"
13          android:text="@string/thcourse"
14          app:layout_constraintBottom_toTopOf="@+id/cbDSP"
15          app:layout_constraintTop_toTopOf="@+id/widget0"
16          android:layout_marginStart="16dp"
17          app:layout_constraintLeft_toLeftOf="@+id/widget0"
18          android:layout_marginEnd="16dp"
19          app:layout_constraintRight_toRightOf="@+id/widget0"
20          app:layout_constraintHorizontal_bias="0.15"
21          app:layout_constraintVertical_bias="0.42">
22      </TextView>
23      <CheckBox
24          android:id="@+id/cbDSP"
25          android:layout_width="150dp"
26          android:layout_height="40dp"
```

```
27              android:text = "@string/mydsp"
28              android:textSize = "14sp"
29              android:layout_marginStart = "16dp"
30              app:layout_constraintLeft_toLeftOf = "@ + id/widget0"
31              android:layout_marginEnd = "16dp"
32              app:layout_constraintRight_toRightOf = "@ + id/widget0"
33              app:layout_constraintTop_toTopOf = "@ + id/widget0"
34              app:layout_constraintBottom_toBottomOf = "@ + id/widget0"
35              app:layout_constraintHorizontal_bias = "0.17"
36              app:layout_constraintVertical_bias = "0.15">
37          </CheckBox>
38          <CheckBox
39              android:id = "@ + id/cbEOS"
40              android:layout_width = "150dp"
41              android:layout_height = "40dp"
42              android:text = "@string/myeos"
43              android:textSize = "14sp"
44              app:layout_constraintTop_toBottomOf = "@ + id/cbDSP"
45              app:layout_constraintBottom_toBottomOf = "@ + id/widget0"
46              app:layout_constraintVertical_bias = "0.04"
47              android:layout_marginEnd = "16dp"
48              app:layout_constraintRight_toRightOf = "@ + id/widget0"
49              app:layout_constraintLeft_toLeftOf = "@ + id/cbDSP"
50              app:layout_constraintHorizontal_bias = "0.0">
51          </CheckBox>
52          <CheckBox
53              android:id = "@ + id/cbEDA"
54              android:layout_width = "150dp"
55              android:layout_height = "40dp"
56              android:text = "@string/myeda"
57              android:textSize = "14sp"
58              app:layout_constraintTop_toBottomOf = "@ + id/cbEOS"
59              app:layout_constraintBottom_toBottomOf = "@ + id/widget0"
60              android:layout_marginEnd = "16dp"
61              app:layout_constraintRight_toRightOf = "@ + id/widget0"
62              app:layout_constraintLeft_toLeftOf = "@ + id/cbEOS"
63              app:layout_constraintHorizontal_bias = "0.0"
64              app:layout_constraintVertical_bias = "0.04">
65          </CheckBox>
66          <Button
67              android:id = "@ + id/btback"
68              android:layout_width = "88dp"
69              android:layout_height = "48dp"
70              android:text = "@string/myback"
71              android:onClick = "myBackMD"
72              app:layout_constraintTop_toBottomOf = "@ + id/cbEDA"
73              app:layout_constraintBottom_toBottomOf = "@ + id/widget0"
74              android:layout_marginEnd = "16dp"
75              app:layout_constraintRight_toRightOf = "@ + id/widget0"
76              android:layout_marginStart = "16dp"
77              app:layout_constraintLeft_toLeftOf = "@ + id/widget0"
```

```
78                app:layout_constraintVertical_bias = "0.13">
79            </Button>
80    </android.support.constraint.ConstraintLayout>
```

从上述代码中可知，mycourse.xml 对应的界面上有一个 TextView 静态文本框(第 9～22 行)和三个复选框(第 23～37 行、第 38～51 行、第 52～65 行)以及一个命令按钮(第 66～79 行)，命令按钮的单击事件为 myBackMD(第 71 行)，如图 5-13(c)所示。

布局文件 myschedule.xml 的内容如下：

```
1   <?xml version = "1.0" encoding = "utf-8"?>
2   <android.support.constraint.ConstraintLayout xmlns:app = "http://schemas.android.com/apk/res-auto"
3       xmlns:tools = "http://schemas.android.com/tools"
4       android:id = "@+id/widget0"
5       android:layout_width = "fill_parent"
6       android:layout_height = "fill_parent"
7       xmlns:android = "http://schemas.android.com/apk/res/android"
8       android:background = "@drawable/darkgray">
9       <TextView
10          android:id = "@+id/tvdispsch"
11          android:layout_width = "135dp"
12          android:layout_height = "33dp"
13          android:text = "@string/thschedule"
14          app:layout_constraintBottom_toTopOf = "@+id/rgschedule"
15          app:layout_constraintTop_toTopOf = "@+id/widget0"
16          android:layout_marginStart = "16dp"
17          app:layout_constraintLeft_toLeftOf = "@+id/widget0"
18          android:layout_marginEnd = "16dp"
19          app:layout_constraintRight_toRightOf = "@+id/widget0"
20          app:layout_constraintHorizontal_bias = "0.18">
21      </TextView>
22      <RadioGroup
23          android:id = "@+id/rgschedule"
24          android:layout_width = "157dp"
25          android:layout_height = "78dp"
26          app:layout_constraintTop_toTopOf = "@+id/widget0"
27          app:layout_constraintBottom_toBottomOf = "@+id/widget0"
28          android:layout_marginStart = "16dp"
29          app:layout_constraintLeft_toLeftOf = "@+id/widget0"
30          android:layout_marginEnd = "16dp"
31          app:layout_constraintRight_toRightOf = "@+id/widget0"
32          app:layout_constraintHorizontal_bias = "0.2"
33          app:layout_constraintVertical_bias = "0.19">
34          <RadioButton
35              android:id = "@+id/rbnomeeting"
36              android:layout_width = "wrap_content"
37              android:layout_height = "wrap_content"
38              android:text = "@string/mywh"
```

```
39              tools:layout_editor_absoluteY = "105dp"
40              tools:layout_editor_absoluteX = "158dp">
41          </RadioButton>
42          <RadioButton
43              android:id = "@ + id/rbmeeting"
44              android:layout_width = "wrap_content"
45              android:layout_height = "wrap_content"
46              android:text = "@string/myyh"
47              android:checked = "true"
48              tools:layout_editor_absoluteY = "153dp"
49              tools:layout_editor_absoluteX = "65dp">
50          </RadioButton>
51      </RadioGroup>
52      <Button
53          android:id = "@ + id/btback"
54          android:layout_width = "88dp"
55          android:layout_height = "48dp"
56          android:text = "@string/myback"
57          android:onClick = "myBackMD"
58          app:layout_constraintTop_toBottomOf = "@ + id/rgschedule"
59          app:layout_constraintBottom_toBottomOf = "@ + id/widget0"
60          android:layout_marginEnd = "16dp"
61          app:layout_constraintRight_toRightOf = "@ + id/widget0"
62          app:layout_constraintLeft_toLeftOf = "@ + id/rgschedule"
63          app:layout_constraintHorizontal_bias = "0.41"
64          app:layout_constraintVertical_bias = "0.21">
65      </Button>
66  </android.support.constraint.ConstraintLayout>
```

由上述代码可知，myschedule.xml 对应的布局如图 5-13(e)所示，即包含一个静态文本框(第 9~21 行)、一个包含两个单选钮的单选钮组(第 22~51 行)和一个命令按钮(第 52~65 行)，命令按钮的单击事件方法名为 myBackMD(第 57 行)。

应用 MyMGUIDCommApp 的执行结果如图 5-13 所示。在图 5-13(a)中输入用户和密码分别为"张勇"和"123456"，输入的密码显示为"……"，如图 5-13(b)所示，选中"查看课表"单选按钮，然后单击"登录"按钮，进入图 5-13(c)所示的界面。图 5-13(c)中的"张勇"是由界面图 5-13(b)传递过来的，在图 5-13(c)中选中"DSP 技术"和"EDA 技术"(可以任意选择)，然后单击"返回"按钮，进入图 5-13(d)所示的界面。在图 5-13(d)下面显示了文本"您刚查看了课表，今天有课：DSP 技术 EDA 技术"，这些信息是由图 5-13(c)所示的界面传递过来的。如果在图 5-13(b)中选中"查看日程"，单击"登录"按钮，则进入图 5-13(e)所示的界面。在图 5-13(e)中选中"今天没有会"单选钮(可以随意选)，然后单击"返回"按钮，进入图 5-13(f)所示的界面。在图 5-13(f)下面显示文本"您刚查看了日程表，今天没有会！"，这些信息是由图 5-13(e)所示界面传递过来的。

综上所述，图 5-13(b)所示界面向图 5-13(c)和图 5-13(e)传递了数据"张勇"，而图 5-13(c)和图 5-13(e)所示界面向图 5-13(b)所示界面传递了上课和开会的信息，从而实现了界面间的双向数据传递。

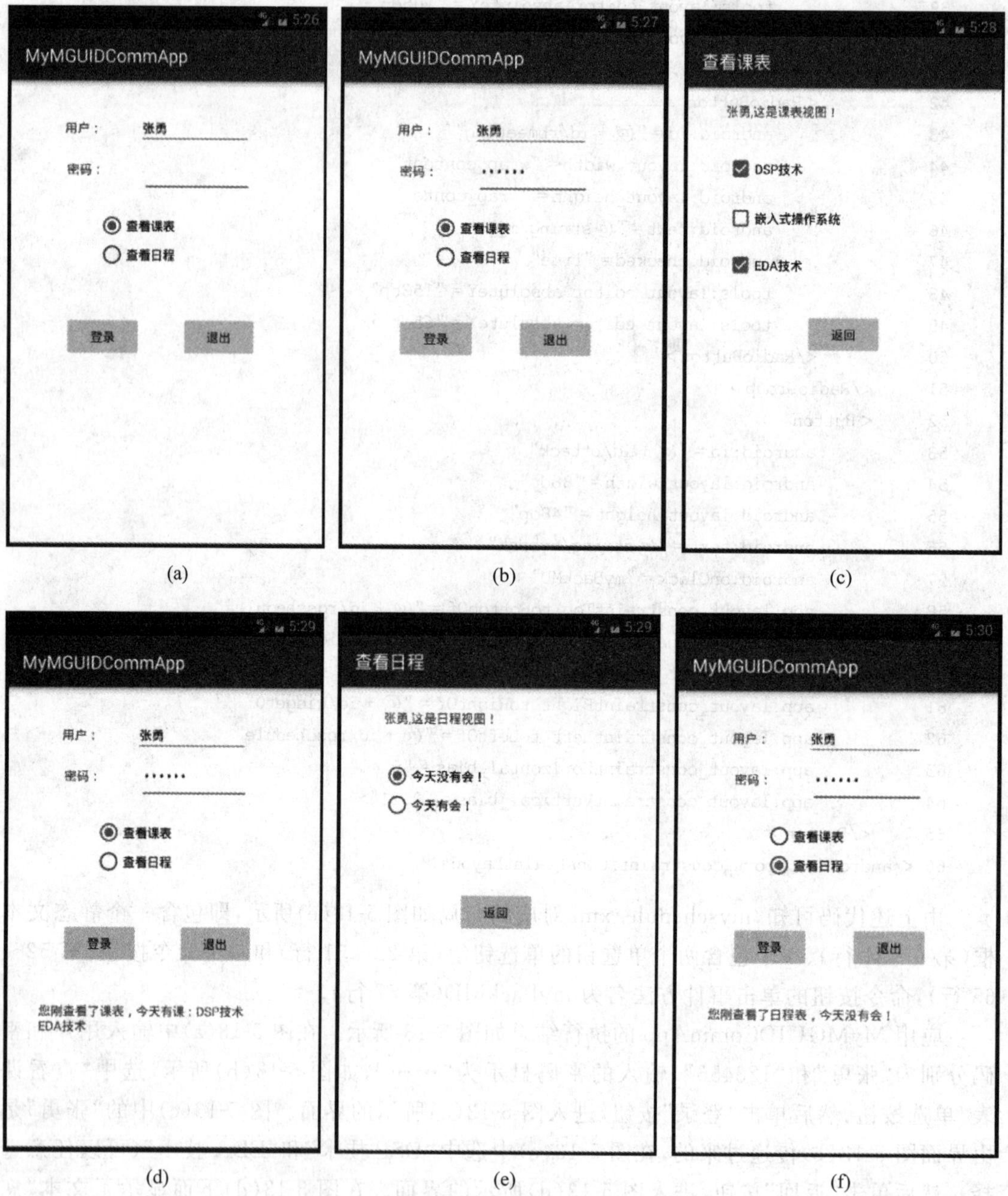

图 5-13 应用 MyMGUIDCommApp 执行结果

源文件 MyMGUIDCommAct.java 的内容如下：

```
1    package cn.edu.jxufe.zhangyong.mymguidcommapp;
2
3    import android.support.v7.app.AppCompatActivity;
4    import android.content.Intent;
5    import android.os.Bundle;
6    import android.view.Gravity;
7    import android.view.View;
```

```
8   import android.widget.EditText;
9   import android.widget.RadioButton;
10  import android.widget.TextView;
11  import android.widget.Toast;
12
13  public class MyMGUIDCommAct extends AppCompatActivity {
14      public RadioButton rbcourse,rbschedule;
15      public EditText etuser,etsecret;
16      public TextView tvatten;
17      private final int COU_REQUEST = 1;
18      private final int SCH_REQUEST = 2;
```

第17、18行的 COU_REQUEST 和 SCH_REQUEST 为自定义常量,用于表示请求码,即表示从哪个显示界面中返回数据。结合第43~45、49~51行就会明白,向 MyCourse 对象请求数据使用请求码 COU_REQUEST,而向 MySchedule 对象请求数据使用请求码 SCH_REQUEST。

```
19
20      @Override
21      protected void onCreate(Bundle savedInstanceState) {
22          super.onCreate(savedInstanceState);
23          setContentView(R.layout.activity_my_mguidcomm);
24
25          myInitGUI();
26      }
27      private void myInitGUI(){
28          rbcourse = (RadioButton)findViewById(R.id.rbcourse);
29          rbschedule = (RadioButton)findViewById(R.id.rbschedule);
30          etuser = (EditText)findViewById(R.id.etuser);
31          etsecret = (EditText)findViewById(R.id.etsecret);
32          tvatten = (TextView)findViewById(R.id.tvatten);
33      }
34      public void myLoginMD(View v){
35          if(getResources().getString(R.string.myfullname)
36                  .equals(etuser.getText().toString())
37                  && "123456".equals(etsecret.getText().toString())){
```

第35~37行为一条语句,即判断用户名是否为字符串资源 R.string.myfullname 指向的字符串(这里是"张勇"),并且密码为"123456"。

```
38
39      Bundle bd = new Bundle();
40      bd.putString("USER", etuser.getText().toString());
```

第39行创建一个 Bundle 对象 bd,第40行将 etuser(编辑框对象)中的字符串写入 bd 对象的数据表中,其键值为"USER",键值就是数据在数据表中的索引值。

```
41      if(rbcourse.isChecked()){
42          Intent it = new Intent();
43          it.setClass(this, MyCourse.class);
44          it.putExtras(bd); //it.putExtra("USER",etuser.getText().toString());
```

```
45        startActivityForResult(it,COU_REQUEST);
46    }
```

第41行判断单选按钮"查看课表"是否选中,如果选中,则rbcourse.isChecked()返回ture(即真),则第42~45行得到执行。第42行创建Intent对象it,第43行将MyCourse类与it对象链接起来,第44行将bd对象存入it对象的数据区,第45行调用startActivityForResult方法而不是startActivity方法启动MyCourse类定义的活动界面对象,从MyCourse界面返回的值的请求号将为COU_REQUEST。

```
47    if(rbschedule.isChecked()){
48        Intent it = new Intent();
49        it.setClass(this, MySchedule.class);
50        it.putExtras(bd);
51        startActivityForResult(it,SCH_REQUEST);
52    }
```

第47行判断"查看日程"单选钮是否选中,如果选中,第48~51行得到执行,第51行调用startActivityForResult方法显示MySchedule类定义的活动界面对象,从MySchedule界面返回的值的请求号为SCH_REQUEST。需要注意的是,这里不能使用finish方法关闭当前的界面。

```
53        }
54        else{
55            Toast toa = Toast.makeText(this, "Input right code!", Toast.LENGTH_LONG);
56            toa.setGravity(Gravity.TOP, 0, 440);
57            toa.show();
58        }
59    }
60    public void myExitMD(View v){
61        this.finish();
62    }
63    @Override
64    protected void onActivityResult(int requestCode, int resultCode, Intent data){
65        if(resultCode == RESULT_OK){
66            switch(requestCode){
67                case COU_REQUEST:
68                    String str_c = data.getExtras().getString("MYCOURSE");
69                    tvatten.setText(str_c);
70                    break;
71                case SCH_REQUEST:
72                    String str_s = data.getExtras().getString("MYSCHEDULE");
73                    tvatten.setText(str_s);
74                    break;
75            }
76        }
77    }
78 }
```

上面第64~77行为onActivityResult方法,当从MyCourse或MySchedule界面返回时,该方法自动执行。如果返回结果码resultCode为RESULT_OK(第65行为真),则第 66～75 行得到执行。返回的结果码可在下面文件 MyCourse.java 中的第 50 行或文件 MySchedule.java 中的第 39 行查到,由那里的方法 setResult 设置,这里的 RESULT_OK 是 Android 系统常数。第 66 行根据请求码 requestCode 的值跳转到相应的分支去执行。如果请求码为 COU_REQUEST,说明是从 MyCourse 界面返回的,然后,第 68～79 行代码得到执行。第 68 行代码得到自 MyCourse 界面返回的数据,这些数据赋给字符串变量 str_c,然后,第 69 行设置静态文本框 tvatten 显示字符串 str_c 的值。需要注意的是:第 68 行获取字符串的键值为"MYCOURSE",与文件 MyCourse.java 中的第 49 行的"MYCOURSE"是对应的。

源文件 MyCourse.java 的内容如下:

```
1   package cn.edu.jxufe.zhangyong.mymguidcommapp;
2
3   import android.content.Intent;
4   import android.os.Bundle;
5   import android.support.v7.app.AppCompatActivity;
6   import android.view.View;
7   import android.widget.CheckBox;
8   import android.widget.TextView;
9
10  public class MyCourse extends AppCompatActivity{
11      private TextView tv;
12      private CheckBox cbdsp,cbeos,cbeda;
13      @Override
14      public void onCreate(Bundle savedInstanceState) {
15          super.onCreate(savedInstanceState);
16          setContentView(R.layout.mycourse);
17          myInitGUI();
18      }
19      private void myInitGUI(){
20          tv = (TextView)findViewById(R.id.tvhint);
21          Bundle bd = getIntent().getExtras();
22          String struser = bd.getString("USER");
23          tv.setText(struser + "," + getResources().getString(R.string.thcourse));
```

第 20～23 行为从 Intent 对象中获取数据并在静态文本框对象 tv 中显示出来。这里的方法 myInitGUI 是在 onCreate 方法中调用的(第 17 行),因此,MyCourse 界面一启动就要接收启动该界面 Intent 中的数据,因此,必须在该界面不显示时调用 finish 方法(第 51 行)关闭它,保证它每次启动都会执行 onCreate 方法。如果没有第 51 行,就应把接收 Intent 数据的程序代码放置在 onResume 方法中。

```
24
25          cbdsp = (CheckBox)findViewById(R.id.cbDSP);
26          cbeos = (CheckBox)findViewById(R.id.cbEOS);
27          cbeda = (CheckBox)findViewById(R.id.cbEDA);
28      }
29      public void myBackMD(View v){
```

```
30      String str = getResources().getString(R.string.myckkb);
31      if(cbdsp.isChecked() || cbeos.isChecked() || cbeda.isChecked()){
32          str = str + getResources().getString(R.string.myyk);
33          if(cbdsp.isChecked()){
34              str = str + getResources().getString(R.string.mydsp) + " ";
35          }
36          if(cbeos.isChecked()){
37              str = str + getResources().getString(R.string.myeos) + " ";
38          }
39          if(cbeda.isChecked()){
40              str = str + getResources().getString(R.string.myeda);
41          }
42      }
43      else{
44          str = str + getResources().getString(R.string.mymk);
45      }
46
47      Intent it_course = new Intent();
48      it_course.setClass(this, MyMGUIDCommAct.class);
49      it_course.putExtra("MYCOURSE", str);
50      setResult(RESULT_OK, it_course);
51      this.finish();
52  }
53 }
```

第 29~52 行代码为单击"返回"按钮时执行的代码，第 30 行 str 的值为"您刚查看了课表，"（即 R.string.myckkb 的值）；当图 5-13(c)中至少有一个复选框选中时（第 31 行为真），第 32 行 str 的值为"您刚查看了课表，今天有课："（即追加上 R.string.myyk 的值）；当选中图 5-13(c)所示中的两个复选框时，第 33 行和第 39 行为真，第 34 和 40 行得到执行，执行完成后，str 的值为"您刚查看了课表，今天有课：DSP 技术 EDA 技术"，即显示在图 5-13(d)中的结果。第 49 行将 str 的值写入 Intent 对象的数据表中，其键值为"MYCOURSE"。然后，调用 setResult 方法返回到 it_course 对象链接的 MyMGUIDCommAct 界面（第 48 行）。最后，第 51 行调用 finish 方法关闭该界面。

源文件 MySchedule.java 的内容如下：

```
1  package cn.edu.jxufe.zhangyong.mymguidcommapp;
2
3  import android.content.Intent;
4  import android.os.Bundle;
5  import android.support.v7.app.AppCompatActivity;
6  import android.view.View;
7  import android.widget.RadioButton;
8  import android.widget.TextView;
9
10 public class MySchedule extends AppCompatActivity{
11     private TextView tv;
12     private RadioButton rbyes,rbno;
```

```
13    @Override
14    public void onCreate(Bundle savedInstanceState) {
15        super.onCreate(savedInstanceState);
16        setContentView(R.layout.myschedule);
17        myInitGUI();
18    }
19    private void myInitGUI(){
20        tv = (TextView)findViewById(R.id.tvdispsch);
21        Bundle bd = getIntent().getExtras();
22        String struser = bd.getString("USER");
23        tv.setText(struser + "," + getResources().getString(R.string.thschedule));
24
25        rbyes = (RadioButton)findViewById(R.id.rbmeeting);
26        rbno = (RadioButton)findViewById(R.id.rbnomeeting);
27    }
28    public void myBackMD(View v){
29        String str = getResources().getString(R.string.myckrc);
30        if(rbyes.isChecked()){
31            str = str + getResources().getString(R.string.myyh);
32        }
33        if(rbno.isChecked()){
34            str = str + getResources().getString(R.string.mywh);
35        }
36        Intent it_schedule = new Intent();
37        it_schedule.setClass(this, MyMGUIDCommAct.class);
38        it_schedule.putExtra("MYSCHEDULE", str);
39        setResult(RESULT_OK, it_schedule);
40        this.finish();
41    }
42 }
```

上述代码与 MyCourse.java 文件的代码工作原理相同，不同的是，第 38 行将字符串 str 的值写入到 Intent 对象的数据表时，使用了键值 "MYSCHEDULE"。这里 str 的值是由第 29～35 行的代码产生的，如图 5-13（e）所示，则 str 的值为"您刚查看了日程表，今天没有会！"。

5.5 本章小结

Android 应用程序中，最常用的创建对话框方法是借助 AlertDialog 类实现的，程序员还可以直接使用 Dialog 类创建对话框。此外，Android 系统还集成了时间选择对话框、日期选择对话框、进度条对话框等基本对话框类，创建这些对话框直接使用这些对话框类即可。

Android 系统支持两种菜单，即顶部弹出菜单和上下文菜单。Android 系统最多支持两级菜单，即顶级菜单和一级子菜单，子菜单的显示样式与上下文菜单相同。可以借助 XML 布局文件创建静态菜单，也可以使用程序代码在程序执行过程中创建动态菜单。Android 系统的菜单显示风格与 Windows CE 完全不同，更适合屏幕较小的智能设备。

Activity 和 Intent 是 Android 应用程序的重要组件，Activity（活动界面）是显示在屏幕上的用户界面，一般地，一个 Activity 对应于一个布局文件，多个 Activity 对应于多个布局文件。一个 Android 应用程序可以拥有多个 Activity，各个 Activity 之间借助于 Intent 对象进行数据通信。方法 startActivity 用于启动（或显示）另一个 Activity，方法 startActivityForResult 用于启动一个需要有返回值的 Activity，当一个 Activity 与多个 Activity 间进行数据通信时，需要指定请求码以区分从哪个 Activity 得到了返回值。Intent 对象本身包含了数据区（或称数据表），使用 putExtra 方法将要传递的数据存入 Intent 对象的数据表中，或者借助于 Bundle 对象将要传递的数据存入 Intent 对象中。

第 6 章

数据访问技术

Android 系统提供了访问磁盘文件、数据库和应用程序间数据共享的方法。通过接口 SharedPreferences 可以实现 XML 格式文件的访问,在这类文件中,每个数据项具有一个对应的键值。Android 系统支持 Java 语言的标准文件输入/输出流操作以及 SQLite 关系查询数据库访问技术,通过 Content Provider 组件可以实现应用程序间数据的共享。本章首先阐述 SharedPreferences(译为共享参考脚本文件)存储 XML 文件的方法,然后介绍文件操作和数据库操作方法,最后列举一个应用实例。

6.1 SharedPreferences 文件访问

SharedPreferences 文件是 XML 格式的文件,由字段组成,每个字段包括一个键值和字段值,键值是自定义常数,用于检索该字段的值。调用如下语句

```
SharedPreferences sharedPref = getPreferences(MODE_PRIVATE);
```

可创建 SharedPreferences 对象 sharedPref,方法 getPreferences 是当前 Activity 对象的方法,MODE_PRIVATE 是 Android 系统定义的常数(其值为 0),表示该 SharedPreferences 文件被该应用程序私有,其他应用程序不能访问。方法 getPreferences 还可以传递参数 MODE_WORLD_READABLE(值为 1)或 MODE_WORLD_WRITEABLE(值为 2),分别表示其他应用程序可以读或写该文件。方法 getPreferences 无法指定文件名,其返回的文件名与其所在的 Activity 界面名相同,如果 Activity 界面文件名为 MyPreferencesAct.java,那么 SharedPreferences 文件名为 MyPreferencesAct.xml。

调用 sharedPref 对象的 getString、getInt、getFloat、getBoolean、getLong 和 getStringSet 等方法,可从 SharedPreferences 文件中读出相应类型的数据。例如,getString 从 SharedPreferences

文件读出字符串数据，读取数据时，需要指定键值，例如，

```
String str1 = sharedPref.getString("USER1","USER");
```

读取键值为"USER1"的字符串值，如果读取失败，则将"USER"字符串赋给字符串变量 str1。

向 SharedPreferences 文件中写入数据，需要借助 SharedPreferences.Editor 接口，定义接口对象如下：

```
SharedPreferences.Editor editor = sharedPref.edit();
```

调用对象 editor 的 putString、putInt、putFloat、putBoolean、putLong 和 putStringSet 等方法向 SharedPreferences 文件中写入相应数型的数据。例如，putString 是写入字符串数据，写入字符串时需要指定键值，例如，

```
editor.putString("USER1","张勇");
editor.commit();      //editor.apply();
```

将键值"USER1"和其对应的字符串值"张勇"写入 SharedPreferences 文件中。写入操作完成后，需要调用 apply 或 commit 方法"提交"本次写入操作，执行方法 commit 或 apply 后数据才真正写入 SharedPreferences 文件中。

例 6-1　SharedPreferences 文件访问实例。

新建应用 MyPreferencesApp，应用名为 MyPreferencesApp，活动界面名为 MyPreferencesAct。应用 MyPreferencesApp 实现的功能如图 6-1 所示。应用 MyPreferencesApp 启动后显示如图 6-1(a)所示界面，该界面对应着布局文件 activity_my_preferences.xml。在图 6-1(a)中输入用户名"ZhangYong"和密码"123456"，如图 6-1(b)所示。

在图 6-1(b)中，先单击"添加"按钮，将输入的数据（这里的用户名"ZhangYong"和密码"123456"）写入 SharedPreferences 文件中（这里是 MyPreferencesAct.xml 文件）；然后单击"查看用户"按钮，进入图 6-1(c)所示界面。在图 6-1(c)中显示了图 6-1(b)中输入的用户信息。用同样的方法再输入两组用户信息，即"HouWengang"和"123554"以及"ZhangQiang"和"654321"，然后，单击"查看用户"按钮将显示图 6-1(d)所示的界面。

在图 6-1(e)中任意单击列表中的项，其用户名和密码都将显示在图中的编辑框中，例如，单击列表框中的"HouWengang　123554"，则其上方的用户名和密码编辑框中将显示"HouWengang 和 123554"，这时单击"删除"按钮，将该项从文件中删除。在图 6-1(f)中单击"查看用户"，将失去删除掉的用户信息，即仅显示文件中的用户信息。

在 Android Studio 软件的 DDMS 环境下（由 Tools|Android|Android Device Monitor 菜单项打开），从 File Explorer 窗口中可以查看到 SharedPreferences 文件的存储位置，如图 6-2 所示。这里生成的 SharedPreferences 文件为 MyPreferencesAct.xml，即图 6-2 中圈住的文件。将该文件通过图 6-2 右上角的快捷按钮（即右上角圈住的快捷钮）将该文件下载到计算机中，用 UltraEdit 等文本编辑软件可查看其内容如下：

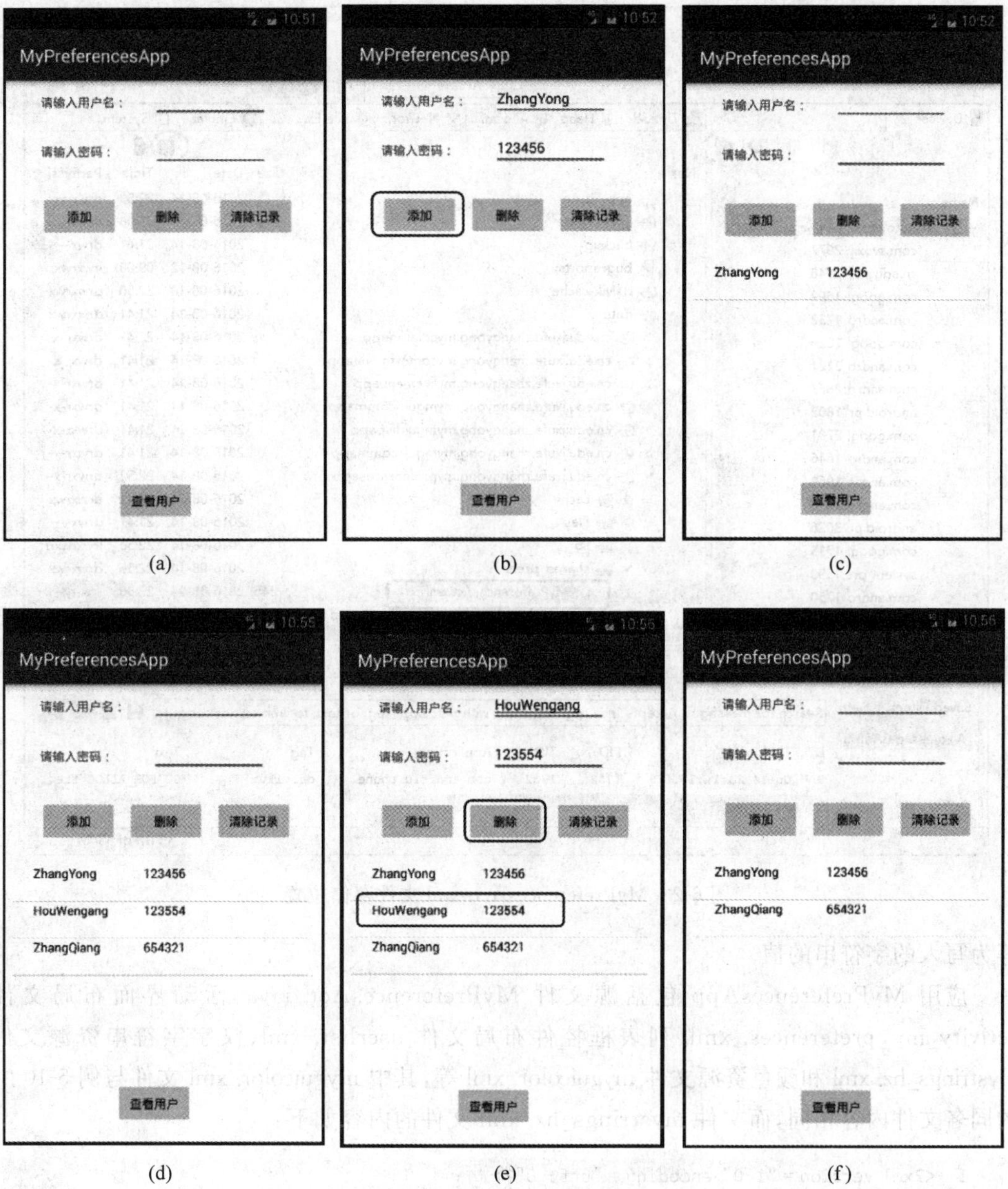

图 6-1　应用 MyPreferencesApp 执行结果（模拟器无中文输入法）

```
1    <?xml version = '1.0' encoding = 'utf-8' standalone = 'yes' ?>
2    <map>
3        <string name = "SECRET1">123456</string>
4        <string name = "SECRET2">654321</string>
5        <string name = "USER2">ZhangQiang</string>
6        <string name = "USER1">ZhangYong</string>
7    </map>
```

可见，MyPreferencesAct.xml 文件类似于字符串资源文件（这是因为程序中仅进行了字符串的写入操作），其中的 name 项为键值常量，用引号包括；而"123456"或"ZhangYong"

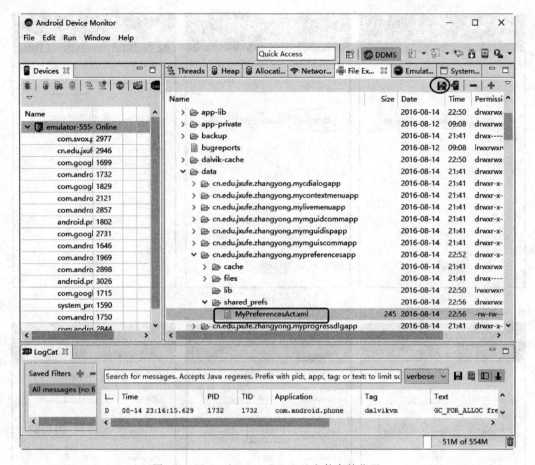

图 6-2 MyPreferencesAct.xml 文件存储位置

等为写入的字符串的值。

应用 MyPreferencesApp 包括源文件 MyPreferencesAct.java、活动界面布局文件 activity_my_preferences.xml、列表框控件布局文件 userlist.xml、汉字字符串资源文件 mystrings_hz.xml 和颜色资源文件 myguicolor.xml 等，其中 myguicolor.xml 文件与例 5-10 中的同名文件内容相同，而文件 mystrings_hz.xml 文件的内容如下：

```
1   <?xml version = "1.0" encoding = "utf - 8"?>
2   < resources >
3       < string name = "myuser">请输入用户名：</string >
4       < string name = "mysecret">请输入密码：</string >
5       < string name = "myadd">添加</string >
6       < string name = "mydel">删除</string >
7       < string name = "mydisp">查看用户</string >
8       < string name = "mydelall">清除记录</string >
9   </resources >
```

活动界面布局文件 activity_my_preferences.xml 对应于图 6-1 所示的界面，其代码如下：

```
1    <?xml version = "1.0" encoding = "utf-8"?>
2    <android.support.constraint.ConstraintLayout xmlns:android = "http://schemas.android.com/apk/res/android"
3        xmlns:app = "http://schemas.android.com/apk/res-auto"
4        xmlns:tools = "http://schemas.android.com/tools"
5        android:id = "@+id/activity_my_preferences"
6        android:layout_width = "match_parent"
7        android:layout_height = "match_parent"
8        tools:context = "cn.edu.jxufe.zhangyong.mypreferencesapp.MyPreferencesAct">
9        <TextView
10           android:id = "@+id/widget29"
11           android:layout_width = "120dp"
12           android:text = "@string/myuser"
13           android:layout_height = "40dp"
14           app:layout_constraintBottom_toTopOf = "@+id/widget30"
15           app:layout_constraintTop_toTopOf = "@+id/activity_my_preferences"
16           android:layout_marginStart = "40dp"
17           app:layout_constraintLeft_toLeftOf = "@+id/activity_my_preferences">
18       </TextView>
```

第9～18行为静态文本框控件，对应于图6-1(a)中的"请输入用户名："。

```
19       <EditText
20           android:id = "@+id/etuser"
21           android:text = ""
22           android:textSize = "16sp"
23           android:layout_width = "130dp"
24           android:layout_height = "40dp"
25           app:layout_constraintBaseline_toBaselineOf = "@+id/widget29"
26           android:layout_marginStart = "8dp"
27           app:layout_constraintLeft_toRightOf = "@+id/widget29"
28           android:layout_marginEnd = "16dp"
29           app:layout_constraintRight_toRightOf = "@+id/activity_my_preferences"
30           app:layout_constraintHorizontal_bias = "0.0">
31       </EditText>
```

第19～31行的编辑框对应于图6-1(b)中显示了"ZhangYong"的编辑框。

```
32       <TextView
33           android:id = "@+id/widget30"
34           android:layout_width = "120dp"
35           android:text = "@string/mysecret"
36           android:layout_height = "40dp"
37           app:layout_constraintBottom_toTopOf = "@+id/btadd"
38           app:layout_constraintTop_toTopOf = "@+id/activity_my_preferences"
39           app:layout_constraintVertical_bias = "0.8"
40           android:layout_marginStart = "40dp"
41           app:layout_constraintLeft_toLeftOf = "@+id/activity_my_preferences">
42       </TextView>
```

第32～42行的静态文本框对应于图6-1(a)中的"请输入密码："。

```
43    <EditText
44        android:id = "@ + id/etsecret"
45        android:text = ""
46        android:textSize = "16sp"
47        android:layout_width = "130dp"
48        android:layout_height = "40dp"
49        app:layout_constraintBaseline_toBaselineOf = "@ + id/widget30"
50        android:layout_marginStart = "8dp"
51        app:layout_constraintLeft_toRightOf = "@ + id/widget30"
52        android:layout_marginEnd = "16dp"
53        app:layout_constraintRight_toRightOf = "@ + id/activity_my_preferences"
54        app:layout_constraintHorizontal_bias = "0.0">
55    </EditText>
```

第43~55行的编辑框对应于图6-1(b)中输入了"123456"的编辑框。

```
56    <Button
57        android:id = "@ + id/btadd"
58        android:text = "@string/myadd"
59        android:onClick = "myAddMD"
60        android:layout_width = "88dp"
61        android:layout_height = "48dp"
62        app:layout_constraintBottom_toTopOf = "@ + id/lvcont"
63        app:layout_constraintTop_toTopOf = "@ + id/activity_my_preferences"
64        app:layout_constraintVertical_bias = "0.84"
65        android:layout_marginStart = "40dp"
66        app:layout_constraintLeft_toLeftOf = "@ + id/activity_my_preferences">
67    </Button>
```

第56~67行的命令按钮对应于图6-1(a)中的"添加"按钮,其单击事件方法名为"myAddMD"。

```
68    <Button
69        android:id = "@ + id/btdel"
70        android:text = "@string/mydel"
71        android:onClick = "myDelMD"
72        android:layout_width = "88dp"
73        android:layout_height = "48dp"
74        app:layout_constraintVertical_bias = "0.85"
75        app:layout_constraintBaseline_toBaselineOf = "@ + id/btadd"
76        android:layout_marginStart = "8dp"
77        app:layout_constraintLeft_toRightOf = "@ + id/btadd"
78        app:layout_constraintRight_toLeftOf = "@ + id/btdelall"
79        android:layout_marginEnd = "8dp"
80        app:layout_constraintHorizontal_bias = "0.47">
81    </Button>
```

第68~81行的命令按钮对应于图6-1(a)中的"删除"按钮,其单击事件方法名为"myDelMD"。

```
82      <Button
83          android:id = "@ + id/btdelall"
84          android:text = "@string/mydelall"
85          android:onClick = "myDelallMD"
86          android:layout_width = "88dp"
87          android:layout_height = "48dp"
88          app:layout_constraintVertical_bias = "0.82"
89          app:layout_constraintBaseline_toBaselineOf = "@ + id/btdel"
90          android:layout_marginEnd = "32dp"
91          app:layout_constraintRight_toRightOf = "@ + id/activity_my_preferences">
92      </Button>
```

第 82～92 行的命令按钮对应于图 6-1(a)中的"清除记录"按钮,其单击事件方法名为"myDelallMD"。

```
93      <Button
94          android:id = "@ + id/btdisp"
95          android:text = "@string/mydisp"
96          android:onClick = "myDispMD"
97          android:layout_width = "88dp"
98          android:layout_height = "48dp"
99          app:layout_constraintTop_toBottomOf = "@ + id/lvcont"
100         app:layout_constraintBottom_toBottomOf = "@ + id/activity_my_preferences"
101         android:layout_marginStart = "16dp"
102         app:layout_constraintLeft_toLeftOf = "@ + id/activity_my_preferences"
103         android:layout_marginEnd = "16dp"
104         app:layout_constraintRight_toRightOf = "@ + id/activity_my_preferences"
105         app:layout_constraintHorizontal_bias = "0.46">
106     </Button>
```

第 93～106 行的命令按钮对应于图 6-1(a)中的"查看用户"按钮,其单击事件方法名为"myDispMD"。

```
107     <ListView
108         android:id = "@ + id/lvcont"
109         android:scrollbars = "vertical"
110         android:layout_width = "337dp"
111         android:layout_height = "226dp"
112         app:layout_constraintTop_toTopOf = "@ + id/activity_my_preferences"
113         app:layout_constraintBottom_toBottomOf = "@ + id/activity_my_preferences"
114         app:layout_constraintLeft_toLeftOf = "@ + id/activity_my_preferences"
115         app:layout_constraintRight_toRightOf = "@ + id/activity_my_preferences"
116         app:layout_constraintHorizontal_bias = "0.36"
117         app:layout_constraintVertical_bias = "0.7">
118     </ListView>
119 </android.support.constraint.ConstraintLayout>
```

第 107～118 行的列表框控件对应于图 6-1(d)中显示了"ZhangYong 123456"等列表。

主活动界面 activity_my_preferences.xml 中有一个列表框控件 lvcont,如上述代码的第 107～118 行所示,该控件需要一个布局文件,由于这里在列表框的每一行中显示了两个文本信息,所以列表框的布局文件中需放置两个 TextView 控件。列表框 lvcont 的布局文件 userlist.xml 内容如下:

```
1   <?xml version = "1.0" encoding = "utf-8"?>
2   <android.support.constraint.ConstraintLayout xmlns:app = "http://schemas.android.com/apk/res-auto"
3       xmlns:tools = "http://schemas.android.com/tools"
4       android:id = "@+id/widget0"
5       android:layout_width = "fill_parent"
6       android:layout_height = "fill_parent"
7       xmlns:android = "http://schemas.android.com/apk/res/android">
8       <TextView
9           android:id = "@+id/user_name"
10          android:layout_width = "100dp"
11          android:layout_height = "40dp"
12          android:text = "TextView"
13          android:textColor = "@drawable/black"
14          app:layout_constraintTop_toTopOf = "@+id/widget0"
15          app:layout_constraintBottom_toBottomOf = "@+id/widget0"
16          android:layout_marginEnd = "16dp"
17          app:layout_constraintRight_toRightOf = "@+id/widget0"
18          android:layout_marginStart = "16dp"
19          app:layout_constraintLeft_toLeftOf = "@+id/widget0"
20          app:layout_constraintHorizontal_bias = "0.04"
21          app:layout_constraintVertical_bias = "0.07">
22      </TextView>
23      <TextView
24          android:id = "@+id/user_secret"
25          android:layout_width = "100dp"
26          android:layout_height = "40dp"
27          android:text = "TextView"
28          android:textColor = "@drawable/black"
29          app:layout_constraintBaseline_toBaselineOf = "@+id/user_name"
30          android:layout_marginStart = "8dp"
31          app:layout_constraintLeft_toRightOf = "@+id/user_name"
32          android:layout_marginEnd = "16dp"
33          app:layout_constraintRight_toRightOf = "@+id/widget0"
34          app:layout_constraintHorizontal_bias = "0.2">
35      </TextView>
36  </android.support.constraint.ConstraintLayout>
```

第 8~22 行和第 23~35 行分别表示两个静态文本框,其 ID 号分别为 user_name 和 user_secret,第 13 和第 28 行设置静态文本框显示字符的颜色为黑色。

Android 系统中列表控件本质上是一种布局控件,即在列表框中显示内容,需要首先向其中放置相应的控件,这里用列表框显示用户名和密码信息,所以需要向列表框中添加静态文本控件;如果向列表框中放置图像显示控件,则列表框可以显示图像信息。

源程序文件 MyPreferencesAct.java 可分为 5 个部分,即 onCreate 方法(或 myInitGUI 方法)、单击图 6-1(a)中的"添加"按钮的方法 myAddMD、单击"删除"按钮的方法 myDelMD、单击"清除记录"按钮的方法 myDelallMD 以及单击"查看用户"按钮的方法

myDispMD。文件 MyPreferencesAct.java 的代码如下：

```
1   package cn.edu.jxufe.zhangyong.mypreferencesapp;
2
3   import android.support.v7.app.AppCompatActivity;
4   import java.util.ArrayList;
5   import java.util.HashMap;
6   import android.content.SharedPreferences;
7   import android.os.Bundle;
8   import android.view.View;
9   import android.widget.AdapterView;
10  import android.widget.AdapterView.OnItemClickListener;
11  import android.widget.EditText;
12  import android.widget.ListView;
13  import android.widget.SimpleAdapter;
14
15  public class MyPreferencesAct extends AppCompatActivity {
16      private int number;
17      private final int MAXREC = 1000;
18      private ListView lvUsers;
19      private EditText etUser,etSecret;
20      private HashMap<String,String> map;
```

第 16 行定义了整型变量 number，用于存储文件中的记录数，每个记录包括一个用户名和一个密码值。第 17 行的整型常量 MAXREC 表示文件中的最大记录数。第 18 和第 19 行定义了一个 ListView 控件对象 lvUser 和两个编辑框对象 etUser、etSecret，对象 lvUser 用于显示文件中的记录数，而两个编辑框对象分别用于输入用户名和密码。第 20 行定义了一个 HashMap 表，该表中的每个记录包括一个字符串型的键值和相应的字符串型数据。

```
21      @Override
22      protected void onCreate(Bundle savedInstanceState) {
23          super.onCreate(savedInstanceState);
24          setContentView(R.layout.activity_my_preferences);
25
26          myInitGUI();
27      }
28      private void myInitGUI(){
29          etUser = (EditText)findViewById(R.id.etuser);
30          etSecret = (EditText)findViewById(R.id.etsecret);
31          lvUsers = (ListView)findViewById(R.id.lvcont);
32          lvUsers.setOnItemClickListener(new OnItemClickListener(){
33              @Override
34              public void onItemClick(AdapterView<?> arg0, View arg1, int arg2,
35                                     long arg3) {
36                  // TODO Auto-generated method stub
37                  SharedPreferences sharedPref = getPreferences(MODE_PRIVATE);
38                  for(int i = 1;i <= arg2 + 1;i++){
39                      String str1 = sharedPref.getString("USER" + String.valueOf(i),
40                              "USER");
41                      String str2 = sharedPref.getString("SECRET" + String.valueOf(i),
```

```
42                        "SECRET");
43                if(i == arg2 + 1){
44                        etUser.setText(str1);
45                        etSecret.setText(str2);
46                }
47            }
48        }
49    });
50    number = 0;
51    SharedPreferences sharedPref = getPreferences(MODE_PRIVATE);
52    map = new HashMap<String,String>();
53    for(int i = 1;i < MAXREC;i++){
54        if(sharedPref.getString("USER" + String.valueOf(i), "USER")
55                .equals("USER")){
56            break;
57        }
58        else{
59            number = number + 1;
60            String str1 = sharedPref.getString("USER" + String.valueOf(i),
61                    "USER");
62            map.put("USER" + String.valueOf(i), str1);
63            String str2 = sharedPref.getString("SECRET" + String.valueOf(i),
64                    "SECRET");
65            map.put("SECRET" + String.valueOf(i), str2);
66        }
67    }
68 }
```

第 22~27 行为 onCreate 方法,第 26 行调用 myInitGUI 方法,即第 28~68 行的 myInitGUI 方法为程序启动时调用的方法。第 29~30 行获得编辑框对象 etUser 和 etSecret；第 31 行得到列表框对象 lvUsers；第 32~49 行为列表框 lvUsers 的列表项单击事件响应方法的事件监听器,当单击列表框中的列表项时,将执行第 34~48 行的代码。第 34 行的参数 arg2 返回列表项的行号,由于是从 0 开始计数,因此,单击第一项时,arg2 为 0；单击第二项时,arg2 为 1,依次类推。第 37 行定义 SharedPreferences 对象 sharedPref,并调用 Activity 的方法 getPreferences 得到 SharedPreferences 文件(其文件名与活动界面名相同,这里是 MyPreferencesAct.xml)；第 38~47 行依次读取第 1~arg2+1 个数据,且只将第 arg2+1 个数据显示在编辑框对象 etUser 和 etSecret 中(第 43 为真),即单击的列表项显示在对象 etUser 和 etSecret 中。

第 50 行将 number 置为 0；第 51 行获得 SharedPreferences 对象 sharedPref；第 52 行得到空的 HashMap 对象；第 53~67 的 for 循环将对象 sharedPref 中的数据读出,并放入 HashMap 表中,即文件 MyPreferencesAct.xml 中的全部内容读出且存放在 HashMap 表中。第 54~57 行判断读出的记录值是否为"USER",如果为"USER"说明已读取了全部记录,第 56 行调用 break 退出循环体；否则,第 59~65 行的代码得到执行,首先记录数 number 加 1(第 59 行),然后,第 60 和 61 行读出记录中的用户名,第 62 行将用户名写入 HashMap 表中,其键值为"USER"加上一个序号,第 63、64 行读出记录中与用户名对应的密码,第 65 行将密码存入 HashMap 表中,其键值为"SECRET"加上同一个序号。

```
69    public void myAddMD(View v){
70        if(etSecret.getText().toString().equals("")){
71        }
72        else{
73            SharedPreferences sharedPref = getPreferences(MODE_PRIVATE);
74            SharedPreferences.Editor editor = sharedPref.edit();
75            number = number + 1;
76            editor.putString("USER" + String.valueOf(number),
77                    etUser.getText().toString());
78            editor.commit();
79            editor.putString("SECRET" + String.valueOf(number),
80                    etSecret.getText().toString());
81            editor.commit();
82            etUser.setText("");
83            etSecret.setText("");
84        }
85    }
```

第69～85行为单击图6-1(a)中的"添加"按钮的事件响应方法。如果输入密码的编辑框为空字符串(第70行为真),则没有操作(第70、71行为空);否则,说明输入有效,第73行获得SharedPreferences对象sharedPref;第74行得到SharedPreferences.Editor接口对象editor;第75行将记录数number加1;第76和第77行将用户名编辑框中的字符串写入editor中,第78行调用commit方法将editor中的数据写入对象sharedPref所表示的文件中;第79～81行将密码编辑框中的内容写入对象sharedPref所表示的文件中;第82和第83行清空用户名和密码输入编辑框中的显示。因此,方法myAddMD实现的功能为:将输入的用户名和密码添加到SharedPreferences文件中,该文件的记录数加1,写入的数据的键值为"USER"和"SECRET"加上记录数。需要注意的是,写入记录的编号必须是从小到大的,否则与列表项的值不对应。

```
86    public void myDelMD(View v){
87        if(etSecret.getText().toString().equals("")){
88        }
89        else{
90            int loc = 0;
91            SharedPreferences sharedPref = getPreferences(MODE_PRIVATE);
92            SharedPreferences.Editor editor = sharedPref.edit();
93            for(int i = 1;i <= number;i++){
94                String str1 = sharedPref.getString("USER" + String.valueOf(i),
95                        "USER");
96                String str2 = sharedPref.getString("SECRET" + String.valueOf(i),
97                        "SECRET");
98                if(str1.equals(etUser.getText().toString())
99                        && str2.equals(etSecret.getText().toString())){
100                   editor.remove("USER" + String.valueOf(i));
101                   editor.apply();
102                   editor.remove("SECRET" + String.valueOf(i));
103                   editor.apply();
104                   loc = i;
```

```
105                    etUser.setText("");
106                    etSecret.setText("");
107                    break;
108                }
109            }
110            if(loc > 0){
111                for(int i = loc + 1; i <= number; i++){
112                    String str1 = sharedPref.getString("USER" + String.valueOf(i),
113                            "USER");
114                    String str2 = sharedPref.getString("SECRET" + String.valueOf(i),
115                            "SECRET");
116                    editor.putString("USER" + String.valueOf(i-1), str1);
117                    editor.commit();
118                    editor.putString("SECRET" + String.valueOf(i-1), str2);
119                    editor.commit();
120                    editor.remove("USER" + String.valueOf(i));
121                    editor.commit();
122                    editor.remove("SECRET" + String.valueOf(i));
123                    editor.commit();
124                }
125                number = number - 1;
126            }
127        }
128    }
```

第 86~128 行为删除一条记录的方法 myDelMD，该方法删除一条记录后，需要将其后的所有记录项的编号都减小 1，即保证记录编号的连续性。删除一条记录要进行的操作为：①判断输入密码的编辑框中的输入是否有效（第 87 行是否为真），如果有效，则执行第 90~127 行代码；②循环遍历文件中的记录数，定位要删除的记录位置 loc，并将记录删除。第 91 和第 92 行得到 SharedPreferences 对象 sharedPref 及 SharedPreferences.Editor 接口对象 editor；第 93~109 行的 for 循环，依次访问每条记录，第 98 行判断读出的记录是否为要删除的记录，如果是，则第 100~107 行得到执行，第 100 和 101 行删除用户名，第 102 和第 103 行删除该用户对应的密码，第 104 行得到删除掉的位置号，第 105 和第 106 行清空用户名和密码输入编辑框的内容，第 107 行跳出 for 循环，转到第 110 行代码执行。③将删除掉的记录后面的所有记录的编号提前 1 位，第 111~124 行的 for 循环实现这个功能，即先把数据读出来（第 112~115 行），然后修改其编号值后再写回去（第 116~119 行），最后，将修改了编号的记录删除（第 120~123 行）。④设置文件中的记录数 number 减少 1。

```
129    public void myDelallMD(View v){
130        SharedPreferences sharedPref = getPreferences(MODE_PRIVATE);
131        SharedPreferences.Editor editor = sharedPref.edit();
132        editor.clear();
133        editor.apply();
134        number = 0;
135        etUser.setText("");
136        etSecret.setText("");
137    }
```

第 129~137 行为图 6-1(a)中"清除记录"按钮的单击事件,该方法清除文件中的所有记录(第 132、133 行),并设置记录数 number 为 0(第 134 行)。

```
138     public void myDispMD(View v){
139         SharedPreferences sharedPref = getPreferences(MODE_PRIVATE);
140         map = new HashMap<String,String>();
141         for(int i = 1;i<MAXREC;i++){
142             if(sharedPref.getString("USER" + String.valueOf(i), "USER").
143                     equals("USER")){
144                 break;
145             }
146             else{
147                 String str1 = sharedPref.getString("USER" + String.valueOf(i),
148                         "USER");
149                 map.put("USER" + String.valueOf(i), str1);
150                 String str2 = sharedPref.getString("SECRET" + String.valueOf(i),
151                         "SECRET");
152                 map.put("SECRET" + String.valueOf(i), str2);
153             }
154         }
155         ArrayList<HashMap<String, String>> arrayList =
156                 new ArrayList<HashMap<String,String>>();
157         for(int i = 1;i<=number;i++){
158             HashMap<String, String> hashMap = new HashMap<String, String>();
159             hashMap.put("user_name", map.get("USER" + String.valueOf(i)));
160             hashMap.put("user_secret", map.get("SECRET" + String.valueOf(i)));
161             arrayList.add(hashMap);
162         }
163         SimpleAdapter listAdapter = new SimpleAdapter(this, arrayList,
164                 R.layout.userlist,
165                 new String[]{"user_name" , "user_secret"},
166                 new int[]{R.id.user_name , R.id.user_secret});
167         lvUsers.setAdapter(listAdapter);
168     }
169 }
```

第 138~169 行为图 6-1(a)中单击"查看用户"按钮的事件响应方法,该方法包括 4 部分处理:①将文件中的数据全部读到一个 HashMap 表中(第 139~154 行)。②将 HashMap 表中的数据存入 ArrayList 对象中(第 155~162 行)。第 155、156 行定义 ArrayList 对象 arrayList;第 157~162 行的 for 循环,每循环一次,通过 get 方法在 map 表中取出数据,再使用 put 方法写入到 hashMap 表中,将每个 hashMap 表作为一个链表节点存放在 arrayList 中(第 161 行)。因此,这里的 for 循环将一个大的 HashMap 表分解成一个个小的 HashMap 表,且键值都相同,再将这些小 HashMap 表作为节点存入 arrayList 表中。③将 ArrayList 对象和列表框布局通过 SimpleAdatper 对象链接起来,构成一个 SimpleAdatper 实例(对象)listAdaper。构造方法 SimpleAdapter 有 5 个参数(第 163~166 行),依次为当前的 Activity、arrayList 对象、ListView 布局 ID 号、arrayList 对象节点键值以及 ListView 布局控件的 ID 号。④调用列表框对象的 setAdapter 方法显示 listAdaper 实例中的数据

(第 167 行)。

例 6-2　SharedPreferences 指定存储文件名访问实例。

例 6-1 中访问的 SharedPreferences 文件与活动界面名同名,而且无法指定其文件名,因为其调用的方法为活动界面对象的 getPreferences 方法,该方法又调用 getSharedPreferences 方法创建文件对象。程序员可以直接调用 getSharedPreferences 方法创建 SharedPreferences 文件,如下所示:

```
SharedPreferences sharedPref = getApplicationContext()
                         .getSharedPreferences("userinfo",MODE_PRIVATE);
```

上述语句创建的 SharedPreferences 文件名为 userinfo,不需要指定扩展名,Android 系统自动添加.xml 扩展名。

新建应用 MyPreferencesExApp,应用名为 MyPreferencesExApp,活动界面名为 MyPreferencesExAct,这样应用 MyPreferencesExApp 包含源文件 MyPreferencesExAct.java、布局文件 activity_my_preferences_ex.xml、列表框布局文件 userlist.xml、汉字字符串文件 mystrings_hz.xml 和颜色资源文件 myguicolor.xml 等,其中,后面三个文件与例 6-1 中的同名文件完全相同,而且,除了活动界面名不同外,布局文件 activity_my_preferences_ex.xml 与例 6-1 中的 activity_my_preferences.xml 内容相同。与例 6-1 的源文件 MyPreferencesAct.java 内容相比,这里的 MyPreferencesExAct.java 做出的变化有:①把文件内容中的包名由 cn.edu.jxufe.zhangyong.mypreferencesapp 变更为 cn.edu.jxufe.zhangyong.mypreferencesexapp;②活动界面名由 MyPreferecesAct 变更为 MyPreferencesExAct;③将文件中所有以下的代码

```
SharedPreferences sharedPref = getPreferences(MODE_PRIVATE);
```

更换为

```
SharedPreferences sharedPref = getApplicationContext()
                         .getSharedPreferences("userinfo",MODE_PRIVATE);
```

接着,执行应用 MyPreferencesExApp,其执行情况与例 6-1 完全相同,如图 6-3 所示,可以在图 6-4 中找到自定义文件名的文件 userinfo.xml,假设此时文件中包含有如图 6-3 所示的两条记录,则文件 userinfor.xml 的内容如下:

```
1   <?xml version = '1.0' encoding = 'utf-8' standalone = 'yes' ?>
2   <map>
3       <string name = "SECRET1">124578</string>
4       <string name = "SECRET2">987654</string>
5       <string name = "USER2">Hou Wen-gang</string>
6       <string name = "USER1">Zhang Yong</string>
7   </map>
```

其中,引号包括的字符串为键值,而"124578"、"987654"、"Zhang Yong"和"Hou Wen-gang"为这些键值对应的字符串值。在例 6-2 中,默认"USER1"和"SECRET1"是第一条记录,而"USER2"和"SECRET2"是第二条记录,依次类推。

第6章 数据访问技术 223

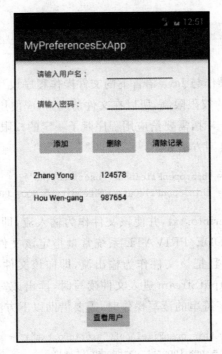

图 6-3 应用 MyPreferencesExApp 执行结果

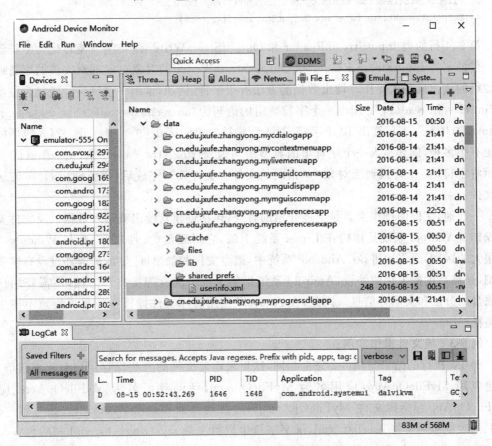

图 6-4 userinfo.xml 文件存储位置

6.2 流文件操作

Android 系统的文件操作与 Java 语言下的文件操作程序设计方法相同。由于 Android 系统对资源的访问有一定的权限限制,所以在文件访问时,应使用 Activity 的创建文件方法创建文件,在访问外部 SD 卡时,需要给应用程序赋予一定的权限。

Activity 创建文件的典型方法如下:

```
1  FileInputStream fis = this.openFileInput("userinfo.txt");
2  FileOutputStream fos = this.openFileOutput("temp.txt",MODE_PRIVATE);
```

第 1 行为打开文件 userinfo.txt,并使该文件作为输入流,即从该文件中读取数据。第 2 行打开文件 temp.txt,MODE_PRIVATE 系统常量指定该文件的打开方法为私有的,即只能在该活动界面内访问,且把该文件作为输出流,即向该文件中写入数据。当直接使用 FileInputStream 或 FileOutputStream 进入文件读写时,读出或写入的数据为字节型数据。当需要对文件进行字符或字符串的读写操作时,需要借助以下方法:

```
1  FileInputStream fis = this.openFileInput("userinfo.txt");
2  InputStreamReader fr = new InputStreamReader (fis);
3  FileOutputStream fos = this.openFileOutput("temp.txt",MODE_PRIVATE);
4  OutputStreamWriter fw = new OutputStreamWriter (fos);
```

第 2、4 行的类 InputStreamReader 和 OutputSteamWriter 是将字节流转化为字符流的类。借助于类 InputStreamReader 和 OutputSteamWriter 的对象可实现对文件的字符或字符数组的读出或写入操作。读出数据的方法有很多种,例如,读出一个字符并转化为整型值为"int ch=fr.read();";读出一个字符数组的语句为"char chusr=new char[30]; fr.read (chusr,0,10);",表示读出 10 个字符,写入到字符数组的第 0~9 个脚标处。写入数据的方法也有很多种,例如写入一个字符串的典型语句为"String str="User"; fw.write(str);",表示将字符串 str 写入到流对象 fw 指向的文件。文件读写完成后,需要调用 close 方法关闭文件。

一般地,文件的操作主要是上述介绍的打开(包括创建)、读出或写入数据以及文件关闭等操作。Android 系统是建构在 Linux 系统上的,它对磁盘(文件)的管理与 Windows CE 有很大的区别,通俗地讲,在 Android 系统中,磁盘文件和触摸屏、显示屏等外设没有本质的区别,是被高度抽象化的外设,Android 系统提供了统一的管理接口和方法,不需要用户关心磁盘的分区和大小等信息。因此,在 Android 应用程序文件操作中应尽可能地不使用绝对路径。

例 6-3 流文件操作实例。

例 6-3 执行的结果与例 6-2 完全相同,这里采用流文件的方法保存用户名和密码信息。新建应用 MyFileOpApp,应用名为 MyFileOpApp,活动界面名为 MyFileOpAct。应用 MyFileOpApp 包括源文件 MyFileOpAct.java、活动界面布局文件 activity_my_file_op.xml、列表框布局文件 userlist.xml、汉字字符串资源文件 mystrings_hz.xml 和颜色资源文件 myguicolor.xml,其中,除了文件 MyFileOpAct.java 外,其余文件均与例 6-2 中的同类型文件或同名文件内容相同（包名变更为 cn. edu. jxufe. zhangyong. myfileopapp）。

文件 MyFileOpAct. java 与例 6-2 中的 MyPreferencesAct. java 相似，主要包括活动界面启动方法 onCreate 和图 6-1(a)中按钮"添加"、"删除""清除记录"与"查看用户"的单击事件方法。下面通过文件 MyFileOpAct. java 的代码依次介绍各个方法的工作原理。

```java
1   package cn. edu. jxufe. zhangyong. myfileopapp;
2
3   import android. support. v7. app. AppCompatActivity;
4   import java. io. FileInputStream;
5   import java. io. FileNotFoundException;
6   import java. io. FileOutputStream;
7   import java. io. IOException;
8   import java. io. InputStreamReader;
9   import java. io. OutputStreamWriter;
10  import java. util. ArrayList;
11  import java. util. HashMap;
12  import android. os. Bundle;
13  import android. view. View;
14  import android. widget. AdapterView;
15  import android. widget. AdapterView. OnItemClickListener;
16  import android. widget. EditText;
17  import android. widget. ListView;
18  import android. widget. SimpleAdapter;
19
20  public class MyFileOpAct extends AppCompatActivity {
21      private ListView lvUsers;
22      private EditText etUser, etSecret;
23      private HashMap < String, String > myMap;
24      @Override
25      protected void onCreate(Bundle savedInstanceState) {
26          super. onCreate(savedInstanceState);
27          setContentView(R. layout. activity_my_file_op);
28          myInitGUI();
29      }
30      private void myInitGUI(){
31          etUser = (EditText)findViewById(R. id. etuser);
32          etSecret = (EditText)findViewById(R. id. etsecret);
33          lvUsers = (ListView)findViewById(R. id. lvcont);
34          lvUsers. setOnItemClickListener(new OnItemClickListener(){
35              @Override
36              public void onItemClick(AdapterView<?> arg0, View arg1, int arg2,
37                      long arg3) {
38                  try {
39                      int ch;
40                      char[] chusr = new char[30];
41                      char[] chsec = new char[30];
42                      FileInputStream fis = MyFileOpAct. this
43                              . openFileInput("userinfo. txt");
44                      InputStreamReader fr = new InputStreamReader (fis);
45                      try {
```

```
46                  int j = 0;
47                  int k = 0;
48                  int m = 0;
49                  int recnum = 1;
50                  while((ch = fr.read())!= -1){
51                      if(ch!= 32){
52                          if(j == 0){
53                              chusr[k++] = (char)ch;
54                          }
55                          else{
56                              chsec[m++] = (char)ch;
57                          }
58                      }
59                      if(ch == 32){
60                          if(j == 0){
61                              j = 1;
62                              String str = new String(chusr,0,k);
63                              k = 0;
64                              if(arg2 + 1 == recnum){
65                                  etUser.setText(str);
66                              }
67                          }
68                          else{
69                              j = 0;
70                              String str = new String(chsec,0,m);
71                              m = 0;
72                              recnum = recnum + 1;
73                              if(arg2 + 2 == recnum){
74                                  etSecret.setText(str);
75                                  break;
76                              }
77                          }
78                      }
79                  }
80              } catch (IOException e) {
81                  e.printStackTrace();
82              }
83              finally{
84                  try {
85                      fr.close();
86                      fis.close();
87                  } catch (IOException e) {
88                      e.printStackTrace();
89                  }
90              }
91          } catch (FileNotFoundException e1) {
92              e1.printStackTrace();
93          }
94      }
95  });
96  }
```

第25~29行为活动界面启动时调用的方法onCreate,第28行调用myInitGUI方法对界面进行初始化。第30~95行为方法myInitGUI,第31和第32行依次获得活动界面上显示的两个编辑框对象,即 etUser 和 etSecret,用于输入用户名和密码。第 33 行获得列表框对象 lvUser,第 34 行设置列表框的事件监听方法,第 36～93 行为单击列表框中的一项时的响应方法 onItemClick,该方法的参数 arg2 返回被单击列表项的行号,从 0 开始索引,即单击第一项,返回 0;单击第 2 项,返回 1,依次类推。

方法 onItemClick 实现的功能为:当第 arg2 项被单击时,从列表框对应的文件中读出相应位置的用户名和密码信息,并将这些信息显示在两个编辑框(etUser 和 etSecret)中。第 42～44 行以流输出方式打开文件 userinfo.txt;第 50～78 行的 while 循环为方法 onItemClick 的主体。通过下文第 97～126 行的 myAddMD 方法可知,文件 userinfo.txt 中每条记录的存储结构为"用户名+空格+密码+空格"的形式,例如,对于用户名为"张勇",密码为"123456"的记录,其存储结构为"张勇 123456"。第 50～79 行读出文件内容的算法如图 6-5 所示。

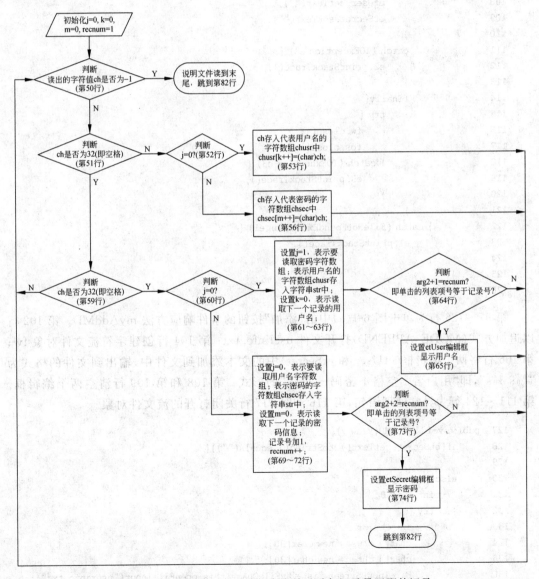

图 6-5　第 50～79 行读出文件内容并显示与记录号匹配的记录

第 80～93 行为异常处理语句,如果在文件创建、打开或读写访问的过程中出现错误,则跳转到这些语句组执行,第 85 和第 86 行为关闭打开的流文件对象。

```
97    public void myAddMD(View v){
98        if(etSecret.getText().toString().equals("")){
99        }
100       else{
101           try {
102               FileOutputStream fos = this.openFileOutput("userinfo.txt",
103                   MODE_APPEND);
104               OutputStreamWriter fw = new OutputStreamWriter (fos);
105               try {
106                   fw.append(String.format("%s %s ",etUser.getText().toString(),
107                       etSecret.getText().toString() ));
108                   etUser.setText("");
109                   etSecret.setText("");
110               }
111               catch (IOException e2) {
112                   e2.printStackTrace();
113               }
114               finally{
115                   try {
116                       fw.close();
117                       fos.close();
118                   } catch (IOException e3) {
119                       e3.printStackTrace();
120                   }
121               }
122           } catch (FileNotFoundException e1) {
123               e1.printStackTrace();
124           }
125       }
126   }
```

第 97～126 行为单击图 6-1(a)中的"添加"按钮的事件响应方法 myAddMD。第 102 行以追加方式(MODE_APPEND)打开文件 userinfo.txt;第 104 行创建字符流文件对象 fw;第 106 行将两个编辑框(etUser 和 etSecret)中的文本添加到文件中,输出到文件的格式为"%s %s",即"用户名+空格+密码+空格"的形式;第 108 和第 109 行清空两个编辑框。第 111～124 行为异常处理语句,第 116 和第 117 行关闭打开的流文件对象。

```
127   public void myDelMD(View v){
128       if(etSecret.getText().toString().equals("")){
129       }
130       else{
131           int loc = 0;
132           try {
133               int ch;
134               char[] chusr = new char[30];
135               char[] chsec = new char[30];
136               FileInputStream fis = MyFileOpAct.this.openFileInput("userinfo.txt");
```

```
137             InputStreamReader fr = new InputStreamReader (fis);
138             try {
139                 int j = 0;
140                 int k = 0;
141                 int m = 0;
142                 int recnum = 1;
143                 int jump = 0;
144                 while((ch = fr.read())!= -1){
145                     if(ch!= 32){
146                         if(j == 0){
147                             chusr[k++] = (char)ch;
148                         }
149                         else{
150                             chsec[m++] = (char)ch;
151                         }
152                     }
153                     if(ch == 32){
154                         if(j == 0){
155                             j = 1;
156                             String str = new String(chusr,0,k);
157                             k = 0;
158                             if(str.equals(etUser.getText().toString())){
159                                 etUser.setText(str);
160                                 jump = 1;
161                             }
162                         }
163                         else{
164                             j = 0;
165                             String str = new String(chsec,0,m);
166                             m = 0;
167                             recnum = recnum + 1;
168                             if(str.equals(etSecret.getText().toString())){
169                                 if(jump == 1){
170                                     loc = recnum - 1;
171                                     etUser.setText("");
172                                     etSecret.setText("");
173                                     break;
174                                 }
175                             }
176                             else{
177                                 jump = 0;
178                             }
179                         }
180                     }
181                 }
182             } catch (IOException e) {
183                 e.printStackTrace();
184             }
185             finally{
186                 try {
187                     fr.close();
```

```
188                    fis.close();
189                } catch (IOException e) {
190                    e.printStackTrace();
191                }
192            }
193        } catch (FileNotFoundException e1) {
194            e1.printStackTrace();
195        }
196
197        if(loc > 0){
198            try {
199                int ch;
200                char[] chusr = new char[30];
201                char[] chsec = new char[30];
202                FileInputStream fis = MyFileOpAct.this
203                        .openFileInput("userinfo.txt");
204                InputStreamReader fr = new InputStreamReader(fis);
205                FileOutputStream fos = this.openFileOutput("temp.txt",
206                        MODE_PRIVATE);
207                OutputStreamWriter fw = new OutputStreamWriter(fos);
208                try {
209                    int j = 0;
210                    int k = 0;
211                    int m = 0;
212                    int recnum = 1;
213                    while((ch = fr.read())!= -1){
214                        if(ch!= 32){
215                            if(j == 0){
216                                chusr[k++] = (char)ch;
217                            }
218                            else{
219                                chsec[m++] = (char)ch;
220                            }
221                        }
222                        if(ch == 32){
223                            if(j == 0){
224                                j = 1;
225                                String str = new String(chusr,0,k);
226                                k = 0;
227                                if(loc == recnum){
228                                }
229                                else{
230                                    fw.write(str + " ");
231                                }
232                            }
233                            else{
234                                j = 0;
235                                String str = new String(chsec,0,m);
236                                m = 0;
237                                recnum = recnum + 1;
238                                if(loc + 1 == recnum){
```

```
239                             }
240                         else{
241                             fw.write(str+" ");
242                         }
243                     }
244                 }
245             }
246         } catch (IOException e) {
247             e.printStackTrace();
248         }
249         finally{
250             try {
251                 fr.close();
252                 fis.close();
253                 fw.close();
254                 fos.close();
255             } catch (IOException e) {
256                 e.printStackTrace();
257             }
258         }
259     } catch (FileNotFoundException e1) {
260         e1.printStackTrace();
261     }
262
263     try {
264         int ch;
265         char[] chusr = new char[30];
266         char[] chsec = new char[30];
267         FileInputStream fis = MyFileOpAct.this.openFileInput("temp.txt");
268         InputStreamReader fr = new InputStreamReader (fis);
269         FileOutputStream fos = this.openFileOutput("userinfo.txt",
270                 MODE_PRIVATE);
271         OutputStreamWriter fw = new OutputStreamWriter (fos);
272         try {
273             int j = 0;
274             int k = 0;
275             int m = 0;
276             while((ch = fr.read())!= -1){
277                 if(ch!= 32){
278                     if(j == 0){
279                         chusr[k++] = (char)ch;
280                     }
281                     else{
282                         chsec[m++] = (char)ch;
283                     }
284                 }
285                 if(ch == 32){
286                     if(j == 0){
287                         j = 1;
288                         String str = new String(chusr,0,k);
289                         k = 0;
```

```
290                              fw.append(str + " ");
291                          }
292                          else{
293                              j = 0;
294                              String str = new String(chsec,0,m);
295                              m = 0;
296                              fw.append(str + " ");
297                          }
298                      }
299                  }
300              } catch (IOException e) {
301                  e.printStackTrace();
302              }
303              finally{
304                  try {
305                      fr.close();
306                      fis.close();
307                      fw.close();
308                      fos.close();
309                  } catch (IOException e) {
310                      e.printStackTrace();
311                  }
312              }
313          } catch (FileNotFoundException e1) {
314              e1.printStackTrace();
315          }
316      }
317  }
318 }
```

第 127～318 行为图 6-1(a)中"删除"按钮的事件响应方法 myDelMD, 当单击"删除"按钮时,将从文件中删除与两个编辑框(etUser 和 etSecret)中匹配的记录。方法 myDelMD 的执行过程分为三步:①第 131～195 行,依次从文件 userinfo.txt 读出各条记录,并与编辑框中的内容比较,找到与编辑框中内容匹配的记录位置,存在 loc 变量中。②如果 loc 大于 0,说明该记录存在于文件 userinfo.txt。第 198～261 行将文件 userinfo.txt 中的记录,除去 loc 位置的记录外,都复制到另一个临时文件 temp.txt 中。③第 263～315 行将 temp.txt 的文件写回文件 userinfo.txt 中,覆盖掉文件 userinfo.txt 中原来的记录,这样,就把原来位于文件 userinfo.txt 中第 loc 位置的记录删除了。上述的每一步中,都使用了类似图 6-5 的方法对文件中的记录进行操作,这里不再赘述。

```
319  public void myDelallMD(View v){
320      try {
321          FileOutputStream fos = this.openFileOutput("userinfo.txt",
322                  MODE_PRIVATE);
323          OutputStreamWriter fw = new OutputStreamWriter (fos);
324          try {
325              fw.write("");
326          }
327          catch (IOException e) {
```

```
328                e.printStackTrace();
329            }
330            finally{
331                try {
332                    fw.close();
333                    fos.close();
334                }
335                catch (IOException e) {
336                    e.printStackTrace();
337                }
338            }
339        }
340        catch (FileNotFoundException e1) {
341            e1.printStackTrace();
342        }
343        etUser.setText("");
344        etSecret.setText("");
345    }
```

第 319~345 行为图 6-1(a)中单击"清除记录"按钮的事件响应方法,该方法清除文件 userinfo.txt 中所有的记录。只需要以 MODE_PRIVATE 方式打开文件 userinfo.txt,然后,向文件中写入一个空格即可清除全部记录,如第 325 行所示。

```
346  public void myDispMD(View v) throws IOException{
347      FileInputStream fis;
348      InputStreamReader fr = null;
349      int ch;
350      char[] chusr = new char[30];
351      char[] chsec = new char[30];
352      myMap = new HashMap<String,String>();
353      try {
354          fis = this.openFileInput("userinfo.txt");
355          fr = new InputStreamReader (fis);
356          try {
357              int j = 0;
358              int k = 0;
359              int m = 0;
360              int recnum = 1;
361              while((ch = fr.read())!= -1){
362                  if(ch!= 32){
363                      if(j == 0){
364                          chusr[k++] = (char)ch;
365                      }
366                      else{
367                          chsec[m++] = (char)ch;
368                      }
369                  }
370                  if(ch == 32){
371                      if(j == 0){
```

```
372                    j = 1;
373                    String str = new String(chusr,0,k);
374                    myMap.put("USER" + String.valueOf(recnum), str);
375                    k = 0;
376                }
377                else{
378                    j = 0;
379                    String str = new String(chsec,0,m);
380                    myMap.put("SECRET" + String.valueOf(recnum), str);
381                    m = 0;
382                    recnum = recnum + 1;
383                }
384            }
385        }
386        ArrayList<HashMap<String, String>> arrayList =
387                new ArrayList<HashMap<String,String>>();
388        for(int i = 1;i < recnum;i++){
389            HashMap<String, String> hashMap = new HashMap<String,
390                String>();
391            hashMap.put("user_name", myMap.get("USER"
392                + String.valueOf(i)));
393            hashMap.put("user_secret", myMap.get("SECRET"
394                + String.valueOf(i)));
395            arrayList.add(hashMap);
396        }
397        SimpleAdapter listAdapter = new SimpleAdapter(this, arrayList,
398            R.layout.userlist,
399            new String[]{"user_name" , "user_secret"},
400            new int[]{R.id.user_name , R.id.user_secret});
401        lvUsers.setAdapter(listAdapter);
402    }
403    finally{
404        try {
405            fr.close();
406            fis.close();
407        } catch (IOException e) {
408            e.printStackTrace();
409        }
410    }
411  } catch (FileNotFoundException e1) {
412      e1.printStackTrace();
413  }
414 }
415 }
```

第 346~414 行为图 6-1(a) 中按钮 "查看用户" 的单击事件响应方法。第 347~385 行代码采用图 6-5 类似的方法从文件 userinfo.txt 中读出全部记录，并将这些记录写入 HashMap 表对象 myMap 中。第 386~396 行将 myMap 中的记录写入到 ArrayList 对象

arrayList 中。第 397～400 行将 arrayList 和列表框布局资源关联起来，生成 SimpleAdapter 类的适配器对象 listAdapter，第 401 行调用列表框控件对象的 setAdapter 方法显示文件 userinfo.txt 中的全部记录。

若执行应用 MyFileOpApp，并在图 6-6 中输入 4 条记录，文件 userinfo.txt 将位于图 6-7 所示的目录中，即位于"/data/data/cn.edu.jxufe.zhangyong.myfileopapp/files/"目录中。将文件传输到 Windows 系统中，并使用 UltraEdit 等文本编辑软件打开，其内容为"张勇 123456 侯文刚 123654 张琼 987621 李雪倩 654321"，即每条记录都采取"用户名＋空格＋密码＋空格"的形式保存。

在打开文件或读写文件时，需要添加必要的异常处理，在 Android 应用程序中如果不添加这些异常处理，程序将提示出错。尽管例 6-3 与例 6-2 实现的功能完全相同，但是使用文件流方法进行记录的处理，显然比使用 SharedPreferences 文件更复杂一些。

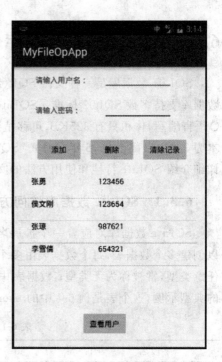

图 6-6 应用 MyFileOpApp 执行结果

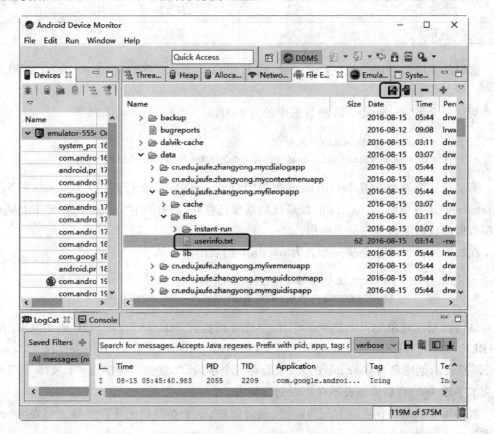

图 6-7 文件 userinfo.txt 存储位置

6.3 SQLite 关系数据库

SQLite 数据库是一种支持 SQL(结构化查询语言)的关系型数据库,严格地讲,SQLite 数据库支持多数 SQL92 标准。SQLite 的最大优势在于它是开源的、免费的数据库引擎,用 C 语言编写,体积只有 325KB,可移植到所有主流的操作系统和嵌入式操作系统上,最新版本为 3.14.1(截至 2016 年 9 月)。SQLite 网址为 http://www.sqlite.org,包括了近 80 篇详细介绍 SQLite 特性和使用方法的文档。Android 系统支持 SQLite 数据库。

6.3.1 SQLite 数据库访问方法

SQLite 数据库对应着一个扩展名为.db 的文件,该文件名称为数据库名,数据库中可以创建多个数据表,每个数据表由多个字段构成,每个字段对应于数据表的列,字段可以为任意类型,常被称为无类型;数据表中的每一行是各个字段具体的取值。表 6-1 为数据表的典型示例,这个表是例 6-4 中的 userinfo 表。

表 6-1 userinfo 表的典型结构

字段名	_id	user	secret
字段值	1	张勇	123456
字段值	2	侯文刚	654321
字段值	3	张琼	123789

这里的字段名_id 被称为主索引键,由系统自动添加;字段 user 和 secret 是用户数据列。

新建一个名为 myuser.db 的数据库的方法为:

```
1   private SQLiteDatabase myDB = null;
2   myDB = this.openOrCreateDatabase("myuser.db", MODE_PRIVATE, null);
```

第 1 行定义一个 SQLiteDatabase 数据库对象 myDB;第 2 行调用方法 openOrCreate-Database 创建数据库 myuser.db,如果该数据库已存在,则打开它,常量 MODE_PRIVATE 说明数据库仅限于打开它的类调用。

在数据库 myDB 中创建一个名为 userinfo 的表的方法如下:

```
1   myDB.execSQL("CREATE TABLE IF NOT EXISTS userinfo" +
2              "(_id INTEGER PRIMARY KEY, user TEXT,secret TEXT)");
```

调用数据库对象 myDB 的 execSQL 方法可以执行 SQL 语句,而 SQL 语句"CREATE TABLE IF NOT EXISTS userinfo(_id INTEGER PRIMARY KEY, user TEXT, secret TEXT)"表示创建名为 userinfo 的表,该表有三个字段,即整型主键字段_id、文本字段 user 和文本字段 secret,如果该表不存在则创建该表。

数据库中表的操作主要有记录添加、记录删除、记录查询和记录清空等。向表 userinfo 中添加一条记录的方法为:

```
1   ContentValues cv = new ContentValues();
2   cv.put("user", "张勇");
3   cv.put("secret", "123456");
4   myDB.insert("userinfo",null,cv);
```

第 1 行创建一个 ContentValues 对象 cv，ContentValues 类是一个类似于 HashMap 的数据表类，具有"键值＋数据"的存储结构；第 2 和第 3 行分别向对象 cv 中存入键值为 "user"和"secret"的值"张勇"和"123456"，这里的键值为字段名；第 4 行调用数据库对象 myDB 的 insert 方法将 cv 的值插入到数据表"userinfo"中，将产生表 6-1 中的第二行数据。

查询表 userinfo 中字段 user 和 secret 为"张勇"和"123456"的记录的方法如下：

```
1   Cursor cursor = myDB.query("userinfo",
2                              null,
3                              " user = '张勇'" +
4                              " AND secret = '123456'",
5                              null, null, null, null);
6   if(cursor!= null){
7        cursor.moveToFirst();
8        int loc = cursor.getColumnIndex("secret");
9        String str = cursor.getString(loc);
10  }
```

第 1 行调用数据库对象 myDB 的 query 方法从表 userinfo 中查询 WHERE 条件为第 3、4 行的记录，这里的 WHERE 条件是指 SQL 语句的 WHERE 条件式，即" user = '张勇' AND secret = '123456'"，查询出所有 user 字段为"张勇"且 secret 字段为"123456"的记录，放在 Cursor 类的对象 cursor 中，cursor 对象类似于 HashMap 表，第 7 行将 cursor 对象中的数据指针指向其第一个记录，第 8 行得到该记录的列号，第 9 行得到该列的值，这里是"123456"。由此可见，查询数据表的记录需要借助于 query 方法和 WHERE 条件式。如果要读取表中所有的记录，则使用以下语句：

```
1   Cursor cursor = myDB.query("userinfo",
2                              null,
3                              null, null, null, null, "_id");
```

第 1～3 行表示读出数据表 userinfo 中的所有记录，并存入 cursor 对象中；第 3 行中的 "_id"表示读出的记录按主键_id 的值从小到大排序。

删除数据表 userinfo 中的字段 user 和 secret 为"张勇"和"123456"的记录的方法为：

```
1   myDB.delete("userinfo", " user = '张勇'" +
2                           " AND secret = '123456'", null);
```

借助于数据库对象 myDB 的 delete 方法和 WHERE 条件"" user = '张勇' AND secret = '123456'""，可以删除满足条件的记录。这里常犯的错误是忽视了字段值需要用单引号括起来，即"'张勇'"，如果字段值是整型，则不需要用单引号括起来。

删除数据表 userinfo 中的所有记录值的方法为：

```
1   myDB.delete("userinfo", null, null);
```

在 delete 方法中只指定表名为"userinfo",其他两个参数赋为 null,即可清除表中所有的记录值,注意,这时字段名和表的结构不发生变化,数据表仍然存在于数据库中。从数据库中删除数据表需要调用 execSQL 方法执行 SQL 语句"DROP TABLE userinfo",如下所示:

```
1    myDB.execSQL("DROP TABLE userinfo");
```

删除数据库文件 myuser.db 的方法为:

```
1    this.deleteDatabase("myuser.db");
```

调用 Acitivity 对象的 deleteDatabase 方法,其参数为数据库名。

例 6-4　SQLite 数据库访问技术。

例 6-4 执行的结果与例 6-3 相同,在例 6-4 中,"查看用户"按钮(见图 6-1(a))的作用被集成在"添加"、"删除"和"清除记录"按钮(见图 6-1(a))的单击事件中了,即单击"添加"、"删除"和"清除记录"按钮后,自动显示数据库中的记录数。因此,例 6-4 中可以删去"查看用户"按钮。

新建应用 MySQLDBApp,应用名为 MySQLDBApp,活动界面名为 MySQLDBAct。应用 MySQLDBApp 包括源文件 MySQLDBAct.java、汉字字符串资源文件 mystrings_hz.xml、颜色资源文件 myguicolor.xml、活动界面布局文件 activity_my_sqldb.xml 和 ListView 控件布局文件 userlist.xml 等。除了源文件 MySQLDBAct.java 之外,其他文件与例 6-3 中的同类型文件或同名文件内容相同(包更新为 cn.edu.jxufe.zhangyong.mysqldbapp)。

源文件 MySQLDBAct.java 的内容如下:

```
1    package cn.edu.jxufe.zhangyong.mysqldbapp;
2    import android.support.v7.app.AppCompatActivity;
3    import android.content.ContentValues;
4    import android.database.Cursor;
5    import android.database.sqlite.SQLiteDatabase;
6    import android.os.Bundle;
7    import android.view.View;
8    import android.widget.AdapterView;
9    import android.widget.AdapterView.OnItemClickListener;
10   import android.widget.CursorAdapter;
11   import android.widget.EditText;
12   import android.widget.ListAdapter;
13   import android.widget.ListView;
14   import android.widget.SimpleCursorAdapter;
15   import android.widget.TextView;
16
17   public class MySQLDBAct extends AppCompatActivity {
18       private ListView lvUsers;
19       private EditText etUser,etSecret;
20       private SQLiteDatabase myDB = null;
21       private final static String MY_DB_NAME = "myuser.db";
22       private final static String MY_TB_NAME = "userinfo";
```

```
23        private final static String MY_FD_USER = "user";
24        private final static String MY_FD_SECRET = "secret";
```

第 18 行定义 ListView 控件对象 lvUsers，用于显示数据表中的记录；第 19 行定义 EditText 控件的对象 etUser 和 etSecret，用于输入用户名和密码，当单击 ListView 列表项时用于输出单击的用户名和密码；第 20 行定义 SQLite 数据库对象 myDB；第 21 行定义常量字符串 myuser.db，用作数据库名；第 22 行定义常量字符串 userinfo，用作数据表名；第 23、24 行定义常量字符串 user 和 secret，用作表中的字段名。

```
25    @Override
26    protected void onCreate(Bundle savedInstanceState) {
27        super.onCreate(savedInstanceState);
28        setContentView(R.layout.activity_my_sqldb);
29        myInitGUI();
30    }
31    public void onDestroy(){
32        super.onDestroy();
33        myDB.close();
34    }
```

第 31～34 行说明当应用程序退出时，调用数据库对象 myDB 的 close 方法关闭数据库。

```
35    private void myInitGUI(){
36        etUser = (EditText)findViewById(R.id.etuser);
37        etSecret = (EditText)findViewById(R.id.etsecret);
38        lvUsers = (ListView)findViewById(R.id.lvcont);
39        lvUsers.setOnItemClickListener(new OnItemClickListener(){
40            @Override
41            public void onItemClick(AdapterView<?> arg0, View arg1, int arg2,
42                                   long arg3) {
43                TextView tv_usr = (TextView)arg1.findViewById(R.id.user_name);
44                TextView tv_sec = (TextView)arg1.findViewById(R.id.user_secret);
45                Cursor cursor = myDB.query(MY_TB_NAME,
46                    null,//new String[]{"_id",MY_FD_USER,MY_FD_SECRET},
47                    " user = '" + tv_usr.getText().toString() + "'" +
48                        " AND secret = '" + tv_sec.getText().toString() + "'",
49                    null, null, null, null);
50                if(cursor!= null){
51                    cursor.moveToFirst();
52                    int loc = cursor.getColumnIndex(MY_FD_USER);
53                    etUser.setText(cursor.getString(loc));
54                    loc = cursor.getColumnIndex(MY_FD_SECRET);
55                    etSecret.setText(cursor.getString(loc));
56                }
57            }
58        });
59        myDB = this.openOrCreateDatabase(MY_DB_NAME, MODE_PRIVATE, null);
60        myDB.execSQL("CREATE TABLE IF NOT EXISTS " + MY_TB_NAME +
61            "(_id INTEGER PRIMARY KEY, user TEXT,secret TEXT)");
```

```
62        myDispRec();
63    }
```

第35～63为myInitGUI方法,该方法被onCreate方法调用(第29行),用于程序启动时执行初始化工作。第36和第37行得到编辑框对象etUser和etSecret,即将这两个对象实例化。第38行得到ListView对象lvUsers,第39行设置该对象的列表项单击事件。第41～57行为列表项的单击事件方法,其中参数arg1表示列表框的视图,根据列表框布局文件userlist.xml可知,该列表框中有两个静态文件框,其ID号分别为user_name和user_secret,第43和第44行得到这两个静态文本框的对象;第45～49行从数据表userinfo中查询user和secret字段分别等于单击的列表项中两个静态文本框内容的记录,并将这些满足条件的记录存放在cursor对象中;第50～56行先判断如果cursor中有记录,将其数据指针移动到第一个记录(第51行),然后得到user字段的列号(第52行),第53行得到user字段的值,并将其赋给编辑框etUser,第54和第55行取得secret字段的值,并赋给etSecret编辑框。即第41～57行实现的功能为:单击列表框中的某项,将该项中的用户名和密码赋给两个编辑框。第62行调用myDispRec自定义方法刷新列表显示(第97～107行)。

```
64    public void myAddMD(View v){
65        if(etSecret.getText().toString().equals("")){
66        }
67        else{
68            ContentValues cv = new ContentValues();
69            cv.put(MY_FD_USER, etUser.getText().toString());
70            cv.put(MY_FD_SECRET, etSecret.getText().toString());
71            myDB.insert(MY_TB_NAME,null,cv);
72            etUser.setText("");
73            etSecret.setText("");
74            myDispRec();
75        }
76    }
```

第64～76行的方法myAddMD为"添加"按钮(见图6-1)的单击事件,将两个编辑框etUser和etSecret中的数据添加到数据表中。当etSecret编辑框不为空格时,则执行第68～74行的代码。第68～70行借助ContentValues类的对象将两个编辑框中的数据格式化为一条记录,第71行调用insert方法将该条记录插入数据表userinfo中。第72和第73行清除两个编辑框etUser和etSecret中的数据。第74行调用myDispRec自定义方法刷新列表显示(第97～107行)。

```
77    public void myDelMD(View v){
78        if(etSecret.getText().toString().equals("")){
79        }
80        else{
81            myDB.delete(MY_TB_NAME, " user = '" + etUser.getText().toString() + "'" +
82                " AND secret = '" + etSecret.getText().toString() + "'", null);
83            myDispRec();
84            etUser.setText("");
85            etSecret.setText("");
86        }
87    }
```

第 77～87 行的方法 myDelMD 为 "删除" 按钮（见图 6-1）的单击事件方法，将删除数据表 userinfo 中 user 和 secret 字段分别等于两个编辑框 etUser 和 etSecret 中的数据的记录。如果编辑框 etSecret 非空，则执行第 81～85 行的代码。第 81 行调用 delete 方法执行删除操作；第 83 行调用 myDispRec 方法刷新列表框显示；第 84 和第 85 行清空两个编辑框 etUser 和 etSecret 中的显示。

```
88    public void myDelallMD(View v){
89        myDB.delete(MY_TB_NAME, null, null);
90        etUser.setText("");
91        etSecret.setText("");
92        myDispRec();
93    }
```

第 88～93 的方法 myDelallMD 是 "清除记录" 按钮（见图 6-1）的单击事件方法，用于清除数据表中的全部记录。第 89 行执行清除数据表 userinfo 中的全部记录的操作；第 92 行调用 myDispRec 方法刷新列表框显示；第 90 和第 91 行清空两个编辑框 etUser 和 etSecret 中的显示。

```
94    public void myDispMD(View v){
95        myDispRec();
96    }
```

第 94～96 行的方法 myDsipMD 调用 myDispRec 方法刷新列表框显示，为 "显示记录" 按钮（见图 6-1）的单击事件方法。

```
97     private void myDispRec(){
98         Cursor cursor = myDB.query(MY_TB_NAME,
99             null,//new String[]{"_id",MY_FD_USER,MY_FD_SECRET},
100            null, null, null, null, "_id");
101        ListAdapter listAdapter = new SimpleCursorAdapter(this,
102            R.layout.userlist,cursor,
103            new String[]{MY_FD_USER,MY_FD_SECRET},
104            new int[]{R.id.user_name, R.id.user_secret},
105            CursorAdapter.FLAG_REGISTER_CONTENT_OBSERVER);
106        lvUsers.setAdapter(listAdapter);
107    }
108 }
```

第 97～107 行为自定义方法 myDispRec，用于刷新列表框的显示。第 98～100 行从数据表 userinfo 中读出全部的数据表记录，存放在 cursor 对象中；第 101～105 行创建一个 listAdapter 适配器对象，简单地说，就是按 ListView 每一行（即每个列表项）的显示格式将 cursor 中的数据调整为相应的显示格式，从而形成一个多行的列表格式；第 106 行调用 setAdapter 方法显示列表框中的记录数。

例 6-4 实现了例 6-1～例 6-3 的所有功能，而且代码比前面三个工程的代码大大减少，这体现了使用数据库进行数据管理的优势。假设数据表中的内容如图 6-8 所示，例 6-4 创建的数据库文件 myuser.db 将位于图 6-9 所示的目录下，在图 6-9 中通过单击右上角框住的快捷按钮可把数据库 myuser.db 导入计算机中，然后，使用 SQLiteManager 软件查看 myuser.db 的内容将如图 6-10 所示，通过这三个图的内容对比，可知程序运行正常。

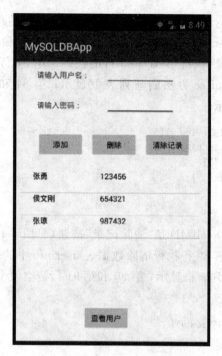

图 6-8　例 6-4 执行结果

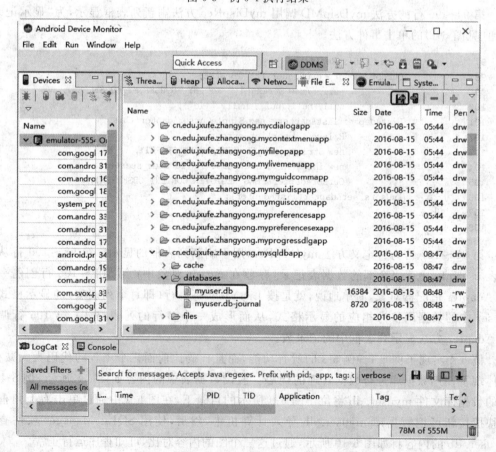

图 6-9　数据库文件 myuser.db 的存储位置

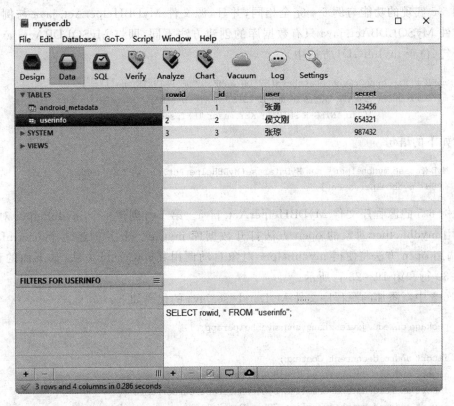

图 6-10　SQLiteManager 软件显示数据库 myuser.db 中的内容

在图 6-10 中清晰地看出，数据库 myuser.db 的表 userinfo 中有三个字段，即"_id"、"user"和"secret"，图 6-10 显示了表 userinfo 中的三个记录的情况。此外，可以通过 SQLiteManager 软件学习 SQL 语言。

6.3.2　SQLiteOpenHelper 类

第 6.3.1 节详细介绍了 SQLite 数据库的使用方法，例 6-4 中直接使用数据库类 SQLiteDatabase 的对象方法创建和访问数据库中的数据表，而这些对象方法的参数较多，另一种有效的方法是将数据库操作封装在一个用户定义的类中，数据库中数据表记录的添加、删除和查询等方法用自定义的方法实现，这样可以简化数据库的设计和维护工作。SQLiteOpenHelper 类是 Android 系统提供的数据库操作类，封装了数据库的创建和版本维护功能，因此，一般地，程序员自定义的数据库管理类都继承了 SQLiteOpenHelper 类。

例 6-5　SQLiteOpenHelper 类用法实例一。

新建应用 MyDBHelperApp，应用名为 MyDBHelperApp，活动界面名为 MyDBHelperAct。应用 MyDBHelperApp 包括源文件 MyDBHelperAct.java、MyDatabase.java、汉字符串资源文件 mystrings_hz.xml、颜色资源文件 myguicolor.xml、活动界面布局文件 activity_my_dbhelper.xml 和 ListView 控件布局文件 userlist.xml 等。除了源文件 MyDBHelperAct.java 和 MyDatabase.java 之外，其他文件与例 6-4 中的同类型文件或同名文件内容相同（包名为 cn.）。

例 6-5 实现的功能与例 6-4 完全相同,并且,源文件 MyDBHelperAct.java 与例 6-4 中的源文件 MySQLDBAct.java 只有数据库的创建方法不同,即将 MySQLDBAct.java 中的以下语句(位于 myInitGUI 方法中)

```
1  myDB = this.openOrCreateDatabase(MY_DB_NAME, MODE_PRIVATE, null);
2  myDB.execSQL("CREATE TABLE IF NOT EXISTS " + MY_TB_NAME +
3           "(_id INTEGER PRIMARY KEY, user TEXT, secret TEXT)");
```

修改为如下的语句

```
1  MyDatabase mydbhelper = new MyDatabase(MyDBHelperAct.this,MY_DB_NAME);
2  myDB = mydbhelper.open();
```

即得到例 6-5 的源程序文件 MyDBHelperAct.java。第 1 行创建一个 mydbhelper 对象,第 2 行调用 mydbhelper 对象的 open 方法打开数据库 myuser.db 并创建一个 userinfo 数据表,即调用 open 方法将使得 mydbhelper 对象自动调用其 onCreate 方法,如下面的程序段第 28~30 行和第 19~23 行所示。

例 6-5 中新添加的源程序文件 MyDatabase.java 的代码如下:

```
1  package cn.edu.jxufe.zhangyong.mydbhelperapp;
2
3  import android.content.Context;
4  import android.database.sqlite.SQLiteDatabase;
5  import android.database.sqlite.SQLiteDatabase.CursorFactory;
6  import android.database.sqlite.SQLiteOpenHelper;
7
8  public class MyDatabase extends SQLiteOpenHelper {
9      private static final int VERSION = 1;
10     public MyDatabase(Context context, String name, CursorFactory factory,
11             int version) {
12         super(context, name, factory, version);
13         // TODO Auto-generated constructor stub
14     }
```

第 8 行自定义公有类 MyDatabase,继承了类 SQLiteOpenHelper。第 10~14 行为类 MyDatabase 的构造方法,由 Android 系统自动添加。下面第 15~17 行添加了一个自定义的构造方法,该方法中只有两个参数,即 Context 类对象和表示数据库名的字符串变量。

```
15     public MyDatabase(Context context, String name){
16         this(context,name,null,VERSION);
17     }
18     @Override
19     public void onCreate(SQLiteDatabase db) {
20         // TODO Auto-generated method stub
21         db.execSQL("CREATE TABLE IF NOT EXISTS userinfo" +
22                 "(_id INTEGER PRIMARY KEY,user TEXT,secret TEXT)");
23     }
```

第 19~23 行为 onCreate 方法,调用自定义类 MyDatabase 的对象方法 getWritableDatabase 或 getReadableDatabase 时自动执行,第 21 和第 22 行为创建名为 userinfo 的数据表。

```
24      @Override
25      public void onUpgrade(SQLiteDatabase db, int oldVersion, int newVersion) {
26          // TODO Auto-generated method stub
27      }
28      public SQLiteDatabase open(){
29          return this.getWritableDatabase();
30      }
31  }
```

第 25~27 行的方法 onUpgrade 用于更新版本号,这里是空方法。第 28~30 行的 open 方法调用 getWritableDatabase 方法打开一个可读可写的数据库(打开只读数据库的方法为 getReadableDatabase)。

例 6-6　SQLiteOpenHelper 类用法实例二。

例 6-5 仅是借助 SQLiteOpenHelper 类封装了数据库的打开和数据表的创建方法,并没有将数据库对象封装起来。而例 6-6 将数据库对象作为自定义类 MyDatabase 的私有成员数据,即封装了数据库对象和其操作方法。例 6-6 实现的功能与例 6-5 完全相同。

新建应用 MyDBHelperExApp,应用 MyDBHelperExApp 包含的文件有源代码文件 MyDBHelperExAct.java、MyDatabase.java、布局文件 activity_my_dbhelper_ex.xml、ListView 布局文件 userlist.xml 和资源文件 myguicolor.xml、mystrings_hz.xml 等。除了源代码文件 MyDBHelperExAct.java 和 MyDatabase.java 之外,其余文件与例 6-5 中的同类型文件或同名文件内容相同(包名为 cn.edu.jxufe.zhangyong.mydbhelperexapp),例 6-6 实现的功能与例 6-5 完全相同。也可直接在例 6-5 的基础上,通过修改例 6-5 中的源文件 MyDatabase.java 和 MyDBHelperAct.java 得到例 6-6。

源文件 MyDatabase.java 的代码如下:

```
1   package cn.edu.jxufe.zhangyong.mydbhelperexapp;
2
3   import android.content.ContentValues;
4   import android.content.Context;
5   import android.database.Cursor;
6   import android.database.sqlite.SQLiteDatabase;
7   import android.database.sqlite.SQLiteDatabase.CursorFactory;
8   import android.database.sqlite.SQLiteOpenHelper;
9
10  public class MyDatabase extends SQLiteOpenHelper {
11      private SQLiteDatabase mydb = null;
12      private final static String MY_TB_NAME = "userinfo";
13      private final static String MY_FD_USER = "user";
14      private final static String MY_FD_SECRET = "secret";
```

第 11 行定义了私有数据库成员 mydb,第 12 行定义该数据库中的表名常量,第 13、14 行定义了数据表中的两个字段常量。

```
15      private static final int VERSION = 1;
16      public MyDatabase(Context context, String name, CursorFactory factory,
17              int version) {
18          super(context, name, factory, version);
```

```
19        // TODO Auto-generated constructor stub
20    }
21    public MyDatabase(Context context, String name){
22        this(context,name,null,VERSION);
23    }
24    @Override
25    public void onCreate(SQLiteDatabase db) {
26        // TODO Auto-generated method stub
27        db.execSQL("CREATE TABLE IF NOT EXISTS " + MY_TB_NAME +
28                "(_id INTEGER PRIMARY KEY,user TEXT,secret TEXT)");
29    }
```

第 25~29 行的 onCreate 方法创建数据表 userinfo，包含三个字段，即主键_id、用户名 user 和密码 secret。

```
30    @Override
31    public void onUpgrade(SQLiteDatabase db, int oldVersion, int newVersion) {
32        // TODO Auto-generated method stub
33
34    }
35    //My Database method
36    public void open(){
37        mydb = this.getWritableDatabase();
38    }
```

第 36~38 行的方法 open 创建或打开数据库。

```
39    public void close(){
40        mydb.close();
41    }
```

第 39~41 行的方法 close 关闭数据库。

```
42    public void add(String name,String secret){
43        ContentValues cv = new ContentValues();
44        cv.put(MY_FD_USER, name);
45        cv.put(MY_FD_SECRET, secret);
46        mydb.insert(MY_TB_NAME,null,cv);
47    }
```

第 42~47 行的 add 方法向数据表中添加记录，其中记录中的 user 字段为 name 参数、secret 字段为 secret 参数。

```
48    public void del(String where){
49        mydb.delete(MY_TB_NAME,where,null);
50    }
51    public void delall(){
52        mydb.delete(MY_TB_NAME, null, null);
53    }
```

第 48~50 行的 del 方法删除满足 where 条件的记录；第 51~53 行的 delall 方法删除数据表中的全部记录。

```
54  public Cursor query(){
55      return mydb.query(MY_TB_NAME,
56              null,//new String[]{"_id",MY_FD_USER,MY_FD_SECRET},
57              null, null, null, null, "_id");
58  }
```

第 54～58 行的 query 方法得到数据表中的全部记录，赋给一个 Cursor 类的对象并返回它。

```
59  public Cursor query(String where){
60      return mydb.query(MY_TB_NAME,
61              null,
62              where,
63              null, null, null, null);
64  }
65  }
```

第 59～64 行的 query 方法得到满足 where 条件的记录，赋给一个 Cursor 类的对象并返回它，这里的方法 query 是第 54～58 行的 query 方法的一个重载方法。

因此，在文件 MyDatabase.java 中的类 MyDatabase 中封装了数据库对象和数据库常用的添加、删除和查询记录的操作。在文件 MyDBHelperExAct.java 中只需要调用 MyDatabase 类的对象方法即可完成数据库的操作，用户无须关心数据库对象的具体操作方法。

源文件 MyDBHelperExAct.java 的内容如下（这里重点介绍与工程 ex06_05 中的 MyDBHelperAct.java 文件的区别之处）：

```
1   package cn.edu.jxufe.zhangyong.mydbhelperexapp;
2
3   import android.support.v7.app.AppCompatActivity;
4   import android.database.Cursor;
5   import android.os.Bundle;
6   import android.view.View;
7   import android.widget.AdapterView;
8   import android.widget.AdapterView.OnItemClickListener;
9   import android.widget.EditText;
10  import android.widget.ListAdapter;
11  import android.widget.ListView;
12  import android.widget.SimpleCursorAdapter;
13  import android.widget.TextView;
14  import android.widget.CursorAdapter;
15  public class MyDBHelperExAct extends AppCompatActivity {
16      private MyDatabase mydb;
```

第 16 行定义 MyDatabase 类的对象 mydb。

```
17      private ListView lvUsers;
18      private EditText etUser,etSecret;
19      private final static String MY_DB_NAME = "myuser.db";
20      private final static String MY_FD_USER = "user";
21      private final static String MY_FD_SECRET = "secret";
```

```
22      @Override
23      protected void onCreate(Bundle savedInstanceState) {
24          super.onCreate(savedInstanceState);
25          setContentView(R.layout.activity_my_dbhelper_ex);
26          myInitGUI();
27      }
28      public void onDestroy(){
29          super.onDestroy();
30          mydb.close();
31      }
```

第 30 行调用对象 mydb 的 close 方法关闭数据库。

```
32      private void myInitGUI(){
33          etUser = (EditText)findViewById(R.id.etuser);
34          etSecret = (EditText)findViewById(R.id.etsecret);
35          lvUsers = (ListView)findViewById(R.id.lvcont);
36          lvUsers.setOnItemClickListener(new OnItemClickListener(){
37              @Override
38              public void onItemClick(AdapterView<?> arg0, View arg1, int arg2,
39                                     long arg3) {
40                  TextView tv_usr = (TextView)arg1.findViewById(R.id.user_name);
41                  TextView tv_sec = (TextView)arg1.findViewById(R.id.user_secret);
42                  String where = " user = '" + tv_usr.getText().toString() + "'" +
43                          " AND secret = '" + tv_sec.getText().toString() + "'";
44                  Cursor cursor = mydb.query(where);
```

第 44 行调用对象 mydb 的 query 方法查询满足 where 条件的记录。

```
45                  if(cursor!= null){
46                      cursor.moveToFirst();
47                      int loc = cursor.getColumnIndex(MY_FD_USER);
48                      etUser.setText(cursor.getString(loc));
49                      loc = cursor.getColumnIndex(MY_FD_SECRET);
50                      etSecret.setText(cursor.getString(loc));
51                  }
52              }
53          });
54          mydb = new MyDatabase(MyDBHelperExAct.this,MY_DB_NAME);
55          mydb.open();
```

第 54 和第 55 行使用构造方法 MyDatabase 传递数据库名,第 55 行创建并打开数据库,同时创建或打开一个名为 userinfo 的数据表(见文件 MyDatabase.java 的第 25~29 行)。

```
56          myDispRec();
57      }
58      public void myAddMD(View v){
59          if(etSecret.getText().toString().equals("")){
60          }
61          else{
62              mydb.add(etUser.getText().toString(), etSecret.getText().toString());
```

第 62 行调用 mydb 对象的方法 add 添加记录。

```
63          etUser.setText("");
64          etSecret.setText("");
65          myDispRec();
66      }
67  }
68  public void myDelMD(View v){
69      if(etSecret.getText().toString().equals("")){
70      }
71      else{
72          String where = " user = '" + etUser.getText().toString() + "'" +
73              " AND secret = '" + etSecret.getText().toString() + "'";
74          mydb.del(where);
```

第 74 行调用对象 mydb 的 del 方法删除满足 where 条件的记录。

```
75          myDispRec();
76          etUser.setText("");
77          etSecret.setText("");
78      }
79  }
80  public void myDelallMD(View v){
81      mydb.delall();
```

第 81 行调用对象 mydb 的 delall 方法删除全部记录。

```
82      etUser.setText("");
83      etSecret.setText("");
84      myDispRec();
85  }
86  public void myDispMD(View v){
87      myDispRec();
88  }
89  private void myDispRec(){
90      Cursor cursor = mydb.query();
```

第 90 行调用 mydb 对象的 query 方法得到数据表中的全部记录。

```
91          ListAdapter listAdapter = new SimpleCursorAdapter(this,
92              R.layout.userlist,cursor,
93              new String[]{MY_FD_USER,MY_FD_SECRET},
94              new int[]{R.id.user_name , R.id.user_secret},
95              CursorAdapter.FLAG_REGISTER_CONTENT_OBSERVER);
96          lvUsers.setAdapter(listAdapter);
97      }
98  }
```

从例 6-6 中可以看出，当封装了数据库对象和数据库操作方法后，在文件 MyDBHelperAct.java 中使用数据库更加方便，只需要了解 MyDatabase 类的公有方法（即操作接口），例如 open、close、query、del 和 delall 等，不需要关心数据库的细节。

6.4 内容提供者

Android 系统中不同应用程序间共享数据需要借助 Content Provider(内容提供者)。Android 系统是多任务实时操作系统,可同时运行多个应用程序,这些运行的应用程序可能同时访问同一个文件或同一个数据库等,为了防止数据访问冲突或死锁,必须借助 Content Provider 提供的接口进行访问,如图 6-11 所示。

图 6-11　应用程序间的文件和数据库共享

从图 6-11 可以看出,Content Provider 提供了一种数据共享的接口,即访问机制,这种访问机制封装了文件或数据库的访问操作。Content Provider 直接管理数据共享的方法由 Android 系统实现,程序员只需要操作 ContentProvider 类提供的接口方法,例如,插入数据 insert、查询数据 query、删除数据 delete、更新数据 update 和返回数据类型 getType 等。Content Provider 为它管理的数据指定一个唯一的 Uri 标识符,例如

```
1   public static final Uri myUri = Uri.parse("content://cn.edu.jxufe.zhangyong.users/userinfo");
```

其中,content://是固定前缀,cn.edu.jxufe.zhangyong.users 必须是唯一的,它对应于共享的文件或数据库,而 userinfo 对应于数据库中的某个数据表,当有多个数据表时,应该为每个数据表指定一个标识符。

假设需要共享数据库"myuser.db",则创建一个继承 ContentProvider 的类 MyDBProvider(该类名随意指定),其对应的文件 MyDBProvider.java 的内容如下:

```
1   package cn.edu.jxufe.zhangyong.mydbshareapp;
2
3   import android.content.ContentProvider;
4   import android.content.ContentUris;
5   import android.content.ContentValues;
6   import android.content.UriMatcher;
7   import android.database.Cursor;
8   import android.net.Uri;
9
10  public class MyDBProvider extends ContentProvider {
11      private MyDatabase mydb;
```

第 11 行的 MyDatabase 是工程 ex06_06 中继承 SQLiteOpenHelper 类的数据库操作自定义类,该类在本节实例中增加了两个方法。

```
12    private static final String MY_DB_NAME = "myuser.db";
13    public static final String myAuthority = "cn.edu.jxufe.zhangyong.users";
```

第12行定义了代表数据库名的字符串常量；第13行为Content Provider的权限名称，即Content Provider封装的数据集的名称。

```
14    public static final Uri myUri = Uri.parse("content://cn.edu.jxufe.zhangyong.users/userinfo");
```

第14行定义了Uri常量，对应于数据库中的数据表"userinfo"，这个常量是任意指定的，常量中的userinfo也是任意指定的，只需要保证它的唯一性。

```
15    private static final int USERINFO = 1;
16    private static final UriMatcher sURIMatcher =
17                       new UriMatcher(UriMatcher.NO_MATCH);
```

第16和第17行定义UriMatcher类的对象sURIMatcher，并将该对象初始化为空。第19~21行向该对象的myAuthority数据集中添加userinfo，并指定userinfo项在数据集中的索引号为USERINFO（即第15行的常数1）。当访问第14行定义的myUri时，语句sURIMatcher.match(myUri)将返回USERINFO（即1），用于识别对那个Content Provider管理的数据库（表）进行操作，如第30行所示。

```
18    private static final String MY_TB_NAME = "userinfo";
19    static{
20        sURIMatcher.addURI(myAuthority, MY_TB_NAME, USERINFO);
21    }
22    @Override
23    public boolean onCreate() {
24        // TODO Auto-generated method stub
25        mydb = new MyDatabase(getContext(),MY_DB_NAME);
26        return true;
27    }
```

第23~27行为onCreate方法，第25行调用MyDatabase类的构造函数，将数据库名传递给mydb对象（数据表名在MyDatabase类内部指定了）。

```
28    @Override
29    public Uri insert(Uri uri, ContentValues values) {
30        switch(sURIMatcher.match(uri)){
31        case USERINFO:
32            mydb.open();
33            long rowId = mydb.add(values);
34            if(rowId > 0){
35                Uri newUri = ContentUris.withAppendedId(myUri, rowId);
36                getContext().getContentResolver().notifyChange(newUri, null);
37                return newUri;
38            }
39            break;
40        }
41        return null;
42    }
```

第 29～42 行为向 Content Provider 管理的数据库（表）插入数据记录的方法 insert，它具有两个参数，第一个为 Uri 数据集标识符，第二个为 ContentValues 类的对象，即插入的数据。第 30 行判断 Uri 是否为第 14 行定义的 myUri，如果是，则 sURIMatcher.match(uri) 返回值 USERINFO，第 32～39 行得到执行。第 32 行打开数据库（表）；第 33 行将参数 values 添加到数据表中，add 方法是自定义的 MyDatabase 类新添加的方法，返回添加记录的行号（记录位置）。如果行号 rowId 大于 0，将该行号添加到 myUri 中（第 35、36 行）。Content Provider 类的对象不能直接访问它本身的方法，需要借助 ContentResolver 对象访问，第 36 行调用 getContentResolver 方法得到一个 ContentResolver 对象，然后再调用 notifyChange 方法登记添加的记录项。从第 29～42 行可以看出，记录的添加本质上仍然是数据库（表）的记录添加，Content Provider 只是提供了一个访问的封装。

```
43      @Override
44      public Cursor query(Uri uri, String[] projection, String selection,
45              String[] selectionArgs, String sortOrder) {
46          switch(sURIMatcher.match(uri)){
47          case USERINFO:
48              mydb.openRd();
49              Cursor cursor = mydb.query(selection);
50              cursor.setNotificationUri(getContext().getContentResolver(), uri);
51              return cursor;
52          }
53          return null;
54      }
```

第 44～54 行为 Content Provider 对象的查询方法 query，具有五个参数，第一个参数为 Uri 数据集标识符，第二个参数为访问的数据表列（或字段），第三个参数为 where 条件，第四个参数为 where 条件中的参数值，第五个参数为数据排序方法。第 48 行调用对象 mydb 的 openRd 方法打开一个只读的数据库（表），该方法是 MyDatabase 类中新添加的方法。第 49 行将查询到的记录存入 cursor 对象中。第 50 行设置 cursor 对象的数据集对象，第 51 行将该 cursor 对象返回。

```
55      @Override
56      public int delete(Uri uri, String selection, String[] selectionArgs) {
57          switch(sURIMatcher.match(uri)){
58          case USERINFO:
59              mydb.open();
60              if(selection == null){
61                  mydb.delall();
62              }
63              else{
64                  mydb.del(selection);
65              }
66              getContext().getContentResolver().notifyChange(uri, null);
67              return 1;
68          }
69          return 0;
70      }
```

第 56～70 行为 Content Provider 对象的删除方法 delete,有三个参数,依次为 Uri 数据集标识符、SQL 语句的 where 条件和 where 条件值。第 59 行打开数据库(表);如果 selection 为 null,则调用 delall 方法删除数据表中的全部记录,否则删除 selection 中指定条件的记录。第 66 行通知 Content Provider 对象数据集做了修改。

```
71    @Override
72    public String getType(Uri uri) {
73        return null;
74    }
75    @Override
76    public int update(Uri uri, ContentValues values, String selection,
77            String[] selectionArgs) {
78        return 0;
79    }
```

第 72～74 行和第 76～79 行的两个方法为返回数据集类型和更新数据集方法,均没有编写具体的实现代码。

```
80    }
```

通过上述的 MyDBProvider 封装了数据库 myuser.db 的插入、查询、删除和清除全部记录等操作。在应用程序中包含了 MyDBProvider 类后,在活动界面类中创建 ContentResolver 对象,并指定相应的 Uri,即可以实现对 MyDBProvider 类封装的数据库的操作,在活动界面类中不需要创建 MyDBProvider 类的对象,这个对象由 Android 系统管理。

下面创建两个实例,其中例 6-7 使用 Content Provider 提供了一个共享数据库,例 6-8 通过 ContentResolver 对象访问例 6-7 的共享数据库,当例 6-7 和例 6-8 的工程都处于运行状态时,从例 6-7 中添加记录,例 6-8 可以读出其添加的记录,从而实现不同应用程序间数据的共享。

例 6-7　Content Provider 共享数据提供者实例。

例 6-7 使用 Content Provider 技术将其数据库共享给其他应用程序使用,其共享的入口 Uri 为"content://cn.edu.jxufe.zhangyong.users/userinfo"。除此之外,例 6-7 执行的功能与例 6-6 完全相同(例 6-7 将例 6-6 应用程序界面上的"请输入密码"改为"请输入 ID 号")。

新建应用 MyDBShareApp,应用名为 MyDBShareApp,活动界面名为 MyDBShareAct。应用名为 MyDBShareApp 包括源文件 MyDBShareAct.java、MyDBProvider.java、MyDatabase.java、汉字字符串资源文件 mystrings_hz.xml、颜色资源文件 myguicolor.xml、主活动界面布局文件 activity_my_dbshare.xml、列表框布局文件 userlist.xml 以及工程配置文件 AndroidManifest.xml 等。其中,MyDBProvider.java 在前面介绍了,文件 myguicolor.xml、activity_my_dbshare.xml 和 userlist.xml 与例 6-6 中的同类型文件或同名文件相同(包为 cn.edu.jxufe.zhangyong.mydbshareapp);文件 mystrings_hz.xml 与例 6-6 中的同名文件相比,只是将 name 为 mysecret 的字符串值由"请输入密码"改为"请输入 ID 号";工程配置文件 AndroidManifest.xml 比例 6-6 中的同名文件多了如下三行代码:

```
1   <provider android:exported = "true" android:authorities = "cn.edu.jxufe.zhangyong.users"
2             android:name = "MyDBProvider">
3   </provider>
```

这三行代码添加到文件中</application>一行的上面。

文件 MyDatabase.java 与例 6-6 中的同名文件相比,其一,包名为 cn.edu.jxufe. zhangyong.mydbshareapp;其二,多了两个方法,即:

```
1   public void openRd(){
2       mydb = this.getReadableDatabase();
3   }
4   public long add(ContentValues cv){
5       return mydb.insert(MY_TB_NAME, null, cv);
6   }
```

第 1~3 行的 openRd 方法打开一个只读的数据库文件;第 4~6 行的 add 方法向数据库 mydb 的数据表 MY_TB_NAME 中添加记录 cv。

源文件 MyDBShareAct.java 的代码如下(主要说明与例 6-6 中文件 MyDBHelperExAct. java 不同的地方):

```
1   package cn.edu.jxufe.zhangyong.mydbshareapp;
2   import android.support.v7.app.AppCompatActivity;
3   import android.content.ContentValues;
4   import android.database.Cursor;
5   import android.os.Bundle;
6   import android.view.View;
7   import android.widget.AdapterView;
8   import android.widget.AdapterView.OnItemClickListener;
9   import android.widget.CursorAdapter;
10  import android.widget.EditText;
11  import android.widget.ListAdapter;
12  import android.widget.ListView;
13  import android.widget.SimpleCursorAdapter;
14  import android.widget.TextView;
15
16  public class MyDBShareAct extends AppCompatActivity {
17      private ListView lvUsers;
18      private EditText etUser, etSecret;
19      private final static String MY_FD_USER = "user";
20      private final static String MY_FD_SECRET = "secret";
21      @Override
22      protected void onCreate(Bundle savedInstanceState) {
23          super.onCreate(savedInstanceState);
24          setContentView(R.layout.activity_my_dbshare);
25          myInitGUI();
26      }
27      private void myInitGUI(){
28          etUser = (EditText)findViewById(R.id.etuser);
29          etSecret = (EditText)findViewById(R.id.etsecret);
30          lvUsers = (ListView)findViewById(R.id.lvcont);
31          lvUsers.setOnItemClickListener(new OnItemClickListener(){
32              @Override
```

```
33          public void onItemClick(AdapterView<?> arg0, View arg1, int arg2,
34                          long arg3) {
35              TextView tv_usr = (TextView)arg1.findViewById(R.id.user_name);
36              TextView tv_sec = (TextView)arg1.findViewById(R.id.user_secret);
37              String where = " user = '" + tv_usr.getText().toString() + "'" +
38                      " AND secret = '" + tv_sec.getText().toString() + "'";
39              Cursor cursor = getContentResolver().query(MyDBProvider.myUri,
40                      null,where,null,null);
```

第39～40行调用 getContentResolver 方法得到 ContentResolver 对象,通过该对象的 query 方法查询 Content Provider 中 Uri 为 MyDBProvider.myUri 的数据集中满足 where 条件的数据(记录)。

```
41              if(cursor!= null){
42                  cursor.moveToFirst();
43                  int loc = cursor.getColumnIndex(MY_FD_USER);
44                  etUser.setText(cursor.getString(loc));
45                  loc = cursor.getColumnIndex(MY_FD_SECRET);
46                  etSecret.setText(cursor.getString(loc));
47              }
48          }
49      });
50      myDispRec();
51  }
52  public void myAddMD(View v){
53      if(etSecret.getText().toString().equals("")){
54      }
55      else{
56          ContentValues cv = new ContentValues();
57          cv.put(MY_FD_USER, etUser.getText().toString());
58          cv.put(MY_FD_SECRET, etSecret.getText().toString());
59          getContentResolver().insert(MyDBProvider.myUri, cv);
```

第59行调用 getContentResolver 方法得到 ContentResolver 对象,通过该对象的 insert 方法向 Uri 为 MyDBProvider.myUri 的数据集中插入数据记录 cv。

```
60          myDispRec();
61          etUser.setText("");
62          etSecret.setText("");
63      }
64  }
65  public void myDelMD(View v){
66      if(etSecret.getText().toString().equals("")){
67      }
68      else{
69          String where = " user = '" + etUser.getText().toString() + "'" +
70                  " AND secret = '" + etSecret.getText().toString() + "'";
71          getContentResolver().delete(MyDBProvider.myUri, where, null);
72          myDispRec();
73          etUser.setText("");
74          etSecret.setText("");
75      }
76  }
```

```
77    public void myDelallMD(View v){
78        getContentResolver().delete(MyDBProvider.myUri, null, null);
79        etUser.setText("");
80        etSecret.setText("");
81        myDispRec();
82    }
```

第 71、78 行调用 ContentResolver 对象的 delete 方法删除满足 where 条件的记录或删除全部记录。

```
83    public void myDispMD(View v){
84        myDispRec();
85    }
86    private void myDispRec(){
87
88        Cursor cursor = getContentResolver().query(MyDBProvider.myUri,
89            null,null,null,null);
```

第 88、89 行调用 ContentResolver 对象的 query 方法查询数据集中的所有记录。

```
90        ListAdapter listAdapter = new SimpleCursorAdapter(this,
91            R.layout.userlist,cursor,
92            new String[]{MY_FD_USER,MY_FD_SECRET},
93            new int[]{R.id.user_name, R.id.user_secret},
94            CursorAdapter.FLAG_REGISTER_CONTENT_OBSERVER);
95        lvUsers.setAdapter(listAdapter);
96    }
97 }
```

从上述代码中可以看出文件 MyDBShareAct.java 没有调用数据库的任何操作，而是通过 ContentResolver 对象调用 Content Provider 提供的方法对数据库进行操作，即 Content Provider 封装了被共享的数据库的所有操作，形成统一的调用接口方法。下面的例 6-8 也体现了这一点。

例 6-7 的执行结果如图 6-12 所示。对比图 6-1 和图 6-12，可见将其中的"请输入密码"改为"请输入 ID 号"（通过修改 mystrings_hz.xml 资源文件实现），其他功能与例 6-6 完全相同。

通过例 6-7 的方法，即借助于 Content Provider 提供的统一数据访问接口进行数据（共享）访问的方法，创建的数据库（表）可以被其他应用程序共享和访问。下面在例 6-8 中创建了一个不同于例 6-7 的应用程序，为了说明数据共享的特点，例 6-8 使用了不同的包名，即 cn.edu.jxufe.zhangyong.myusershareapp，例 6-8 的应用程序通过 Uri 标识符、字段名和字段类型（必须知道这三个量），可以读取例 6-7 中创建的数

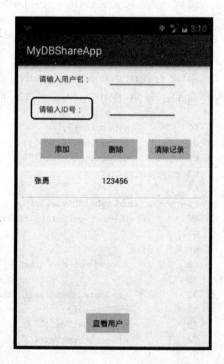

图 6-12　工程 ex06_07 运行界面

据集中的所有数据。

例 6-8 Content Provider 共享数据访问者实例。

新建应用 MyUserShareApp，应用名为 MyUserShareApp，包名为 cn. edu. jxufe. zhangyong. myusershareapp，活动界面名为 MyUserShareAct。应用 MyUserShareApp 包括源文件 MyUserShareAct. java、活动界面布局文件 activity_my_user_share. xml、列表框布局文件 userlist. xml、颜色资源文件 myguicolor. xml 和汉字字符串资源文件 mystrings_hz. xml 等。这里的文件 userlist. xml、myguicolor. xml 和 mystrings_hz. xml 与例 6-7 中的同名文件完全相同。此外，布局文件 activity_my_user_share. xml 仅包含一个列表框和命令按钮，当单击命令按钮时，在列表框中显示例 6-7 中 Content Provider 的数据集，activity_my_user_share. xml 的代码如下：

```
1   <?xml version = "1.0" encoding = "utf - 8"?>
2   < android. support. constraint. ConstraintLayout xmlns: android = "http://schemas. android. com/apk/res/android"
3       xmlns:app = "http://schemas. android. com/apk/res - auto"
4       xmlns:tools = "http://schemas. android. com/tools"
5       android:id = "@ + id/activity_my_user_share"
6       android:layout_width = "match_parent"
7       android:layout_height = "match_parent"
8       tools:context = "earth. china. jiangxi. myusershareapp. MyUserShareAct">
9       < Button
10          android:id = "@ + id/btdisp"
11          android:layout_width = "88dp"
12          android:layout_height = "48dp"
13          android:text = "@string/mydisp"
14          android:onClick = "myDispMD"
15          app:layout_constraintTop_toBottomOf = "@ + id/lvcont"
16          app:layout_constraintBottom_toBottomOf = "@ + id/activity_my_user_share"
17          app:layout_constraintLeft_toLeftOf = "@ + id/lvcont"
18          app:layout_constraintVertical_bias = "0.71000004"
19          app:layout_constraintRight_toRightOf = "@ + id/lvcont">
20      </Button>
21      < ListView
22          android:id = "@ + id/lvcont"
23          android:layout_width = "254dp"
24          android:layout_height = "427dp"
25          android:scrollbars = "vertical"
26          app:layout_constraintTop_toTopOf = "@ + id/activity_my_user_share"
27          app:layout_constraintBottom_toBottomOf = "@ + id/activity_my_user_share"
28          android:layout_marginStart = "16dp"
29          app:layout_constraintLeft_toLeftOf = "@ + id/activity_my_user_share"
30          android:layout_marginEnd = "16dp"
31          app:layout_constraintRight_toRightOf = "@ + id/activity_my_user_share"
32          app:layout_constraintHorizontal_bias = "0.47"
33          app:layout_constraintVertical_bias = "0.110000014">
34      </ListView>
35  </android. support. constraint. ConstraintLayout >
```

文件 MyUserShareAct.java 的代码如下所示：

```
1   package cn.edu.jxfue.zhangyong.myusershareapp;
2   import android.support.v7.app.AppCompatActivity;
3   import android.database.Cursor;
4   import android.net.Uri;
5   import android.os.Bundle;
6   import android.view.View;
7   import android.widget.CursorAdapter;
8   import android.widget.ListAdapter;
9   import android.widget.ListView;
10  import android.widget.SimpleCursorAdapter;
11
12  public class MyUserShareAct extends AppCompatActivity {
13      public static final Uri myUri = Uri.parse("content://cn.edu.jxufe.zhangyong.users/userinfo");
```

第 13 行指定访问的 Uri，通过 Uri 标识符访问 Content Provider 提供的共享数据。

```
14      private ListView lvUsers;
15
16      private final static String MY_FD_USER = "user";
17      private final static String MY_FD_SECRET = "secret";
```

第 16、17 行定义了要访问的数据集中的字段名。

```
18      @Override
19      protected void onCreate(Bundle savedInstanceState) {
20          super.onCreate(savedInstanceState);
21          setContentView(R.layout.activity_my_user_share);
22          myInitGUI();
23      }
24
25      private void myInitGUI(){
26          lvUsers = (ListView)findViewById(R.id.lvcont);
27      }
28      public void myDispMD(View v) {
29          myDispRec();
30      }
```

第 28～30 行为单击"显示用户"命令按钮（如图 6-13(b)）执行的方法 myDispMD，其中调用方法 myDispRec。

```
31      private void myDispRec(){
32          Cursor cursor = getContentResolver().query(myUri,
33                  null,null,null,null);
```

第 32、33 行通过 ContentResolver 对象的 query 方法从 Content Provider 共享数据集 myUri 中读出全部记录，并赋给 cursor 对象。

```
34          ListAdapter listAdapter = new SimpleCursorAdapter(this,
35                  R.layout.userlist,cursor,
36                  new String[]{MY_FD_USER,MY_FD_SECRET},
```

```
37                    new int[]{R.id.user_name, R.id.user_secret},
38                    CursorAdapter.FLAG_REGISTER_CONTENT_OBSERVER);
39         lvUsers.setAdapter(listAdapter);
40     }
41 }
```

例 6-7 和例 6-8 的执行结果如图 6-13 和图 6-14 所示。

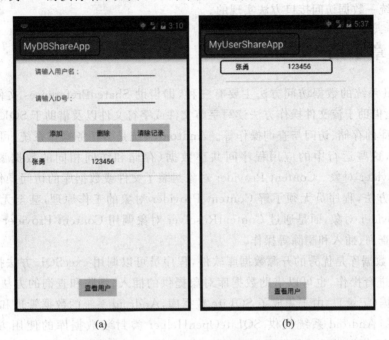

图 6-13　例 6-7 和例 6-8 的执行结果一

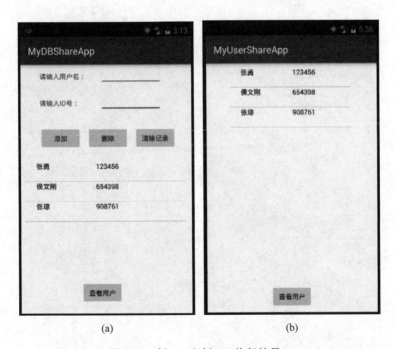

图 6-14　例 6-7 和例 6-8 执行结果二

图 6-13(a)中只有一条记录,此时在图 6-13(b)所示界面下,单击"查看用户",将读出该记录并显示在列表框中。图 6-14(a)中新添加了两条记录,然后,在例 6-8 的执行结果界面中单击"查看用户",将显示所有的记录,如图 6-14(b)所示。从图 6-13 和图 6-14 可以看出,尽管工程 ex06_07 和工程 ex06_08 对应于不同的应用程序(甚至应用程序所在的包都不同),但是这两个不同的应用程序间实现了数据的共享,而这种数据的共享是基于 Content Provider 的统一数据访问接口方法实现的。

6.5 本章小结

Android 系统的数据访问方法主要有三种,即借助 SharedPreferences 文件读写 XML 格式的文件、借助于流文件操作方法读写字节文件或字符文件以及借助于 SQLite 数据库进行关系数据库的存储、访问与查询操作等。Android 系统是多任务操作系统,可同时运行多个应用程序,这些运行中的应用程序间共享数据(存储和访问相同的数据源)需要借助 Content Provider 对象。Content Provider 对象封装了文件或数据库的访问操作,并提供了统一的接口方法,程序员无须了解 Content Provider 对象的工作原理,甚至无法直接使用 Content Provider 对象,而是通过 ContentResolver 对象调用 Content Provider 对象的方法进行数据的查询、插入和删除等操作。

SQLite 数据库是优秀的开源数据库软件,程序员可以调用 execSQL 方法执行 SQL 语句对数据库进行操作,也可以借助数据库对象提供的插入、删除和查询的方法操作数据库(表)中的数据(记录)。由于集成了 SQLite 数据库,Android 系统的数据管理和访问能力异常强大,同时,Android 系统提供 SQLiteOpenHelper 类封装数据库的使用方法,借助于 SQLiteOpenHelper 类进行数据库开发更加安全方便。

图形与动画

Android 系统中与图形相关的类或包有 View 类、SurfaceView 类和 graphics 包等，View 类直接继承类 java.lang.Object，SurfaceView 类是 View 类（即 android.view.View 类）的子类，包 android.graphics 包含了绘图相关的类，例如 Bitmap 类（位图类）、Canvas 类（画布类）、Color 类（颜色类）、Matrix 类（坐标变换类）、Paint 类（绘图类）、Point 类（像素类）和 Rect 类（矩形类）等。Android 系统实现动画相关的类位于包 Android.view.animation 中，该包中包括了 Animation 抽象类、AlphaAnimation 类、RotateAnimation 类、ScaleAnimation 类、TranslateAnimation 类和 RotateAnimation 类等，用于实现图形对象的渐变动画；Android 系统还可实现图像多帧切换形式的动画，相关的类为 android.graphics.drawable.AnimationDrawable。本章介绍与图形和动画相关的类的应用方法，从简单的绘图方法开始。

7.1 绘图

Android 应用程序在视图上绘图，实质上是在 View 类或 SurfaceView 类的屏幕上绘图，这时需要借助四种基本元素：其一，View 类或 SurfaceView 类的对象（一般是位图的形式），用于显示图形；其二，Canvas 类的对象，称为画布，包含绘图方法的调用，即通过画布向视图输出图形；其三，图形元素，即描述图形大小、形状和位置等属性的数据结构；其四，Paint 类的对象，称为画笔，描述绘图的样式和颜色等信息。

Android 系统具有 ImageView 等图像显示控件，可以用于显示多种格式的图像文件，Android 系统通过 ImageView 控件管理这类图像的显示。也就是说，当某个视图覆盖了正在显示的图像控件，当这个视图关闭后，图像控件中的图像仍然是可见的。

如果不借助 ImageView 控件显示图像，而是调用绘图方法直接在窗口（视图）上绘制图形时,绘制图形的视图被其他视图覆盖后,再回到屏幕前方时,绘制的图形就不存在了。Android系统需要重新执行绘图方法将显示的图形再绘制出来,这一过程称为"显示刷新"。与Windows CE相似,当有视图切换或大小改变等事件发生时,Android系统自动向需要重绘的视图发送重绘请求"消息",要求该视图重绘。如果程序员需要该视图重绘,则调用invalidate或postinvalidate方法请求重绘当前视图。

在Android应用程序中,整个绘图过程在View类的onDraw方法中进行,该方法不能在应用程序中直接调用,应用程序可调用invalidate方法请求该视图重绘,然后,Android系统会调用视图的onDraw方法完成视图重绘,即显示刷新工作。

7.1.1 View类绘图程序框架

借助View类绘图,需要在View类的onDraw方法添加绘图语句,然后在应用程序中显示该View视图。如果只是单纯地绘制几个图形,Android系统会管理View视图的显示刷新工作,即程序员无须编写刷新代码,如例7-1所示。

例7-1 View类简单绘图程序示例。

例7-1的运行结果如图7-1所示,即显示一个填充圆和圆圈。

新建应用MyViewDrawApp,应用名为MyViewDrawApp,活动界面名为MyViewDrawAct。应用名为MyViewDrawApp包括源代码文件MyViewDrawAct.java和MySimpleView.java等,其中,文件MySimpleView.java负责绘图工作,而MyViewDrawAct.java负责显示。

MySimpleView.java文件的代码如下:

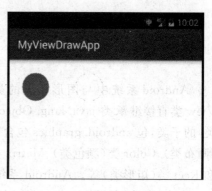

图7-1 例7-1运行结果
(填充圆和圆圈为红色)

```
1   package cn.edu.jxfue.zhangyong.myviewdrawapp;
2
3   import android.content.Context;
4   import android.graphics.Canvas;
5   import android.graphics.Color;
6   import android.graphics.Paint;
7   import android.graphics.Paint.Style;
8   import android.view.View;
9
10  public class MySimpleView extends View {
11      public MySimpleView(Context context) {
12          super(context);
13          // TODO Auto-generated constructor stub
14      }
```

第10行说明MySimpleView类继承了View类。第11~14行为MySimpleView类的构造函数。

```
15      @Override
16      protected void onDraw(Canvas canvas){
```

```
17      Paint paint = new Paint();
18      paint.setColor(Color.RED);
19      paint.setStyle(Style.FILL);
20      canvas.drawColor(Color.WHITE);     //(Color.DKGRAY);
21      canvas.drawCircle(100, 100, 50, paint);
22      for(int theta = 0;theta < 360;theta++){
23          float x = (float) (50 * Math.sin(theta * Math.PI/180.0) + 300);
24          float y = (float) (50 * Math.cos(theta * Math.PI/180.0) + 100);
25          canvas.drawPoint(x, y, paint);
26      }
27  }
28 }
```

第 16～27 行为 MySimpleView 类中的 onDraw 方法,该方法由 Android 系统直接调用,当程序员调用 MySimpleView 类的对象的 invalidate 方法时,将通知 Android 系统调用 onDraw 方法。第 17 行定义画笔;第 18 行设置画笔为红色;第 19 行设置画笔为填充绘制方式;第 20 行设置画布的背景色为白色;第 21 行在点(100,100)处画半径为 50 的圆形并填充;第 22～26 行使用绘点方法 drawPoint 画一个圆形。

文件 MyViewDrawAct.java 的内容如下:

```
1  package cn.edu.jxfue.zhangyong.myviewdrawapp;
2
3  import android.support.v7.app.AppCompatActivity;
4  import android.os.Bundle;
5
6  public class MyViewDrawAct extends AppCompatActivity {
7      private MySimpleView mySimpleView;
8      @Override
9      protected void onCreate(Bundle savedInstanceState) {
10         super.onCreate(savedInstanceState);
11         setContentView(R.layout.activity_my_view_draw);
12         myInitGUI();
13     }
14     private void myInitGUI(){
15         mySimpleView = new MySimpleView(this);
16         setContentView(mySimpleView);
17     }
18 }
```

第 15 行创建一个 MySimpleView 类的对象 mySimpleView,第 16 行将该对象设置为显示界面,从而得到图 7-1 所示结果。

例 7-2 View 类单线程刷新绘图示例。

例 7-1 的 onDraw 方法是不变的,即绘制的是静态图形,这时的应用程序无须程序员编写显示刷新代码,即无须调用 invalidate 方法。例 7-2 给出了一种单线程实现的 View 类绘图操作,在例 7-2 中需要程序员编写显示刷新代码,例 7-2 的执行结果如图 7-2 所示。

在图 7-2 中,选中"填充"或"不填充",然后,单击"绘圆"或"绘矩形"按钮,将显示图 7-2(a)～图 7-2(d)所示的图形,注意各个图形使用灰度显示,图形本身的颜色为红色。

例如，在图 7-2(a)中，选择了"填充"单选钮，然后单击"绘圆"，此时，需要通过调用 invalidate 方法使 Android 系统执行 View 类的 onDraw 方法，对显示屏幕进行刷新，才能正确地显示出一个红色的填充圆。

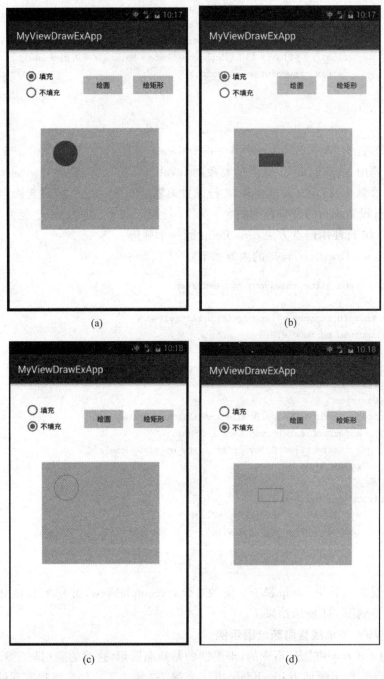

图 7-2 例 7-2 执行结果

新建应用 MyViewDrawExApp，应用名为 MyViewDrawExApp，活动界面名为 MyViewDrawExAct。应用 MyViewDrawExApp 包括源文件 MyViewDrawExAct.java、

MySimpleView.java、布局文件 activity_my_view_draw_ex.xml、颜色资源文件 myguicolor.mxl 和汉字字符串资源文件 mystrings_hz.xml 等。其中，myguicolor.xml 文件与例 6-8 中的同名文件相同。

汉字字符串资源文件 mystrings_hz.xml 的代码如下：

```
1   <?xml version = "1.0" encoding = "utf-8"?>
2   <resources>
3       <string name = "strfilled">填充</string>
4       <string name = "strstroke">不填充</string>
5       <string name = "strcircle">绘圆</string>
6       <string name = "strrect">绘矩形</string>
7   </resources>
```

布局文件 activity_my_view_draw_ex.xml 的代码如下所示，重点需要关注第 64～74 行关于自定义的 View 类的布局 XML 语言写法。

```
1   <?xml version = "1.0" encoding = "utf-8"?>
2   <android.support.constraint.ConstraintLayout xmlns:android = "http://schemas.android.com/apk/res/android"
3       xmlns:app = "http://schemas.android.com/apk/res-auto"
4       xmlns:tools = "http://schemas.android.com/tools"
5       android:id = "@+id/activity_my_view_draw_ex"
6       android:layout_width = "match_parent"
7       android:layout_height = "match_parent"
8       tools:context = "cn.edu.jxfue.zhangyong.myviewdrawexapp.MyViewDrawExAct">
9       <RadioGroup
10          android:id = "@+id/rgshape"
11          android:layout_width = "90dp"
12          android:layout_height = "65dp"
13          android:layout_x = "20px"
14          android:layout_y = "20px"
15          app:layout_constraintBottom_toTopOf = "@+id/myview"
16          app:layout_constraintTop_toTopOf = "@+id/activity_my_view_draw_ex"
17          android:layout_marginStart = "16dp"
18          app:layout_constraintLeft_toLeftOf = "@+id/activity_my_view_draw_ex"
19          android:layout_marginEnd = "16dp"
20          app:layout_constraintRight_toRightOf = "@+id/activity_my_view_draw_ex"
21          app:layout_constraintHorizontal_bias = "0.06"
22          app:layout_constraintVertical_bias = "0.41000003">
23          <RadioButton
24              android:id = "@+id/rbfill"
25              android:layout_width = "wrap_content"
26              android:layout_height = "wrap_content"
27              android:text = "@string/strfilled"
28              android:checked = "true" >
29          </RadioButton>
30          <RadioButton
31              android:id = "@+id/rbstroke"
32              android:layout_width = "wrap_content"
33              android:layout_height = "wrap_content"
```

```xml
34              android:text = "@string/strstroke" >
35          </RadioButton>
36      </RadioGroup>
37      <Button
38          android:id = "@+id/btcirc"
39          android:layout_width = "88dp"
40          android:layout_height = "48dp"
41          android:text = "@string/strcircle"
42          android:onClick = "myShowMD"
43          android:layout_marginStart = "8dp"
44          app:layout_constraintLeft_toRightOf = "@+id/rgshape"
45          android:layout_marginEnd = "16dp"
46          app:layout_constraintRight_toRightOf = "@+id/activity_my_view_draw_ex"
47          app:layout_constraintHorizontal_bias = "0.16"
48          app:layout_constraintBottom_toTopOf = "@+id/myview"
49          app:layout_constraintTop_toTopOf = "@+id/rgshape"
50          app:layout_constraintVertical_bias = "0.13">
51      </Button>
52      <Button
53          android:id = "@+id/btrect"
54          android:layout_width = "88dp"
55          android:layout_height = "48dp"
56          android:text = "@string/strrect"
57          android:onClick = "myShowMD"
58          app:layout_constraintBaseline_toBaselineOf = "@+id/btcirc"
59          android:layout_marginStart = "8dp"
60          app:layout_constraintLeft_toRightOf = "@+id/btcirc"
61          android:layout_marginEnd = "16dp"
62          app:layout_constraintRight_toRightOf = "@+id/activity_my_view_draw_ex">
63      </Button>
64      <cn.edu.jxfue.zhangyong.myviewdrawexapp.MySimpleView
65          android:id = "@+id/myview"
66          android:layout_width = "240dp"
67          android:layout_height = "200dp"
68          android:layout_marginStart = "16dp"
69          app:layout_constraintLeft_toLeftOf = "@+id/activity_my_view_draw_ex"
70          android:layout_marginEnd = "16dp"
71          app:layout_constraintRight_toRightOf = "@+id/activity_my_view_draw_ex"
72          app:layout_constraintTop_toTopOf = "@+id/activity_my_view_draw_ex"
73          app:layout_constraintBottom_toBottomOf = "@+id/activity_my_view_draw_ex"
74          app:layout_constraintHorizontal_bias = "0.56" />
75  </android.support.constraint.ConstraintLayout>
```

上述代码中,第 42 和第 57 行表示单击"绘圆"和"绘矩形"按钮(见图 7-2)的事件响应方法均为 myShowMD。

源文件 MySimpleView.java 的代码如下,重点介绍其与例 7-1 中的同名文件的区别之处。

```
1  package cn.edu.jxfue.zhangyong.myviewdrawexapp;
2
```

```
3   import android.content.Context;
4   import android.graphics.Canvas;
5   import android.graphics.Color;
6   import android.graphics.Paint;
7   import android.graphics.Paint.Style;
8   import android.graphics.Rect;
9   import android.util.AttributeSet;
10  import android.view.View;
11
12  public class MySimpleView extends View {
13      private Style style = Style.STROKE;
14      private int shape = 0;
```

第13和第14行定义了两个变量 style 和 shape, 分别表示填充方法和绘制圆形还是矩形, shape 为0时绘圆, shape 为1时绘矩形。

```
15      public MySimpleView(Context context, AttributeSet attr){
16          super(context,attr);
17      }
```

必须采用第15～17行的构造方法,其中 attr 表示视图的属性。

```
18      public void onDraw(Canvas canvas){
19          canvas.drawColor(Color.LTGRAY);    //(Color.DKGRAY);
20          Paint paint = new Paint();
21          paint.setColor(Color.RED);
22          paint.setStyle(style);
23          switch(shape){
24          case 0:
25              canvas.drawCircle(100, 100, 50, paint);
26              break;
27          case 1:
28              Rect r = new Rect(100,100,200,150);
29              canvas.drawRect(r, paint);
30              break;
31          }
32      }
```

第22行通过变量 style 设置填充方式; 第23～31行根据 shape 的值设置画圆还是画矩形, 当 shape 为0时, 画圆; 当 shape 为1时, 画矩形。

```
33      public void setStyle(Style style, int shape){
34          this.style = style;
35          this.shape = shape;
36      }
37  }
```

第33～36行的方法 setStyle 用于设置表示绘图的填充方法的变量 style 以及画圆还是画矩形的变量 shape。

源文件 MyViewDrawExAct.java 的代码如下所示, 重点分析与例7-1中的文件 MyViewDrawAct.java 不同的地方。

```java
1   package cn.edu.jxfue.zhangyong.myviewdrawexapp;
2
3   import android.support.v7.app.AppCompatActivity;
4   import android.graphics.Paint.Style;
5   import android.os.Bundle;
6   import android.view.View;
7   import android.widget.RadioButton;
8
9   public class MyViewDrawExAct extends AppCompatActivity {
10      private RadioButton rbfilled,rbstroke;
11      private MySimpleView mySimpleView;
12
13      @Override
14      protected void onCreate(Bundle savedInstanceState) {
15          super.onCreate(savedInstanceState);
16          setContentView(R.layout.activity_my_view_draw_ex);
17          myInitGUI();
18      }
```

第 10 行定义两个单选钮对象，表示图 7-2 中的"填充"和"不填充"单选按钮（第 21、22 行）。第 11 行定义 MySimpleView 类的对象 mySimpleView，该对象为布局文件中的 myview（第 20 行）。第 16 行设置显示界面为 activity_my_view_draw_ex.xml 布局文件的内容。

```java
19      private void myInitGUI(){
20          mySimpleView = (MySimpleView)findViewById(R.id.myview);
21          rbfilled = (RadioButton)findViewById(R.id.rbfill);
22          rbstroke = (RadioButton)findViewById(R.id.rbstroke);
23      }
24      public void myShowMD(View v){
25          int shape = 0;
26          switch(v.getId()){
27              case R.id.btcirc:
28                  shape = 0;
29                  break;
30              case R.id.btrect:
31                  shape = 1;
32                  break;
33          }
34          if(rbfilled.isChecked()){
35              mySimpleView.setStyle(Style.FILL,shape);
36          }
37          if(rbstroke.isChecked()){
38              mySimpleView.setStyle(Style.STROKE,shape);
39          }
40          mySimpleView.postInvalidate();//mySimpleView.invalidate();
41      }
42  }
```

第 24~41 行为单击按钮"绘圆"或"绘矩形"（见图 7-2）的事件方法 myShowMD。第

26～33行判断单击的是"绘圆"按钮还是"绘矩形"按钮,如果是"绘圆"按钮,则shape赋为0;如果是"绘矩形"按钮,则shape赋为1。第34～39行根据"填充"和"不填充"单选按钮(见图7-2)的选中情况,调用setStyle方法设置mySimpleView对象的私有成员style和shape(见程序MySimpleView.java的第13和第14行和第33～36行)。

第40行调用postInvalidate方法刷新显示,这里也可以执行"mySimpleView.invalidate();"语句进行显示刷新。

Android系统建构在Linux系统之上,一个Android应用程序对应着一个Linux的进程(或线程),第40行调用postInvalidate方法,相当于一个线程本身告诉Android系统刷新它本身,这在Android系统中被认为是不安全的。因此对于View类绘图方法而言,常常采用多线程显示刷新方法,即新创建一个线程,用它调用invalidate方法请求Android系统刷新显示。例7-3采用了这种显示刷新方法,该实例可作为View类绘图的程序框架,更复杂的绘图工作可以在该程序框架上进一步扩展代码得到。

例7-3 View类多线程刷新绘图示列。

例7-3执行的结果与例7-2完全相同,如图7-2所示。

新建应用MyViewDrawEx1App,应用名为MyViewDrawEx1App,活动界面名为MyViewDrawExAct。应用MyViewDrawEx1App包含源文件MyViewDrawExAct.java、MySimpleView.java、MyThread.java、布局文件activity_my_view_draw_ex.xml、颜色资源文件myguicolor.mxl和汉字字符串资源文件mystrings_hz.xml等。与例7-2相比,新添加了一个文件MyThread.java,并且对MyViewDrawExAct.java做了改动,其余文件与例7-2中的同名文件内容完全相同,这里不再重述。

文件MyThread.java的内容如下:

```
1   package cn.edu.jxfue.zhangyong.myviewdrawex1app;
2
3   import android.os.Bundle;
4   import android.os.Handler;
5   import android.os.Message;
6
7   public class MyThread extends Thread {
8       private Handler h;
9       private Message msg;
10      public MyThread(Handler h) {
11          // TODO Auto-generated constructor stub
12          this.h = h;
13      }
14      @Override
15      public void run(){
16          try{
17              Thread.sleep(100);
18          }
19          catch(InterruptedException e){
20              Thread.currentThread().interrupt();
21          }
22          msg = h.obtainMessage();
23          Bundle bd = new Bundle();
```

```
24          bd.putInt("value", MyViewDrawExAct.VIEW_FRESH);
25          msg.setData(bd);
26          h.sendMessage(msg);
27      }
28  }
```

上述代码自定义了一个线程类 MyThread,继承类 Thread。该线程每次执行将发送消息 msg,包含了一个自定义常量 MyViewDrawAct. VIEW_FRESH,即 1(见下面程序 MyViewDrawAct.java 的第 12 行)。

文件 MyViewDrawExAct.java 的内容如下所示,重点分析与例 7-2 的同名文件中不同的地方。

```
1   package cn.edu.jxfue.zhangyong.myviewdrawex1app;
2
3   import android.support.v7.app.AppCompatActivity;
4   import android.graphics.Paint.Style;
5   import android.os.Bundle;
6   import android.os.Handler;
7   import android.os.Message;
8   import android.view.View;
9   import android.widget.RadioButton;
10
11  public class MyViewDrawExAct extends AppCompatActivity {
12      public static final int VIEW_FRESH = 1;
13      private RadioButton rbfilled,rbstroke;
14      private MySimpleView mySimpleView;
15
16      @Override
17      protected void onCreate(Bundle savedInstanceState) {
18          super.onCreate(savedInstanceState);
19          setContentView(R.layout.activity_my_view_draw_ex);
20          myInitGUI();
21      }
22      private void myInitGUI(){
23          mySimpleView = (MySimpleView)findViewById(R.id.myview);
24          rbfilled = (RadioButton)findViewById(R.id.rbfill);
25          rbstroke = (RadioButton)findViewById(R.id.rbstroke);
26      }
27      public void myShowMD(View v){
28          int shape = 0;
29          switch(v.getId()){
30              case R.id.btcirc:
31                  shape = 0;
32                  break;
33              case R.id.btrect:
34                  shape = 1;
35                  break;
36          }
37          if(rbfilled.isChecked()){
38              mySimpleView.setStyle(Style.FILL,shape);
```

```
39          }
40          if(rbstroke.isChecked()){
41              mySimpleView.setStyle(Style.STROKE,shape);
42          }
43          MyThread myThread = new MyThread(handler);
44          myThread.start();
45      }
46      final Handler handler = new Handler(){
47          @Override
48          public void handleMessage(Message msg){
49              int state = msg.getData().getInt("value");
50              if(state == VIEW_FRESH){
51                  mySimpleView.invalidate();
52              }
53          }
54      };
55  }
```

在第27～45行的"绘圆"或"绘矩形"按钮（见图7-2）的单击事件方法 myShowDM 中，每次执行时都创建一个新的自定义线程对象 myThread，如第43行所示，在第44行启动该线程，该线程将发送消息 msg。第46～54行的 Handler 对象将接收到 msg 消息，并解析出其中的值赋给 state 变量（第49行），如果 state 变量的值为 VIEW_FRESH（即1），那么调用 invalidate 方法刷新界面显示。从第43和第44行可以看出，随着应用程序的运行和用户多次单击"绘圆"或"绘矩形"按钮，例7-3将创建很多自定义线程，而且每个线程运行一次后就再没有用了，Android 系统会管理那些不再使用的线程。相比于单线程绘图而言，这种通过创建新的线程刷新界面显示的方法被认为是安全的。

7.1.2 SurfaceView 类绘图程序框架

SurfaceView 类是专用绘图类，比 View 类更加实用。借助 SurfaceView 类绘图需要通过 SurfaceHolder 接口实现，即需要调用 getHolder 方法得到绘图的容器（设为 myholder），这个容器是不可见的；然后，调用 myholder 对象的 lockCanvas 方法得到绘图的画布（设为 canvas），这个画布对象仍是不可见的，且被"锁定"；在 canvas 画布上绘图；最后，调用 myholder 对象的 unlockCanvasAndPost 方法将画布 canvas 解锁并显示出来。由于 SurfaceView 类绘图需要使用 SurfaceHolder 接口，因此还要实现该接口的三个回调方法，即 surfaceChanged、surfaceCreated 和 surfaceDestroyed 方法，分别表示 surfaceView 绘图对象大小变化、首次创建和被清除时自动调用的方法。

例7-4 SurfaceView 类绘图示例。

例7-4与例7-3执行的功能相同，如图7-2所示。

新建应用 MySurfaceDrawApp，应用名为 MySurfaceDrawApp，活动界面名为 MySurfaceDrawAct。应用 MySurfaceDrawApp 包括源文件 MySurfaceDrawAct.java、MySurfaceView.java、布局文件 activity_my_surface_draw.xml、颜色资源文件 myguicolor.mxl 和汉字字符串资源文件 mystrings_hz.xml 等。其中，文件 myguicolor.mxl 和 mystrings_hz.xml 与例7-4中的同名文件相同。

例 7-4 的布局文件 activity_my_surface_draw.xml 只是将例 7-3 的布局文件 activity_my_view_draw_ex.xml 中第 64～74 行换成以下代码：

```xml
1   <cn.edu.jxfue.zhangyong.mysurfacedrawapp.MySurfaceView
2       android:id = "@ + id/myview"
3       android:layout_width = "240dp"
4       android:layout_height = "200dp"
5       android:layout_marginStart = "16dp"
6       app:layout_constraintLeft_toLeftOf = "@ + id/activity_my_view_draw_ex"
7       android:layout_marginEnd = "16dp"
8       app:layout_constraintRight_toRightOf = "@ + id/activity_my_view_draw_ex"
9       app:layout_constraintTop_toTopOf = "@ + id/activity_my_view_draw_ex"
10      app:layout_constraintBottom_toBottomOf = "@ + id/activity_my_view_draw_ex"
11      app:layout_constraintHorizontal_bias = "0.56" />
```

源文件 MySurfaceView.java 的代码如下：

```java
1   package cn.edu.jxfue.zhangyong.mysurfacedrawapp;
2
3   import android.content.Context;
4   import android.graphics.Canvas;
5   import android.graphics.Color;
6   import android.graphics.Paint;
7   import android.graphics.Rect;
8   import android.graphics.Paint.Style;
9   import android.util.AttributeSet;
10  import android.view.SurfaceHolder;
11  import android.view.SurfaceView;
12
13  public class MySurfaceView extends SurfaceView
14                  implements SurfaceHolder.Callback, Runnable {
15      private SurfaceHolder myholder;
16      private Style style = Style.STROKE;
17      private int shape = 0;
18      public MySurfaceView(Context context, AttributeSet attrs) {
19          super(context, attrs);
20          myholder = this.getHolder();
21          myholder.addCallback(this);
22      }
```

由于使用布局文件，必须使用第 18～22 行的构造方法。第 20 行得到 SurfaceHolder 对象 myholder 作为绘图容器。第 21 行注册 myholder 对象的回调方法，即当发生绘图对象大小变化、首次创建或被清除时 Android 系统将自动调用第 24～26 行、第 28～30 行或第 32、33 行的方法。

```java
23  @Override
24  public void surfaceChanged(SurfaceHolder arg0, int arg1,
25                                  int arg2, int arg3) {
26  }
27  @Override
28  public void surfaceCreated(SurfaceHolder holder) {
```

```
29        new Thread(this).start();
30    }
```

当首次创建绘图对象时,创建并执行 SurfaceView 类的线程(第 29 行)。

```
31    @Override
32    public void surfaceDestroyed(SurfaceHolder holder) {
33    }
34    @Override
35    public void run() {
36        try{
37            Thread.sleep(300);
38        }
39        catch(Exception e){}
40        synchronized(myholder){
41            draw();
42        }
43    }
```

在线程中通过同步调用 draw 方法绘图(第 40~42 行)。

```
44    public void draw(){
45        Canvas canvas = myholder.lockCanvas();
46        canvas.drawColor(Color.DKGRAY);
47        Paint paint = new Paint();
48        paint.setColor(Color.RED);
49        paint.setStyle(style);
50        switch(shape){
51        case 0:
52            canvas.drawCircle(100, 100, 50, paint);
53            break;
54        case 1:
55            Rect r = new Rect(100,100,200,150);
56            canvas.drawRect(r, paint);
57            break;
58        }
59        myholder.unlockCanvasAndPost(canvas);
60    }
61    public void setStyle(Style style, int shape){
62        this.style = style;
63        this.shape = shape;
64    }
65 }
```

第 44~60 行为绘图方法 draw。第 45 行得到锁定的画布对象 canvas;然后,第 46~58 行进行绘图,这一过程与例 7-3 的 onDraw 方法相同;第 59 行调用 unlockCanvasAndPost 方法显示绘图结果。

从第 35~43 行可见 SurfaceView 类绘图是通过其内部的线程方法刷新绘图的,这一点与 View 绘图不同。

源文件 MySurfaceDrawAct.java 的内容如下:

```java
1   package cn.edu.jxfue.zhangyong.mysurfacedrawapp;
2
3   import android.support.v7.app.AppCompatActivity;
4   import android.graphics.Paint.Style;
5   import android.os.Bundle;
6   import android.view.View;
7   import android.widget.RadioButton;
8
9   public class MySurfaceDrawAct extends AppCompatActivity {
10      public static final int VIEW_FRESH = 1;
11      private RadioButton rbfilled,rbstroke;
12      private MySurfaceView mySurfaceView;
13
14      @Override
15      protected void onCreate(Bundle savedInstanceState) {
16          super.onCreate(savedInstanceState);
17          setContentView(R.layout.activity_my_surface_draw);
18          myInitGUI();
19      }
20      private void myInitGUI(){
21          mySurfaceView = (MySurfaceView)findViewById(R.id.myview);
22          rbfilled = (RadioButton)findViewById(R.id.rbfill);
23          rbstroke = (RadioButton)findViewById(R.id.rbstroke);
24      }
25      public void myShowMD(View v){
26          int shape = 0;
27          switch(v.getId()){
28              case R.id.btcirc:
29                  shape = 0;
30                  break;
31              case R.id.btrect:
32                  shape = 1;
33                  break;
34          }
35          if(rbfilled.isChecked()){
36              mySurfaceView.setStyle(Style.FILL,shape);
37          }
38          if(rbstroke.isChecked()){
39              mySurfaceView.setStyle(Style.STROKE,shape);
40          }
41          new Thread(mySurfaceView).start();
42      }
43  }
```

第21行从布局文件中得到mySurfaceView对象；第41行创建mySurfaceView对象的线程并启动该线程,则文件MySurfaceView.java的第35～43行的代码得到执行,在其第41行中调用draw方法刷新绘图界面。

从例7-4可以看出,在SurfaceView视图上绘图,整个绘图过程在"后台"(不可见的)的

SurfaceHolder 接口对象上进行,绘图完成后调用 unlockCanvasAndPost 方法把整个绘图结果显示出来,这样可有效地避免屏幕闪烁,称为双缓冲显示技术。

7.1.3 基本图形与字符串

Android 系统中常用的基本绘图方法有以下几种:

1. 画点

画点方法有三种,即:

public void drawPoint(float x, float y, Paint paint)在点(x,y)使用 paint 画笔绘一个单点。

public void drawPoints(float[]pts, int offset, int count, Paint paint)使用 paint 画笔绘制数组 pts 中的点列,offset 指定跳过开始的点数,count 指定绘制的总点数。pts 的结构为$[x_0,y_0,x_1,y_1,\cdots,x_n,y_n]$。

public void drawPoints(float[]pts, Paint paint)使用画笔 paint 绘制数组 pts 中的所有点。

2. 画线

画线方法有三种,即:

public void drawLine(float startX, float startY, float stopX, float stopY, Paint paint)使用画笔 paint 从起点(startX, startY)画到终点(stopX, stopY)。

public void drawLines(float[]pts, Paint paint)
public void drawLines(float[]pts, int offset, int count, Paint paint)

上面这两种方法使用 paint 画笔以数组 pts 中的点为顶点画线,offet 指定跳过开始的点数,count 指定绘制的点数。pts 的结构为$[x_0,y_0,x_1,y_1,\cdots,x_n,y_n]$,每 4 个数值组成一条线,即$[x_0,y_0,x_1,y_1]$为第一条线的起止点,$[x_2,y_2,x_3,y_3]$为第二条线的起止点。

3. 画复杂图形

public void drawPath(Path path, Paint paint)使用画笔 paint 按 path 指定的绘画路线绘制图形。

4. 画圆

public void drawCircle(float cx, flaot cy, float radius, Paint paint)使用画笔 paint 绘制圆心在(cx,cy)半径为 radius 的圆。

5. 画椭圆

public void drawOval(RectF oval, Paint paint)使用画笔 paint 绘制内接于矩形 oval 的椭圆。

6. 画矩形

public void drwaRect(float left, float top, float right, float bottom, Paint paint)
public void drawRect(RectF rect, Paint paint)
public void drawRect(Rect r, Paint paint)

使用画笔 paint 绘制矩形，其中，矩形的左上角为（left，top），右下角为（right，bottom），RectF 和 Rect 类的对象含义相同，都包含矩形的左上角和右下角的点的坐标值，可以调用 width 或 height 方法得到矩形的宽和高。

7. 画圆角矩形

public void drawRoundRect(RectF rect，float rx，float ry，Paint paint)使用画笔 paint 绘制圆角矩形，rx 和 ry 指定圆角区域的 X 和 Y 轴方向的半径。

8. 字符串

绘制字符串的方法有 6 种，即：

public void drawText(String text, float x, float y, Paint paint)
public void drawText(CharSequence text, int start, int end, float x, float y, Paint paint)
public void drawText(String text, int start, int end, float x, float y, Paint paint)
public void drawText(char[] text, int index, int count, float x, float y, Paint paint)
public void drawTextOnPath(String text, Path path, float hOffset, float vOffset, Paint paint)
public void drawTextOnPath(char[] text, int index, int count, Path path, float hOffset, float vOffset, Paint paint)

上述绘制字符串方法中，点(x,y)表示绘图的起点坐标，start 和 end 表示绘图的字符串中的起止字符位置，index 和 count 表示字符串中要绘制的字符的起始位置和绘制字符的个数，path 表示字符串绘制的路线。hOffset 和 vOffset 表示字符串相对于绘制路线水平和垂直方向上偏移的位置。

在例 7-1 至例 7-4 中演示了画点、画圆和画矩形的方法，其他绘图方法的使用与这三种方法类似，这里不再举例。需要说明的是，除了上述列举的 8 类基本绘图方法外，在 android.graphics.Canvas 类中有大量的绘图方法，请参考 Android 开发者手册。

7.2 动画

Android 系统支持两种动画设计方法，即渐变动画和帧切换动画，这两种动画都是通过改变绘图屏幕的背景显示动画效果。基于绘图屏幕前景的动画设计，需要借助定时器实现。这里首先介绍借助于定时器实现动画的原理，然后再依次阐述渐变动画和帧切换动画。

7.2.1 定时器动画

借助定时器实现动画的基本原理为：在应用程序中添加一个定时器，定时器周期性地产生定时事件（实际上是定时中断），在定时事件中刷新屏幕显示，因此，屏幕被周期性地连续更新显示，表现为动画的效果。

例 7-5　定时器动画示例一。

新建应用 MyTimerAnimApp，应用名为 MyTimerAnimApp，活动界面名为 MyTimerAnimAct。应用 MyTimerAnimApp 包括源文件 MyTimerAnimAct.java、MySurfaceView.java、布局文件 activity_my_timer_anim.xml、汉字字符串资源文件 mystrings_hz.xml 和颜色资源文件 myguicolor.xml 等。其中，文件 myguicolor.xml 与例 7-4 中的同名文件内容相同。

应用 MyTimerAnimApp 的执行结果如图 7-3 所示。

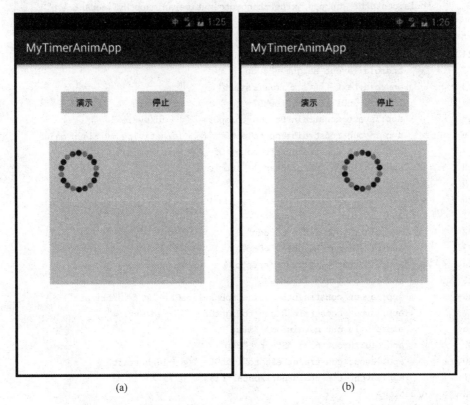

图 7-3　应用 MyTimerAnimApp 执行结果（注：小球有红、绿、蓝三色）

在图 7-3 中，单击"演示"按钮，由红、绿、蓝三色组成的小球（图中以灰度显示）将从屏幕左侧向右侧滚动。单击"停止"按钮，停止滚动。

汉字字符串资源文件 mystrings_hz.xml 定义了两个汉字字符串"演示"和"停止"，其内容如下：

```
1    <?xml version = "1.0" encoding = "utf - 8"?>
2    < resources >
3        < string name = "strstart">演示</string >
4        < string name = "strstop">停止</string >
5    </resources >
```

布局文件 activity_my_timer_anim.xml 中定义了两个命令按钮和一个 MySurfaceView 控件，两个命令按钮的事件单击方法均为"myShowMD"，文件 activity_my_timer_anim.xml 的内容如下：

```
1    <?xml version = "1.0" encoding = "utf - 8"?>
2    < android.support.constraint.ConstraintLayout xmlns:android = "http://schemas.android.com/apk/res/android"
3        xmlns:app = "http://schemas.android.com/apk/res - auto"
4        xmlns:tools = "http://schemas.android.com/tools"
5        android:id = "@ + id/activity_my_timer_anim"
6        android:layout_width = "match_parent"
```

```xml
7      android:layout_height = "match_parent"
8      tools:context = "cn.edu.jxfue.zhangyong.mytimeranimapp.MyTimerAnimAct">
9      <Button
10         android:id = "@+id/btstart"
11         android:layout_width = "88dp"
12         android:layout_height = "48dp"
13         android:text = "@string/strstart"
14         android:onClick = "myShowMD"
15         app:layout_constraintBottom_toTopOf = "@+id/myview"
16         app:layout_constraintTop_toTopOf = "@+id/activity_my_timer_anim"
17         app:layout_constraintLeft_toLeftOf = "@+id/myview"
18         android:layout_marginStart = "16dp">
19     </Button>
20     <Button
21         android:id = "@+id/btstop"
22         android:layout_width = "88dp"
23         android:layout_height = "48dp"
24         android:text = "@string/strstop"
25         android:onClick = "myShowMD"
26         app:layout_constraintBaseline_toBaselineOf = "@+id/btstart"
27         app:layout_constraintRight_toRightOf = "@+id/myview"
28         android:layout_marginEnd = "16dp"
29         android:layout_marginStart = "8dp"
30         app:layout_constraintLeft_toRightOf = "@+id/btstart"
31         app:layout_constraintHorizontal_bias = "0.75">
32     </Button>
33     <cn.edu.jxfue.zhangyong.mytimeranimapp.MySurfaceView
34         android:id = "@+id/myview"
35         android:layout_width = "261dp"
36         android:layout_height = "235dp"
37         android:layout_marginStart = "16dp"
38         app:layout_constraintLeft_toLeftOf = "@+id/activity_my_timer_anim"
39         android:layout_marginEnd = "16dp"
40         app:layout_constraintRight_toRightOf = "@+id/activity_my_timer_anim"
41         app:layout_constraintBottom_toBottomOf = "@+id/activity_my_timer_anim"
42         app:layout_constraintTop_toTopOf = "@+id/activity_my_timer_anim"
43         app:layout_constraintVertical_bias = "0.46"
44         app:layout_constraintHorizontal_bias = "0.58" />
45 </android.support.constraint.ConstraintLayout>
```

源文件 MySurfaceView.java 负责绘图,其代码如下：

```java
1  package cn.edu.jxfue.zhangyong.mytimeranimapp;
2
3  import android.content.Context;
4  import android.graphics.Canvas;
5  import android.graphics.Color;
6  import android.graphics.Paint;
7  import android.graphics.Paint.Style;
8  import android.util.AttributeSet;
9  import android.view.SurfaceHolder;
```

```
10   import android.view.SurfaceView;
11
12   public class MySurfaceView extends SurfaceView
13                       implements SurfaceHolder.Callback,Runnable {
14       private int locx,dir;
15       private SurfaceHolder myholder;
16       public MySurfaceView(Context context, AttributeSet attrs) {
17           super(context, attrs);
18           myholder = this.getHolder();
19           myholder.addCallback(this);
20           locx = 100;
21           dir = 0;
22       }
```

第 14 行定义了两个私有数据成员 locx 和 dir,其中,locx 用于表示图 7-3 中各小球围成的圆形的圆心横坐标,其取值为 100～380；dir 表示小球的颜色切换索引号,取值为 0～2。第 20 和第 21 行为这两个变量赋了初值 100 和 0。

```
23       @Override
24       public void surfaceChanged(SurfaceHolder arg0, int arg1,
25                                               int arg2, int arg3) {
26       }
27       @Override
28       public void surfaceCreated(SurfaceHolder holder) {
29           new Thread(this).start();
30       }
31       @Override
32       public void surfaceDestroyed(SurfaceHolder holder) {
33       }
34       @Override
35       public void run() {
36           try{
37               Thread.sleep(300);
38           }
39           catch(Exception e){}
40           synchronized(myholder){
41               draw();
42           }
43       }
44       public void draw(){
45           int[] color = new int[]{Color.BLUE,Color.RED,Color.GREEN};
46           Canvas canvas = myholder.lockCanvas();
47           canvas.drawColor(Color.LTGRAY);      //(Color.DKGRAY);
48           Paint paint = new Paint();
49           paint.setStyle(Style.FILL);
50           for(int i = 0;i < 360;i += 20){
51               paint.setColor(color[((i/20) % 3 + dir) % 3]);
52               float x0 = (float)(locx + 60 * Math.sin(i * Math.PI/180));
53               float y0 = (float)(100 + 60 * Math.cos(i * Math.PI/180));
54               canvas.drawCircle(x0,y0,10.0f,paint);
55           }
```

```
56          myholder.unlockCanvasAndPost(canvas);
57      }
58      public void setLocation(int locx,int dir){
59          this.locx = locx;
60          this.dir = dir;
61      }
62  }
```

第 44～57 行为绘图方法 draw。第 47 行设置画布背景为浅灰色；第 49 行设置画笔的填充方式；第 50～55 行的 for 循环用于绘制 18 个实心小圆球，这些小圆球围成的圆形的圆心纵坐标为 100，横坐标为 locx，各个小圆球的颜色依次为红、绿、蓝。

第 58～61 行的 setLocation 方法用于更新 locx 和 dir 的值，即改变小圆球围成的圆形的圆心横坐标位置和小圆球的颜色，实现绘图的动画效果。

源文件 MyTimerAnimAct.java 创建了一个定时器，定时周期为 0.2s，每个定时事件中请求刷新绘图，该文件的代码如下：

```
1   package cn.edu.jxfue.zhangyong.mytimeranimapp;
2
3   import android.support.v7.app.AppCompatActivity;
4   import java.util.Timer;
5   import java.util.TimerTask;
6   import android.os.Bundle;
7   import android.os.Handler;
8   import android.os.Message;
9   import android.view.View;
10
11  public class MyTimerAnimAct extends AppCompatActivity {
12      private int locx,dir;
13      private boolean animate = false;
14      private MySurfaceView mySurfaceView;
15      @Override
16      protected void onCreate(Bundle savedInstanceState) {
17          super.onCreate(savedInstanceState);
18          setContentView(R.layout.activity_my_timer_anim);
19          myInitGUI();
20      }
21      private void myInitGUI(){
22          locx = 100;
23          dir = 0;
24          mySurfaceView = (MySurfaceView)findViewById(R.id.myview);
25          timer.schedule(timerTask, 1000, 200);
26      }
27      public void myShowMD(View v){
28          switch(v.getId()){
29              case R.id.btstart:
30                  animate = true;
31                  break;
32              case R.id.btstop:
33                  animate = false;
```

```
34              break;
35          }
36     }
```

第 27～36 行为按钮"演示"和"停止"(见图 7-3)的单击事件方法 myShowMD,如果单击了"演示"按钮,则第 30 行将布尔型变量 animate 设为真;如果单击了"停止"按钮,则第 33 行将布尔型变量 animate 设为假。

```
37   private final Timer timer = new Timer();
38   private TimerTask timerTask = new TimerTask(){
39       @Override
40       public void run(){
41           Message msg = new Message();
42           if(animate){
43               msg.what = 1;
44               handler.sendMessage(msg);
45           }
46           else{
47               msg.what = 2;
48               handler.sendMessage(msg);
49           }
50       }
51   };
```

第 37～51 行为定义定时器和定时器任务,第 25 行打开定时器,定时周期为 0.2s。关于定时器更详细的用法请参考第 4.2.8 节。每个定时事件到来后,如果变量 animate 为真,则发送消息 1(第 42～45 行);如果变量 animate 为假,则发送消息 2(第 46～49 行)。

```
52   private Handler handler = new Handler(){
53       @Override
54       public void handleMessage(Message msg){
55           int msgID = msg.what;
56           switch(msgID){
57               case 1:
58                   locx = locx + 10;
59                   dir = dir + 1;
60                   if(locx > 380){
61                       locx = 100;
62                   }
63                   if(dir > 2){
64                       dir = 0;
65                   }
66                   mySurfaceView.setLocation(locx, dir);
67                   new Thread(mySurfaceView).start();
68                   break;
69               case 2:
70                   break;
71           }
72           super.handleMessage(msg);
73       }
74   };
75   
```

第 52 行的 handler 对象接收消息，如果消息为 1，则更新 locx 和 dir 的值（第 58～65 行），调用 mySurfaceView 对象的 setLocation 方法设置绘图用的 locx 和 dir，第 67 行为 mySurfaceView 开启一个新的线程，该线程启动新的绘图。

例 7-5 的运行过程如框图 7-4 所示。

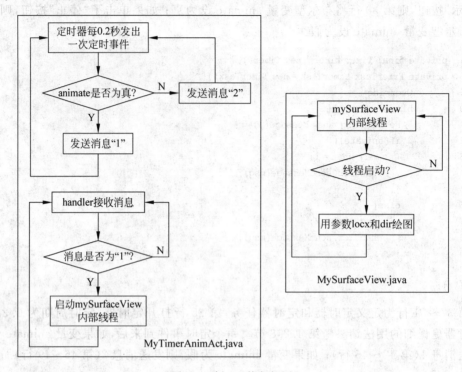

图 7-4　例 7-5 执行框图

从图 7-4 中可以看出，图中三个部分是独立的，事实上，这三部分是有联系的，Android 系统管理着它们之间的通信。

例 7-6　定时器动画示例二。

新建应用 MyTimerAnimExApp，应用名为 MyTimerAnimExApp，活动界面名为 MyTimerAnimAct。应用 MyTimerAnimExApp 包括源文件 MyTimerAnimAct.java、MySurfaceView.java 布局文件 activity_my_timer_anim.xml、汉字字符串资源文件 mystrings_hz.xml、颜色资源文件 myguicolor.xml 和图像文件 gsyg.jpg 等。其中，mystrings_hz.xml 和 myguicolor.xml 文件与例 7-5 中的同名文件内容相同，图像文件 gsyg.jpg 如图 7-5 中的图像所示（保存在 drawable 目录下）。布局文件 activity_my_timer_anim.xml 相对于工程 ex07_05 中的同名文件而言，只是将第 33 行的代码

< cn.edu.jxfue.zhangyong.mytimeranimapp.MySurfaceView

修改为

< cn.edu.jxfue.zhangyong.mytimeranimexapp.MySurfaceView

这是因为例 7-6 的包名为 cn.edu.jxfue.zhangyong.mytimeranimexapp。

例 7-6 实现的功能如图 7-5 所示。当单击"演示"按钮时，上面的图像循环做缩放运动，

而下面的图像循环做旋转运动;当单击"停止"按钮时,两幅图像停止运动。

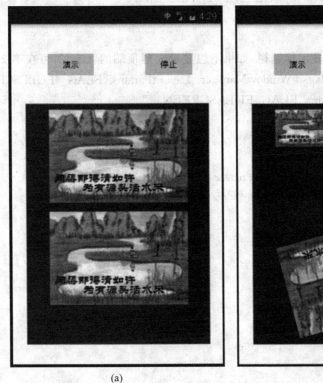

(a)　　　　　　　　　　　(b)

图 7-5　工程 ex07_06 运行结果

文件 MyTimerAnimAct.java 用于设置缩放的步长和旋转的角度,并启动定时器每 0.2 秒刷新绘图一次,该文件的内容如下:

```
 1  package cn.edu.jxfue.zhangyong.mytimeranimexapp;
 2
 3  import android.support.v7.app.AppCompatActivity;
 4  import java.util.Timer;
 5  import java.util.TimerTask;
 6  import android.os.Bundle;
 7  import android.os.Handler;
 8  import android.os.Message;
 9  import android.view.View;
10  import android.view.Window;
11
12  public class MyTimerAnimAct extends AppCompatActivity {
13      private float scale,angle;
14      private boolean animate = false;
15      private MySurfaceView mySurfaceView;
16      @Override
17      protected void onCreate(Bundle savedInstanceState) {
18          super.onCreate(savedInstanceState);
19          if(getSupportActionBar()!= null){
20              getSupportActionBar().hide();
```

```
21          }
22          setContentView(R.layout.activity_my_timer_anim);
23          myInitGUI();
24      }
```

第19～21行设置窗口不显示标题栏,如果不想显示屏幕顶端的状态栏,可在第22行插入语句"getWindow().setFlags(WindowManager.LayoutParams.FLAG_FULLSCREEN, WindowManager.LayoutParams.FLAG_FULLSCREEN);"。

```
25   private void myInitGUI(){
26       scale = 1.0f;
27       angle = 0f;
28       mySurfaceView = (MySurfaceView)findViewById(R.id.myview);
29       timer.schedule(timerTask, 1000, 200);
30   }
31   public void myShowMD(View v){
32       switch(v.getId()){
33           case R.id.btstart:
34               animate = true;
35               break;
36           case R.id.btstop:
37               animate = false;
38               break;
39       }
40   }
```

第31～40行为单击图7-5中"演示"或"停止"按钮的事件响应方法。如果单击"演示"按钮,则animate变量赋为真(第34行),否则赋为假。

```
41   private final Timer timer = new Timer();
42   private TimerTask timerTask = new TimerTask(){
43       @Override
44       public void run(){
45           Message msg = new Message();
46           if(animate){
47               msg.what = 1;
48               handler.sendMessage(msg);
49           }
50           else{
51               msg.what = 2;
52               handler.sendMessage(msg);
53           }
54       }
55   };
```

第41～55行定义了定时器timer和定时器任务timerTask。定时器每0.2秒(第29行)发送一则消息,如果animate为真,则发送消息"1";否则发送消息"2"。

```
56   private Handler handler = new Handler(){
57       @Override
58       public void handleMessage(Message msg){
```

```
59              int msgID = msg.what;
60              switch(msgID){
61                  case 1:
62                      scale = scale – 0.1f;
63                      angle = angle + 5f;
64                      if(scale < 0.1f)
65                          scale = 1f;
66                      if(angle > 360f)
67                          angle = 0f;
68                      mySurfaceView.setShape(scale, angle);
69                      new Thread(mySurfaceView).start();
70                      break;
71                  case 2:
72                      break;
73              }
74              super.handleMessage(msg);
75          }
76      };
77  }
```

第 56～76 行的 handler 对象接收消息，当接收到消息"1"时，设置 scale 和 angle 的值，第 68 行调用 setShape 方法将 scale 和 angle 的值传送给 mySurfaceView 对象，第 69 行运行 mySurfaceView 的线程。

源文件 MySurfaceView.java 负责图像的缩放和旋转，其代码如下：

```
1   package cn.edu.jxfue.zhangyong.mytimeranimexapp;
2
3   import android.content.Context;
4   import android.graphics.Bitmap;
5   import android.graphics.BitmapFactory;
6   import android.graphics.Canvas;
7   import android.graphics.Color;
8   import android.graphics.Matrix;
9   import android.util.AttributeSet;
10  import android.view.SurfaceHolder;
11  import android.view.SurfaceView;
12
13  public class MySurfaceView extends SurfaceView
14                  implements SurfaceHolder.Callback, Runnable {
15      private SurfaceHolder myholder;
16      private Bitmap mybmp;
17      private int bmpWidth, bmpHeight;
18      private Matrix scMatrix, rtMatrix;
19      private float angle, scale;
20      public MySurfaceView(Context context, AttributeSet attrs) {
21          super(context, attrs);
22          myholder = this.getHolder();
23          myholder.addCallback(this);
24          mybmp = BitmapFactory.decodeResource(getResources(), R.drawable.gsyg);
25          bmpWidth = mybmp.getWidth();
```

```
26        bmpHeight = mybmp.getHeight();
27        angle = 0f;
28        scale = 1f;
29        scMatrix = new Matrix();
30        rtMatrix = new Matrix();
31    }
```

第 24 行通过方法 decodeResource 将 JPG 格式的文件转化为 BMP 格式的位图,并赋给对象 mybmp。第 25 和第 26 行得到图像的宽和高。第 27 行设置初始角度,旋转角度为 0°至 360°。第 28 行设置初始缩放系数,小于 1 为缩小,大于 1 为放大。第 29 和第 30 行为两个 Matrix 变量,通过 Matrix 变量设置缩放大小和旋转角度。

```
32    @Override
33    public void surfaceChanged(SurfaceHolder arg0, int arg1, int arg2, int arg3) {
34    }
35    @Override
36    public void surfaceCreated(SurfaceHolder holder) {
37        new Thread(this).start();
38    }
39    @Override
40    public void surfaceDestroyed(SurfaceHolder holder) {
41    }
42    @Override
43    public void run() {
44        try{
45            Thread.sleep(300);
46        }
47        catch(Exception e){}
48        synchronized(myholder){
49            draw();
50        }
51    }
52    public void draw(){
53        Canvas canvas = myholder.lockCanvas();
54        canvas.drawColor(Color.DKGRAY);
55        scMatrix.reset();
56        scMatrix.postScale(scale, scale);
57        Bitmap myBmpScale = Bitmap.createBitmap(mybmp, 0, 0,
58                            bmpWidth, bmpHeight, scMatrix, true);
59        canvas.drawBitmap(myBmpScale, 80, 20, null);
```

第 55 行清空 scMatrix；第 56 行设置缩放大小；第 57、58 行由 mybmp 对象创建一个新的位图对象 myBmpScale,其缩放由 scMatrix 指定；第 59 行绘制对象 myBmpScale,其左上角位置为(80,20)。

```
60        rtMatrix.reset();
61        scMatrix.setRotate(angle);
62        Bitmap myBmpRotate = Bitmap.createBitmap(mybmp, 0, 0,
63                            bmpWidth, bmpHeight, scMatrix, true);
64        canvas.drawBitmap(myBmpRotate, 80, 360, null);
```

第 60 行清空 rtMatrix；第 61 行设置旋转角度；第 62、63 行由 mybmp 对象创建一个新的位图对象 myBmpRotate，其旋转角由 rtMatrix 指定；第 64 行绘制对象 myBmpRotate，其左上角位置为(80,360)。

```
65              myholder.unlockCanvasAndPost(canvas);
66         }
67         public void setShape(float scale,float angle){
68              this.scale = scale;
69              this.angle = angle;
70         }
71    }
```

由例 7-5 和例 7-6 可知，基于定时器的动画设计需要创建一个定时器，该定时器将一直处于工作状态，它不能直接启动 MySurfaceView 线程。定时器通过发送消息，在消息处理对象 handler 中启动 MySurfaceView 线程。由于定时器一直工作，应用程序通过控制发送消息的内容控制绘图动作，而不是控制定时器的启动和停止。当实现复杂的动画时，可以放置多个定时器。

7.2.2 渐变动画

渐变动画包括 4 种，即透明度渐变动画、缩放渐变动画、位置转移渐变动画和旋转渐变动画。渐变动画改变的是屏幕上所有的绘图对象，如果用 SurfaceView 类实现，则是改变它的背景；如果是 View 类实现，将改变其 onDraw 方法中绘制的所有图像。

例 7-7 渐变动画示例。

新建应用 MyTweenApp，应用名为 MyTweenApp，活动界面名为 MyTweenAct。应用 MyTweenApp 包括源文件 MyTweenAct.java、MySimpleView.java、布局文件 activity_my_tween.xml、汉字字符串资源文件 mystrings_hz.xml、颜色资源文件 myguicolor.xml 和图像文件 gsyg.jpg 等。其中，文件 myguicolor.xml 和 gsyg.jpg 与例 7-6 中的同名文件内容相同。

例 7-7 的执行结果如图 7-6 所示。图 7-6(a)为启动应用程序时的界面，当选中"透明变换"单选按钮时，单击"演示"按钮，则如图 7-6(b)所示，图像的透明度渐变；如果选中"大小变换"单选按钮，单击"演示"按钮，则如图 7-6(c)所示，图像的大小渐变；如果选中"平移变换"单选按钮，单击"演示"按钮，则如图 7-6(d)所示，图像的位置从左上角滑动到右下角；当单击"旋转变换"单选按钮时，再单击"演示"按钮，则如图 7-6(e)所示，图像发生旋转。

文件 mystrings_hz.xml 定义了应用程序中使用的汉字字符串，其内容如下：

```
1  <?xml version = "1.0" encoding = "utf-8"?>
2  <resources>
3      <string name = "strstart">演示</string>
4      <string name = "strstop">停止</string>
5      <string name = "stralpha">透明变换</string>
6      <string name = "strscale">大小变换</string>
7      <string name = "strtrans">平移变换</string>
8      <string name = "strrotate">旋转变换</string>
9  </resources>
```

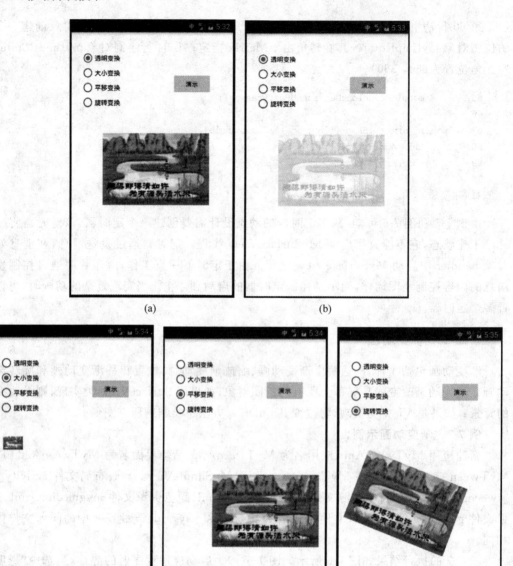

图 7-6 工程 ex07_07 执行结果

例 7-7 的布局结果如图 7-6(a)所示,其布局文件 activity_my_tween.xml 内容如下:

```
1   <?xml version = "1.0" encoding = "utf - 8"?>
2   <android.support.constraint.ConstraintLayout xmlns:android = "http://schemas.android.com/apk/res/android"
3       xmlns:app = "http://schemas.android.com/apk/res - auto"
4       xmlns:tools = "http://schemas.android.com/tools"
5       android:id = "@ + id/activity_my_tween"
6       android:layout_width = "match_parent"
7       android:layout_height = "match_parent"
```

```
8        tools:context = "cn.edu.jxfue.zhangyong.mytweenapp.MyTweenAct">
9        <Button
10           android:id = "@+id/btstart"
11           android:layout_width = "88dp"
12           android:layout_height = "48dp"
13           android:text = "@string/strstart"
14           android:onClick = "myShowMD"
15           android:layout_marginStart = "64dp"
16           app:layout_constraintLeft_toRightOf = "@+id/rgselect"
17           app:layout_constraintRight_toRightOf = "@+id/myview"
18           app:layout_constraintBottom_toTopOf = "@+id/myview"
19           app:layout_constraintTop_toTopOf = "@+id/rgselect"
20           app:layout_constraintHorizontal_bias = "0.56"
21           app:layout_constraintVertical_bias = "0.38">
22       </Button>
23       <cn.edu.jxfue.zhangyong.mytweenapp.MySimpleView
24           android:id = "@+id/myview"
25           android:layout_width = "316dp"
26           android:layout_height = "345dp"
27           android:layout_marginStart = "16dp"
28           app:layout_constraintLeft_toLeftOf = "@+id/activity_my_tween"
29           android:layout_marginEnd = "16dp"
30           app:layout_constraintRight_toRightOf = "@+id/activity_my_tween"
31           app:layout_constraintTop_toTopOf = "@+id/activity_my_tween"
32           app:layout_constraintBottom_toBottomOf = "@+id/activity_my_tween"
33           app:layout_constraintHorizontal_bias = "0.45"
34           app:layout_constraintVertical_bias = "0.96" />
35       <RadioGroup
36           android:id = "@+id/rgselect"
37           android:layout_width = "111dp"
38           android:layout_height = "133dp"
39           app:layout_constraintBottom_toTopOf = "@+id/myview"
40           app:layout_constraintTop_toTopOf = "@+id/activity_my_tween"
41           app:layout_constraintLeft_toLeftOf = "@+id/myview"
42           app:layout_constraintRight_toRightOf = "@+id/myview"
43           app:layout_constraintHorizontal_bias = "0.0">
44           <RadioButton
45               android:id = "@+id/rbalpha"
46               android:layout_width = "wrap_content"
47               android:layout_height = "wrap_content"
48               android:text = "@string/stralpha"
49               android:checked = "true"
50               >
51           </RadioButton>
52           <RadioButton
53               android:id = "@+id/rbscale"
54               android:layout_width = "wrap_content"
55               android:layout_height = "wrap_content"
56               android:text = "@string/strscale"
57               >
58           </RadioButton>
59           <RadioButton
```

```
60              android:id = "@ + id/rbtrans"
61              android:layout_width = "wrap_content"
62              android:layout_height = "wrap_content"
63              android:text = "@string/strtrans"
64              >
65          </RadioButton>
66          <RadioButton
67              android:id = "@ + id/rbrotate"
68              android:layout_width = "wrap_content"
69              android:layout_height = "wrap_content"
70              android:text = "@string/strrotate"
71              >
72          </RadioButton>
73      </RadioGroup>
74  </android.support.constraint.ConstraintLayout>
```

上述代码中,第 35～73 行为包含 4 个单选按钮的单选按钮组控件;第 9～22 行为"演示"命令按钮(见图 7-6(a));第 23、34 行为 MySimpleView 自定义控件,其 ID 号为 myview。

源文件 MySimpleView.java 的内容如下:

```
1   package cn.edu.jxfue.zhangyong.mytweenapp;
2
3   import android.content.Context;
4   import android.graphics.Bitmap;
5   import android.graphics.BitmapFactory;
6   import android.graphics.Canvas;
7   import android.util.AttributeSet;
8   import android.view.View;
9
10  public class MySimpleView extends View {
11      private Bitmap mybmp;
12      public MySimpleView(Context context, AttributeSet attr){
13          super(context,attr);
14          mybmp = BitmapFactory.decodeResource(getResources(),R.drawable.gsyg);
15      }
16      public void onDraw(Canvas canvas){
17          canvas.drawBitmap(mybmp, 100,10, null);
18      }
19  }
```

第 14 行由图像文件 gsyg.jpg 得到位图对象 mybmp,第 16～18 行的 onDraw 方法绘制该位图对象,程序启动后的界面如图 7-6(a)所示。

源文件 MyTweenAct.java 控制图像的渐变效果,其内容如下:

```
1   package cn.edu.jxfue.zhangyong.mytweenapp;
2
3   import android.support.v7.app.AppCompatActivity;
4   import android.os.Bundle;
5   import android.view.View;
6   import android.view.animation.AlphaAnimation;
7   import android.view.animation.Animation;
8   import android.view.animation.RotateAnimation;
9   import android.view.animation.ScaleAnimation;
```

```
10   import android.view.animation.TranslateAnimation;
11   import android.widget.RadioGroup;
12
13   public class MyTweenAct extends AppCompatActivity {
14       private MySimpleView mySimpleView;
15       private Animation myAnim;
16       private RadioGroup rgselect;
17
18       @Override
19       protected void onCreate(Bundle savedInstanceState) {
20           super.onCreate(savedInstanceState);
21           if(getSupportActionBar()!= null){
22               getSupportActionBar().hide();
23           }
24           setContentView(R.layout.activity_my_tween);
25           myInitGUI();
26       }
27       private void myInitGUI(){
28           mySimpleView = (MySimpleView)findViewById(R.id.myview);
29           rgselect = (RadioGroup)findViewById(R.id.rgselect);
30       }
31       public void myShowMD(View v){
32           switch(rgselect.getCheckedRadioButtonId()){
33               case R.id.rbalpha:
34                   myAnim = new AlphaAnimation(0.1f,1f);
35                   myAnim.setDuration(10000l);
36                   mySimpleView.startAnimation(myAnim);
37                   break;
```

第 15 行定义了类 Animation 的对象 myAnim，类 Animation 是个抽象类。第 34 行将对象 myAnim 赋值为 AlphaAnimation 类的对象，两个参数 0.1 和 1 分别表示透明度从 0.1 变为 1（即不透明）。第 35 行设置透明度从 0.1 变为 1 的过程需要花费 10 秒的时间。第 36 行调用对象 mySimpleView 的 startAnimation 方法启动透明渐变动画。

```
38               case R.id.rbscale:
39                   myAnim = new ScaleAnimation(0.1f,1f,0.1f,1f);
40                   myAnim.setDuration(10000l);
41                   mySimpleView.startAnimation(myAnim);
42                   break;
```

第 39 行得到 ScaleAnimation 类的对象 myAnim，设置其 X 和 Y 方向尺度变换均为从 0.1 变为 1，即从缩小 0.01 倍到原图大小。第 40 行设置缩放时间为 10 秒。第 41 行启动缩放动画。

```
43               case R.id.rbrotate:
44                   myAnim = new RotateAnimation(0f,30f);
45                   myAnim.setDuration(10000l);
46                   mySimpleView.startAnimation(myAnim);
47                   break;
```

第 44 行得到 RotatoAnimation 类的对象 myAnim，设置旋转角度为 0°到 30°；第 45 行设置旋转所用时间为 10 秒；第 46 行启动旋转。

```
48              case R.id.rbtrans:
49                  myAnim = new TranslateAnimation(10f,100f,10f,200f);
50                  myAnim.setDuration(100001);
51                  mySimpleView.startAnimation(myAnim);
52                  break;
53          }
54      }
55  }
```

第 49 行得到 TranslateAnimation 类的对象 myAnim，4 个参数表示 X 方向上由 10 移动到 100，Y 方向上由 10 移动到 200；第 50 行设置平移的时间为 10 秒；第 51 行启动平移动画。

从上面的代码可见，渐变动画不需要程序员刷新显示，Android 系统负责刷新显示，因此，渐变动画中无需定时器的帮助。渐变动画是一种简单的动画方法，它主要用于文字图形的透视、缩放、旋转和平移等效果。

7.2.3 帧切换动画

帧切换方式形成动画是动画的最基本形式，即连续地显示一组画面（每幅画面称为一帧）形成动画的形式。帧切换动画需要准备大量的图像素材，Android 应用程序将这些图像素材按照设定的时间连续播放，形成动画效果。

例 7-8 帧切换动画实例。

例 7-8 的执行效果如图 7-7 所示。在图 7-7 中，当单击"演示"按钮时，将显示一个扇动翅膀的蝴蝶，当单击"停止"按钮时，蝴蝶停止扇动翅膀。

图 7-7 例 7-8 执行结果

新建应用MyFrameExApp,应用名为MyFrameExApp,活动界面名为MyFrameAct。应用MyFrameExApp包括源文件MyFrameAct.java、MySimpleView.java、布局文件activity_my_frame.xml、汉字字符串资源文件mystrings_hz.xml、颜色资源文件myguicolor.xml和6个图像资源文件,这6个图像文件名为zeh01.gif、zeh02.gif、zeh03.gif、zeh04.gif、zeh05.gif和zeh06.gif(保存在drawable目录下)。文件myguicolor.xml与例7-7中的同名文件内容相同。

汉字字符串资源文件mystrings_hz.xml的内容如下:

```
1   <?xml version = "1.0" encoding = "utf-8"?>
2   <android.support.constraint.ConstraintLayout xmlns:android = "http://schemas.android.com/apk/res/android"
3       xmlns:app = "http://schemas.android.com/apk/res-auto"
4       xmlns:tools = "http://schemas.android.com/tools"
5       android:id = "@+id/activity_my_frame"
6       android:layout_width = "match_parent"
7       android:layout_height = "match_parent"
8       tools:context = "cn.edu.jxfue.zhangyong.myframeexapp.MyFrameAct">
9       <Button
10          android:id = "@+id/btstart"
11          android:layout_width = "88dp"
12          android:layout_height = "48dp"
13          android:text = "@string/strstart"
14          android:onClick = "myShowMD"
15          app:layout_constraintBottom_toTopOf = "@+id/myview"
16          app:layout_constraintLeft_toLeftOf = "@+id/myview"
17          app:layout_constraintRight_toRightOf = "@+id/myview"
18          app:layout_constraintHorizontal_bias = "0.18"
19          app:layout_constraintTop_toTopOf = "@+id/activity_my_frame"
20          app:layout_constraintVertical_bias = "0.71000004">
21      </Button>
22      <Button
23          android:id = "@+id/btstop"
24          android:layout_width = "88dp"
25          android:layout_height = "48dp"
26          android:text = "@string/strstop"
27          android:onClick = "myShowMD"
28          app:layout_constraintBaseline_toBaselineOf = "@+id/btstart"
29          android:layout_marginStart = "8dp"
30          app:layout_constraintLeft_toRightOf = "@+id/btstart"
31          android:layout_marginEnd = "16dp"
32          app:layout_constraintRight_toRightOf = "@+id/activity_my_frame"
33          app:layout_constraintHorizontal_bias = "0.38">
34      </Button>
35      <cn.edu.jxfue.zhangyong.myframeexapp.MySimpleView
36          android:id = "@+id/myview"
37          android:layout_width = "286dp"
38          android:layout_height = "186dp"
39          app:layout_constraintTop_toTopOf = "@+id/activity_my_frame"
40          app:layout_constraintBottom_toBottomOf = "@+id/activity_my_frame"
```

```
41          android:layout_marginStart = "16dp"
42          app:layout_constraintLeft_toLeftOf = "@ + id/activity_my_frame"
43          android:layout_marginEnd = "16dp"
44          app:layout_constraintRight_toRightOf = "@ + id/activity_my_frame"
45          app:layout_constraintVertical_bias = "0.58000004" />
46 </android.support.constraint.ConstraintLayout>
```

第 9~21 行和第 22~34 行为图 7-7 上的"演示"和"停止"按钮,它们的单击事件方法均为 myShowMD。

源文件 MySimpleView.java 的内容如下:

```
1  package cn.edu.jxfue.zhangyong.myframeexapp;
2  import android.content.Context;
3  import android.graphics.Canvas;
4  import android.graphics.drawable.AnimationDrawable;
5  import android.graphics.drawable.Drawable;
6  import android.support.v4.content.res.ResourcesCompat;
7  import android.util.AttributeSet;
8  import android.view.View;
9
10 public class MySimpleView extends View {
11     private Drawable myframe;
12     private AnimationDrawable myAnim = null;
13     public MySimpleView(Context context, AttributeSet attr){
14         super(context,attr);
15         myAnim = new AnimationDrawable();
16         for(int i = 1;i<=6;i++){
17             int id = getResources().getIdentifier("zeh0" + String.valueOf(i),
18                 "drawable", context.getPackageName());
19             myframe = ResourcesCompat.getDrawable(getResources(),id,null);
20             myAnim.addFrame(myframe,300);
21         }
22         myAnim.setOneShot(false);
23         this.setBackground(myAnim);
24     }
25     @Override
26     public void onDraw(Canvas canvas){
27         super.onDraw(canvas);
28     }
29     public void setAction(boolean b){
30         if(b){
31             myAnim.start();
32         }
33         else{
34             myAnim.stop();
35         }
36     }
37 }
```

第 11 行定义类 Drawable 的对象 myframe,用于存储一幅图像;第 12 行定义帧切换动画的对象 myAnim,它是类 AnimationDrawable 的实例。第 16~21 行将 6 幅图像添加到对象 myAnim 中，其中，第 17 和第 18 行得到资源名为 "zeh0" + 数字 "i" 的图像的 id 号，方法 getIdentifier 的第一个参数为资源名，第二个参数为资源类型，这里是 drawable，第三个参数为所在的包名。第 19 行得到图像并赋给 myframe 对象。第 20 行将 myframe 作为一帧图像添加到动画对象 myAnim 中，每幅图像的停留时间为 0.3 秒。第 22 行的方法 setOneShot 的参数为 flase 时，表示不停地播放动画，如果为 true 时，仅播放一次，即 "One Shot"。

第 29～36 行的方法 setAction 为启动或停止动画播放的方法，如果 b 为真，则启动动画；如果 b 为假，则停止动画。方法 setAction 在文件 MyFrameAct.java 中调用。

源文件 MyFrameAct.java 的代码如下：

```
1   package cn.edu.jxfue.zhangyong.myframeexapp;
2
3   import android.support.v7.app.AppCompatActivity;
4   import android.os.Bundle;
5   import android.view.View;
6
7   public class MyFrameAct extends AppCompatActivity {
8       private MySimpleView mySimpleView;
9
10      @Override
11      protected void onCreate(Bundle savedInstanceState) {
12          super.onCreate(savedInstanceState);
13          if(getSupportActionBar()!= null){
14              getSupportActionBar().hide();
15          }
16          setContentView(R.layout.activity_my_frame);
17          myInitGUI();
18      }
19      private void myInitGUI(){
20          mySimpleView = (MySimpleView)findViewById(R.id.myview);
21      }
22      public void myShowMD(View v){
23          switch(v.getId()){
24              case R.id.btstart:
25                  mySimpleView.setAction(true);
26                  break;
27              case R.id.btstop:
28                  mySimpleView.setAction(false);
29                  break;
30          }
31      }
32  }
```

第 22～31 行的方法 myShowMD 为图 7-7 中 "演示" 和 "停止" 按钮的事件响应方法。当单击 "演示" 按钮时，第 25 行调用带有 true 参数的 setAction 方法启动动画；当单击 "停止" 按钮时，第 28 行调用带有 false 参数的 setAction 方法停止动画。

7.3 本章小结

Android 绘图需要基于 View 类或 SurfaceView 类,使用 graphics 包的绘图方法绘制图形,绘图操作需要 4 个组成部分,即图形显示视图(View 类或 SurfaceView 类的对象)、绘图方法(通过 Canvas 画布调用)、画笔(Paint 对象)和图形属性(描述图形结构和形状的数据结构)。绘制在 View 类中的图形对象,必须通过调用 postInvalidate 或 invalidate 方法刷新后才能显示在屏幕上;而绘制在 SurfaceView 类上的图形对象,则使用 SurfaceView 类内部的线程刷新显示。

常规的动画设计是借助于定时器实现的,即通过定时器的定时事件周期性地绘制变化了位置或形状等属性的图形。基于 View 类的定时器动画,可以在定时器事件中调用 invalidate 更新图形显示;而基于 SurfaceView 类的定时器动画,需要借助于 Handler 类更新图形显示。Android 系统提供了无需程序员直接使用定时器的渐变动画和帧切换动画,一般地,渐变动画在文字或图形的动画显示方面应用较多;而帧动画则可按设定的时间间隔播放一系列图像形成动画效果。

第 8 章

多媒体技术

Android 系统的 MediaPlayer 类集成了播放音乐或视频文件的方法,该类直接继承类 java.lang.Object。借助 MediaPlayer 对象播放音频文件和视频文件的过程相似,都需要依次调用 reset()方法复位 MediaPlayer 对象、setDataSource 方法获得音像文件、prepare 方法准备播放、start 方法开始播放等。随着 Android 系统版本的升级,其多媒体技术也得到了较大的提升,几乎支持现有的所有媒体类型。本章介绍使用 MediaPlayer 类的对象方法播放 MP3 音频文件和 MP4 视频文件的方法。

8.1 音频文件播放

音频文件的格式众多,除了 WAV 格式是没有经过压缩的音频格式外,其余格式都属于压缩存储格式,例如 MP3 格式和 WMA 格式等。其中,WMA 格式(Android 系统支持)的全称为 Windows Media Audio,是微软公司设计的一种优秀音频压缩格式,而目前流行的高压缩比音频格式为 MP3 格式。本节介绍 MP3 格式音频文件的播放技术。

Android 系统的 MediaPlayer 类集成了 MP3 音频解码器,这项技术十分成熟,借助 MediaPlayer 类的对象播放 MP3 文件只需调用播放控制方法即可,其流程如图 8-1 所示。

图 8-1 演示了播放音频文件的整个过程,首先创建 MediaPlayer 对象 myMP3Player(对象名随意取),接着调用 reset 方法复位 myMP3Player 对象(实际上做内存清理工作),然后调用 setDataSource 方法设置音频文件,调用 prepare 方法准备播放,之后调用 start 方法开始播放,播放过程中:一方面会监听 onCompletionListener 接口,如果播放结束,则调用 onCompletion 方法,然后调用 reset 方法复位 myMP3Player 对象;另一方面,可以调用 pause 方法暂停播放,被暂停的播放再次调用 start 方法从暂停处继续播放。如果调用了 stop 方法则停止播放,这时需调用 reset 方法复位 myMP3Player,当调用 release 方法时将

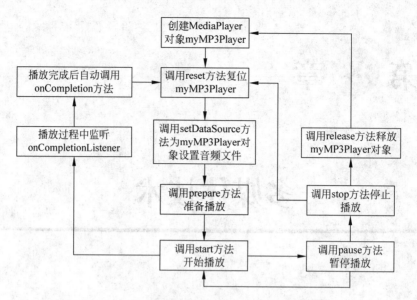

图 8-1　MediaPlayer 对象播放 MP3 文件流程图

myMP3Player 对象从内存中清除，如果需要再次播放音频文件，则需要重新创建 myMP3Player 对象，方法 release 一般在用户程序退出时调用。

本节（甚至是本章）的准备工作是向模拟的 SD 卡中写入音频（和视频）文件，如图 8-2 所

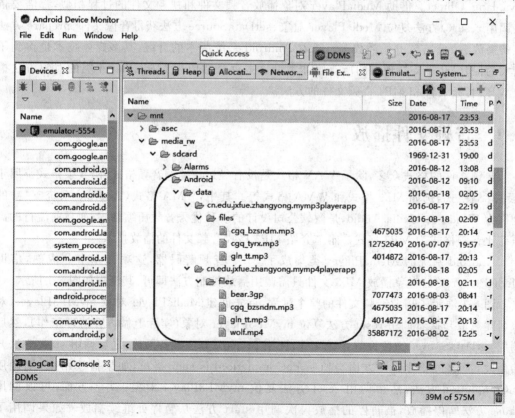

图 8-2　模拟器 SD 卡上的多媒体文件

示。在 Android 版本 4.4 以后,每个应用仅能自由访问外部 SD 卡上其包所在的目录,即"sdcard/Android/data/包名/files",在例 8-1 中的"包名"为"cn.edu.jxfue.zhangyong.mymp3playerapp"。这里向 SD 卡中传送了三个 MP3 文件供本节使用(包为 cn.edu.jxfue.zhangyong.mymp3playerapp),还有另外两个视频文件供第 8.3 节使用(那里使用了包 cn.edu.jxfue.zhangyong.mymp4playerapp)。这里三个音频文件分别为 cgq_bzsndm.mp3、cgq_tyrx.mp3 和 gln_tt.mp3,两个视频文件依次为 bear.3GP 和 wolf.MP4。

例 8-1 播放 MP3 文件工程。

新建应用 MyMP3PlayerApp,应用名为 MyMP3PlayerApp,活动界面名为 MyMP3PlayerAct。应用 MyMP3PlayerApp 包括源文件 MyMP3PlayerAct.java、布局文件 activity_my_mp3_player.xml、列表框布局文件 mp3list.xml、颜色资源文件 myguicolor.xml 以及图像资源文件 play1.png、stop1.png、stop2.png、pause1.png、pause2.png 和 pause3.png 等,其中 6 个图像文件用于表示播放控制按钮的状态,如图 8-3 所示。文件 myguicolor.xml 与例 7-8 中的同名文件内容相同。

列表布局文件 mp3list.xml 的内容如下所示,表示列表框中每行显示一个文本。

```
1   <?xml version = "1.0" encoding = "utf - 8"?>
2   < android.support.constraint.ConstraintLayout xmlns:app = "http://schemas.android.com/apk/res - auto"
3       xmlns:tools = "http://schemas.android.com/tools"
4       android:id = "@ + id/widget0"
5       android:layout_width = "fill_parent"
6       android:layout_height = "fill_parent"
7       xmlns:android = "http://schemas.android.com/apk/res/android">
8       < TextView
9           android:id = "@ + id/mp3name"
10          android:layout_width = "192dp"
11          android:layout_height = "27dp"
12          android:text = "TextView"
13          android:textColor = "@drawable/black"
14          tools:layout_constraintTop_creator = "1"
15          tools:layout_constraintLeft_creator = "1"
16          app:layout_constraintLeft_toLeftOf = "@ + id/widget0"
17          app:layout_constraintTop_toTopOf = "@ + id/widget0"
18          app:layout_constraintBottom_toBottomOf = "@ + id/widget0"
19          app:layout_constraintRight_toRightOf = "@ + id/widget0"
20          app:layout_constraintHorizontal_bias = "0.14"
21          app:layout_constraintVertical_bias = "0.050000012">
22      </TextView>
23  </android.support.constraint.ConstraintLayout>
```

例 8-1 的布局如图 8-3 所示,其布局文件 activity_my_mp3_player.xml.xml 的内容如下:

```
1   <?xml version = "1.0" encoding = "utf - 8"?>
2                       < android.support.constraint.ConstraintLayout
                            xmlns:android = "http://schemas.android.com/apk/res/android"
```

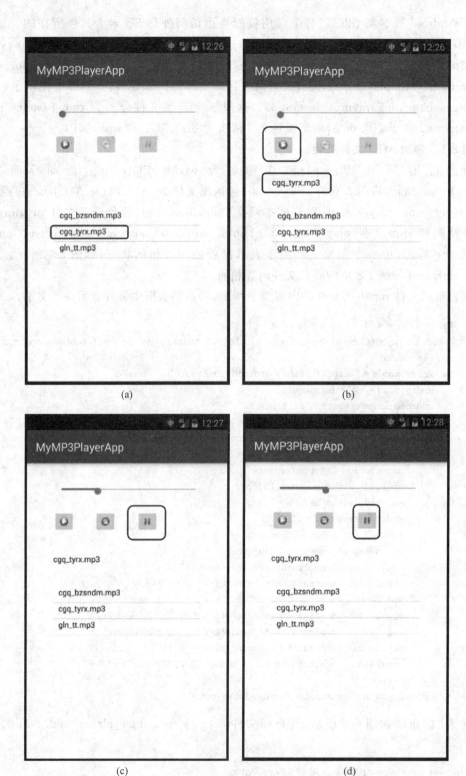

图 8-3 例 8-1 执行结果

```
3       xmlns:app = "http://schemas.android.com/apk/res-auto"
4       xmlns:tools = "http://schemas.android.com/tools"
5       android:id = "@+id/activity_my_mp3_player"
6       android:layout_width = "match_parent"
7       android:layout_height = "match_parent"
8       tools:context = "cn.edu.jxfue.zhangyong.mymp3playerapp.MyMP3PlayerAct">
9       <SeekBar
10          android:id = "@+id/seekbar"
11          android:layout_height = "25dp"
12          android:layout_width = "274dp"
13          android:max = "100"
14          android:progress = "0"
15          app:layout_constraintTop_toTopOf = "@+id/activity_my_mp3_player"
16          app:layout_constraintBottom_toTopOf = "@+id/textview"
17          app:layout_constraintVertical_bias = "0.25"
18          android:layout_marginEnd = "56dp"
19          app:layout_constraintRight_toRightOf = "@+id/activity_my_mp3_player"
20          app:layout_constraintLeft_toLeftOf = "@+id/imbplay"
21          app:layout_constraintHorizontal_bias = "0.05">
22      </SeekBar>
23      <ImageButton
24          android:id = "@+id/imbplay"
25          android:layout_width = "40dp"
26          android:layout_height = "40dp"
27          android:src = "@drawable/play1"
28          android:scaleType = "fitCenter"
29          android:onClick = "myPlayMD"
30          app:layout_constraintLeft_toLeftOf = "@+id/textview"
31          app:layout_constraintBottom_toTopOf = "@+id/textview"
32          app:layout_constraintTop_toBottomOf = "@+id/seekbar"
33          app:layout_constraintVertical_bias = "0.4">
34      </ImageButton>
35      <ImageButton
36          android:id = "@+id/imbpause"
37          android:layout_width = "40dp"
38          android:layout_height = "40dp"
39          android:src = "@drawable/pause1"
40          android:scaleType = "fitCenter"
41          android:onClick = "myPauseMD"
42          app:layout_constraintBottom_toBottomOf = "@+id/imbstop"
43          app:layout_constraintLeft_toRightOf = "@+id/imbstop"
44          app:layout_constraintRight_toRightOf = "@+id/listview"
45          app:layout_constraintHorizontal_bias = "0.33">
46      </ImageButton>
47      <ImageButton
48          android:id = "@+id/imbstop"
49          android:layout_width = "40dp"
50          android:layout_height = "40dp"
51          android:src = "@drawable/stop1"
52          android:scaleType = "fitCenter"
53          android:onClick = "myStopMD"
```

```
54              app:layout_constraintBottom_toBottomOf = "@ + id/imbplay"
55              android:layout_marginStart = "8dp"
56              app:layout_constraintLeft_toRightOf = "@ + id/imbplay"
57              android:layout_marginEnd = "16dp"
58              app:layout_constraintRight_toRightOf = "@ + id/activity_my_mp3_player"
59              app:layout_constraintHorizontal_bias = "0.13">
60          </ImageButton>
61          <TextView
62              android:id = "@ + id/textview"
63              android:layout_width = "176dp"
64              android:layout_height = "36dp"
65              android:text = ""
66              android:textColor = "@drawable/black"
67              app:layout_constraintLeft_toLeftOf = "@ + id/listview"
68              app:layout_constraintRight_toRightOf = "@ + id/listview"
69              app:layout_constraintHorizontal_bias = "0.0"
70              app:layout_constraintBottom_toTopOf = "@ + id/listview"
71              app:layout_constraintTop_toTopOf = "@ + id/activity_my_mp3_player"
72              app:layout_constraintVertical_bias = "0.88">
73          </TextView>
74          <ListView
75              android:id = "@ + id/listview"
76              android:layout_width = "265dp"
77              android:layout_height = "187dp"
78              android:layout_marginStart = "16dp"
79              app:layout_constraintLeft_toLeftOf = "@ + id/activity_my_mp3_player"
80              android:layout_marginEnd = "16dp"
81              app:layout_constraintRight_toRightOf = "@ + id/activity_my_mp3_player"
82              app:layout_constraintBottom_toBottomOf = "@ + id/activity_my_mp3_player"
83              app:layout_constraintTop_toTopOf = "@ + id/activity_my_mp3_player"
84              app:layout_constraintVertical_bias = "0.65999997">
85          </ListView>
86      </android.support.constraint.ConstraintLayout>
```

由上述代码可知,第9～22行为SeekBar控件;第23～60行为三个ImageButton按钮,依次表示播放的开始、暂停和停止,其单击事件响应方法分别为myPlayMD、myPauseMD和myStopMD。第61～73行的静态文本框用于显示正在播放的文件名;第74～85行为显示播放文件列表的列表框控件。

例8-1的执行结果如图8-3所示。应用程序启动后的界面如图8-3(a)所示,在图8-3(a)中选择cgq_tyrx.mp3("谭维维——一生所爱"),此时,图8-3(b)中显示被单击的文件,然后,单击"播放"按钮,将开始播放音频文件,如图8-3(c)所示。播放过程中拖动滑块可以从滑块所在位置开始播放。此外,播放器还具有暂停和停止播放的功能。

源文件MyMP3PlayerAct.java的代码如下:

```
1   package cn.edu.jxfue.zhangyong.mymp3playerapp;
2   import android.support.v4.content.res.ResourcesCompat;
3   import android.support.v7.app.AppCompatActivity;
4   import java.io.File;
5   import java.io.FilenameFilter;
```

```
6   import java.io.IOException;
7   import java.util.ArrayList;
8   import java.util.HashMap;
9   import java.util.Timer;
10  import java.util.TimerTask;
11  import android.media.MediaPlayer;
12  import android.media.MediaPlayer.OnCompletionListener;
13  import android.os.Bundle;
14  import android.view.View;
15  import android.widget.AdapterView;
16  import android.widget.AdapterView.OnItemClickListener;
17  import android.widget.ImageButton;
18  import android.widget.ListView;
19  import android.widget.SeekBar;
20  import android.widget.SimpleAdapter;
21  import android.widget.TextView;
22
23  public class MyMP3PlayerAct extends AppCompatActivity
24          implements SeekBar.OnSeekBarChangeListener{
25      private SeekBar mySeekbar;
26      private TextView mytv;
27      private ImageButton imbplay, imbpause, imbstop;
28      private ListView lvMP3List;
29      private boolean blstop = false;
30      private int blpause = 0;
31      private MediaPlayer myMP3Player;
32      private Timer timer;
33      private TimerTask timerTask;
34      private boolean stopTrack = false;
```

第25行定义了进度条对象mySeekbar；第26行定义了静态文本框对象mytv；第27行定义了三个图像按钮控件对象imbplay、imbpause和imbstop；第28行定义了列表框控件对象lvMP3List；第31行定义了MediaPlayer对象myMP3Player；第32和第33行定义了定时器对象timer和定时器任务timerTask。

```
35      @Override
36      protected void onCreate(Bundle savedInstanceState) {
37          super.onCreate(savedInstanceState);
38          setContentView(R.layout.activity_my_mp3_player);
39
40          myInitGUI();
41      }
42
43      private void myInitGUI(){
44          mySeekbar = (SeekBar)findViewById(R.id.seekbar);
45          mySeekbar.setMax(100);
46          mySeekbar.setProgress(0);
47          mySeekbar.setOnSeekBarChangeListener(this);
48          mytv = (TextView)findViewById(R.id.textview);
49          imbplay = (ImageButton)findViewById(R.id.imbplay);
```

```
50    imbstop = (ImageButton)findViewById(R.id.imbstop);
51    imbpause = (ImageButton)findViewById(R.id.imbpause);
52    imbplay.setImageDrawable(ResourcesCompat.getDrawable(getResources(),R.drawable.
      play1,null));
53    imbstop.setImageDrawable(ResourcesCompat.getDrawable(getResources(),R.drawable.
      stop2,null));
54    imbpause.setImageDrawable(ResourcesCompat.getDrawable(getResources(),R.drawable.
      pause3,null));
55    lvMP3List = (ListView)findViewById(R.id.listview);
56    lvMP3List.setOnItemClickListener(new OnItemClickListener(){
57        @Override
58        public void onItemClick(AdapterView<?> arg0, View arg1, int arg2,
59                                long arg3) {
60            TextView tv_name = (TextView)arg1.findViewById(R.id.mp3name);
61            mytv.setText(tv_name.getText().toString());
62        }
63    });
```

第 56～63 行为 lvMP3List 列表框列表项的单击事件，当某项被单击时，该项的文本（即要播放音频文件的文件名）显示在静态文本框 tv_name 中，如图 8-3(b)所示。

```
64    MP3PlayList();
65    myMP3Player = new MediaPlayer();
```

第 65 行创建 myMP3Player 对象。

```
66    timer = new Timer();
67    timerTask = new TimerTask(){
68        @Override
69        public void run() {
70            if(!stopTrack){
71                if(myMP3Player.isPlaying())
72                    mySeekbar.setProgress(myMP3Player.getCurrentPosition());
73            }
74        }
75    };
76    timer.schedule(timerTask, 100,100);
77 }
```

第 66 行创建定时器对象 timer；第 67～75 行创建定时器任务 timerTask，每次定时事件中，判断 myMP3Player 是否正在播放，如果是，则调整进度条的位置与播放位置同步（第 71 和第 72 行）。第 76 行设定定时间隔为 100ms。

```
78    @Override
79    public void onProgressChanged(SeekBar seekBar,int progress,boolean fromTouch){
80    }
81    @Override
82    public void onStartTrackingTouch(SeekBar seekBar) {
83        stopTrack = true;
84    }
85    @Override
86    public void onStopTrackingTouch(SeekBar seekBar) {
87        stopTrack = false;
```

```
 88         myMP3Player.seekTo(seekBar.getProgress());
 89     }
```

第79～89行的三个覆盖方法依次表示进度条滑块改变时、改变过程中和停止拖动后的事件方法，为了避免拖动过程中与定时器设定滑块位置冲突，通过boolean量stopTrack进行协调，如第70和第83行所示。当滑块停止拖动时，调用seekTo方法将播放文件的位置与滑块的位置同步。

```
 90 public void myPlayMD(View v){
 91     if(playMP3()){
 92         blstop = true;
 93         blpause = 1;
 94         imbstop.setImageDrawable(ResourcesCompat
 95                 .getDrawable(getResources(),R.drawable.stop1,null));
 96         imbpause.setImageDrawable(ResourcesCompat
 97                 .getDrawable(getResources(),R.drawable.pause1,null));
 98     }
 99 }
```

第90～99行为"播放"按钮的单击事件方法。第91行调用playMP3方法开始播放音频文件，如果调用成功，则返回真，此时，设置变量blstop为真（第92行），表示"停止"按钮可用；设置blpause为1（第93行），表示"暂停"按钮进入状态1。"暂停"按钮有三个状态，即不可用、单击暂停和单击继续播放，分别对应于blpause的值为0、1和2。第94～97行设置"停止"和"暂停"按钮的图标。

```
100 public void myPauseMD(View v){
101     switch(blpause){
102         case 1:
103             imbpause.setImageDrawable(ResourcesCompat
104                     .getDrawable(getResources(),R.drawable.pause2,null));
105             myMP3Player.pause();
106             blpause = 2;
107             break;
108         case 2:
109             imbpause.setImageDrawable(ResourcesCompat
110                     .getDrawable(getResources(),R.drawable.pause1,null));
111             myMP3Player.start();
112             blpause = 1;
113             break;
114     }
115 }
```

第100～115行为"暂停"按钮的单击事件方法。当blpause为1时，处于单击暂停状态，此时，单击"暂停"按钮，则调用pause方法（第105行）暂停播放，并设置"暂停"按钮的状态为"单击继续播放"状态，即blpause为2（第106行）。当blpause为2时，处于单击继续播放状态，则调用start方法（第111行）继续播放，并设置blpause为1。

```
116 public void myStopMD(View v){
117     if(blstop){
118         blstop = false;
119         blpause = 0;
```

```
120        imbstop.setImageDrawable(ResourcesCompat
121                .getDrawable(getResources(),R.drawable.stop2,null));
122        imbpause.setImageDrawable(ResourcesCompat
123                .getDrawable(getResources(),R.drawable.pause3,null));
124        myMP3Player.reset();
125        mySeekbar.setProgress(0);
126    }
127 }
```

第116～127行为"停止"按钮的单击事件方法。如果blstop为真,则调用reset方法(第124行)复位播放器,同时,设置blstop为假,blpause为0。

```
128 private boolean playMP3(){
129     myMP3Player.reset();
130     try {
131         File myMp3 = new File(getExternalFilesDir(null).toString() + "/" + mytv.getText().toString());
132         if(myMp3.exists()){
133             myMP3Player.setDataSource(myMp3.getAbsolutePath());
134             myMP3Player.prepare();
135             mySeekbar.setMax(myMP3Player.getDuration());
136             myMP3Player.start();
137             myMP3Player.setOnCompletionListener(new OnCompletionListener(){
138                 @Override
139                 public void onCompletion(MediaPlayer mp) {
140                     // TODO Auto-generated method stub
141                     myMP3Player.reset();
142                     blstop = false;
143                     blpause = 0;
144                     mySeekbar.setProgress(0);
145                     imbstop.setImageDrawable(getResources()
146                             .getDrawable(R.drawable.stop2));
147                     imbpause.setImageDrawable(getResources()
148                             .getDrawable(R.drawable.pause3));
149                 }
150             });
151             return true;
152         }
```

第128行的方法playMP3为播放音频文件的方法。第129行调用reset方法复位播放器;第131行得到播放文件路径;第132行判断文件是否存在;第133行将该文件设为要播放的音频文件;第134行调用prepare方法准备播放;第136行调用start方法开始播放。第137～150行说明播放过程中监听OnCompletionListener接口,当播放完成后,自动调用onCompletion方法,复位myMP3Player对象(第141行)。

```
153     } catch (IllegalArgumentException e) {
154         e.printStackTrace();
155     } catch (IllegalStateException e) {
156         e.printStackTrace();
157     } catch (IOException e) {
158         e.printStackTrace();
```

```
159          }
160          return false;
161    }
162    class MP3Filter implements FilenameFilter{
163          @Override
164          public boolean accept(File dir, String filename) {
165              return filename.endsWith(".mp3");
166          }
167    }
```

第 172 行的 listFiles 方法的参数为 FilenameFilter 接口类型。上述第 162~167 行定义了实现该接口的类 MP3Filter，实现了该接口的抽象方法 accept，用于过滤扩展名为.mp3 的文件。

```
168    private void MP3PlayList(){
169          ArrayList<HashMap<String,String>> mp3NameList =
170                  new ArrayList<HashMap<String,String>>();
171          File file = new File(getExternalFilesDir(null).toString());
172          if(file.listFiles(new MP3Filter()).length>0){
173              for(File mp3file:file.listFiles(new MP3Filter())){
174                  HashMap<String,String> hashMap = new HashMap<String,String>();
175                  hashMap.put("mp3_name", mp3file.getName());
176                  mp3NameList.add(hashMap);
177              }
178          }
179          SimpleAdapter listAdapter = new SimpleAdapter(this, mp3NameList,
180                  R.layout.mp3list,
181                  new String[]{"mp3_name"},
182                  new int[]{R.id.mp3name});
183          lvMP3List.setAdapter(listAdapter);
184    }
185    }
```

第 172 行得到 SD 卡上目录"sdcard/Android/data/ cn. edu. jxfue. zhangyong. mymp3playerapp/files"内扩展名为.mp3 的所有文件；第 173~177 行的 for 循环将文件名通过 HashMap 表赋给 ArrayList 对象 mp3NameList；第 179~182 行设置适配器 listAdapter；第 183 行在列表框 lvMP3List 中显示 SD 卡上的音频文件名。

通过例 8-1 可知，借助 MediaPlayer 类播放音频文件，主要的工作在于调用 MediaPlayer 类的对象方法对音频文件进行控制，除此之外，就是界面设计工作。

8.2 服务

Android 系统支持没有用户界面的应用程序，称之为服务（Service）。嵌入式操作系统的系统程序大都属于服务，这里的服务属于应用程序的范畴，属于应用服务。最容易理解的应用服务是播放背景音乐，例如在阅读小说的过程中，循环播放着轻音乐提高阅读兴趣。

应用服务的特点是没有用户界面，如果服务是由某个界面启动的，该界面可以关闭服务，如果该界面没有关闭服务而退出时（指调用 finish 方法关闭了），服务仍然处于运行状态。下面通过实例阐述服务的程序设计方法。

例 8-2 背景音乐播放实例。

例 8-2 的运行结果如图 8-4 所示,当单击"播放"按钮时,启动服务,服务将播放一段轻音乐"ghsy.mp3",如果单击"暂停"按钮,则停止服务,服务将停止播放音乐。当单击"退出"按钮(图 8-4(b)中最右侧按钮)时,将关闭当前应用程序界面,此时,音乐继续播放,因为服务还在工作。

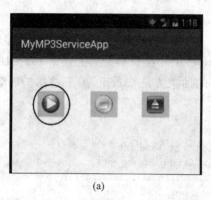

(a)   (b)

图 8-4 例 8-2 执行结果

新建应用 MyMP3ServiceApp,应用名为 MyMP3ServiceApp,活动界面名为 MyMP3ServiceAct。应用 MyMP3ServiceApp 的结构如图 8-5 所示。

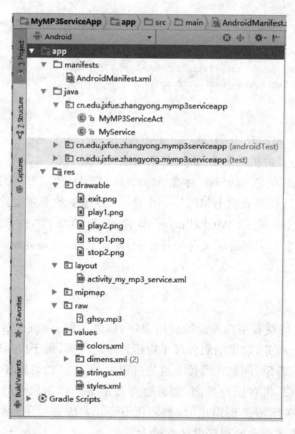

图 8-5 应用 MyMP3ServiceApp 文件目录结构

由图 8-5 可知，应用 MyMP3ServiceApp 包括源文件 MyMP3ServiceAct.java、MyService.java、布局文件 activity_my_mp3_service.xml、音频文件 ghsy.mp3（用作服务播放的背景音乐，保存在 raw 目录下）、配置文件 AndroidManifest.xml 以及图像资源文件 exit.png、play1.png、play2.png、stop1.png 和 stop2.png 等。其中，图像文件用于播放控制按钮的图标显示，如图 8-4 所示。

由于应用 MyMP3ServiceApp 中有服务，其配置文件 AndroidManifest.xml 的内容如下：

```
1  <?xml version = "1.0" encoding = "utf-8"?>
2  <manifest xmlns:android = "http://schemas.android.com/apk/res/android"
3      package = "cn.edu.jxfue.zhangyong.mymp3serviceapp">
4      <application
5          android:allowBackup = "true"
6          android:icon = "@mipmap/ic_launcher"
7          android:label = "@string/app_name"
8          android:supportsRtl = "true"
9          android:theme = "@style/AppTheme">
10         <activity android:name = ".MyMP3ServiceAct">
11             <intent-filter>
12                 <action android:name = "android.intent.action.MAIN" />
13                 <category android:name = "android.intent.category.LAUNCHER" />
14             </intent-filter>
15         </activity>
16         <service android:name = "MyService">
17             <intent-filter>
18                 <action android:name = "myMP3ServIt"/>
19                 <category android:name = "android.intent.category.DEFAULT"/>
20             </intent-filter>
21         </service>
22     </application>
23 </manifest>
```

上述代码中，第 16～21 行为服务配置，第 18 行声明了服务的 Intent 动作名为 "myMP3ServIt"。

应用 MyMP3ServiceApp 的布局如图 8-4 所示，包括三个图像按钮控件，其布局文件 activity_my_mp3_service.xml 代码如下：

```
1  <?xml version = "1.0" encoding = "utf-8"?>
2  <android.support.constraint.ConstraintLayout xmlns:android = "http://schemas.android.com/apk/res/android"
3      xmlns:app = "http://schemas.android.com/apk/res-auto"
4      xmlns:tools = "http://schemas.android.com/tools"
5      android:id = "@+id/activity_my_mp3_service"
6      android:layout_width = "match_parent"
7      android:layout_height = "match_parent"
8      tools:context = "cn.edu.jxfue.zhangyong.mymp3serviceapp.MyMP3ServiceAct">
9      <ImageButton
10         android:id = "@+id/imbplay"
11         android:layout_width = "60dp"
```

```
12            android:layout_height = "60dp"
13            android:src = "@drawable/play1"
14            android:scaleType = "fitCenter"
15            android:onClick = "myPlayMD"
16            app:layout_constraintTop_toTopOf = "@ + id/activity_my_mp3_service"
17            app:layout_constraintBottom_toBottomOf = "@ + id/activity_my_mp3_service"
18            android:layout_marginStart = "16dp"
19            app:layout_constraintLeft_toLeftOf = "@ + id/activity_my_mp3_service"
20            android:layout_marginEnd = "16dp"
21            app:layout_constraintRight_toRightOf = "@ + id/activity_my_mp3_service"
22            app:layout_constraintHorizontal_bias = "0.11"
23            app:layout_constraintVertical_bias = "0.120000005">
24        </ImageButton>
25        < ImageButton
26            android:id = "@ + id/imbstop"
27            android:layout_width = "60dp"
28            android:layout_height = "60dp"
29            android:src = "@drawable/stop1"
30            android:scaleType = "fitCenter"
31            android:onClick = "myStopMD"
32            app:layout_constraintBottom_toBottomOf = "@ + id/imbplay"
33            android:layout_marginStart = "8dp"
34            app:layout_constraintLeft_toRightOf = "@ + id/imbplay"
35            android:layout_marginEnd = "16dp"
36            app:layout_constraintRight_toRightOf = "@ + id/activity_my_mp3_service"
37            app:layout_constraintHorizontal_bias = "0.2">
38        </ImageButton>
39        < ImageButton
40            android:id = "@ + id/imbexit"
41            android:layout_width = "60dp"
42            android:layout_height = "60dp"
43            android:src = "@drawable/exit"
44            android:scaleType = "fitCenter"
45            android:onClick = "myExitMD"
46            app:layout_constraintBottom_toBottomOf = "@ + id/imbstop"
47            android:layout_marginStart = "8dp"
48            app:layout_constraintLeft_toRightOf = "@ + id/imbstop"
49            android:layout_marginEnd = "16dp"
50            app:layout_constraintRight_toRightOf = "@ + id/activity_my_mp3_service"
51            app:layout_constraintHorizontal_bias = "0.44">
52        </ImageButton>
53 </android.support.constraint.ConstraintLayout>
```

上述代码的第15、31和第45行声明了三个图像按钮控件的单击事件方法分别为myPlayMD、myStopMD和myExitMD。

源文件MyMP3ServiceAct.java的内容如下：

```
1  package cn.edu.jxfue.zhangyong.mymp3serviceapp;
2
3  import android.support.v7.app.AppCompatActivity;
4  import android.content.Intent;
5  import android.os.Bundle;
6  import android.view.View;
7  import android.widget.ImageButton;
```

```
8
9   public class MyMP3ServiceAct extends AppCompatActivity {
10      private ImageButton imbplay, imbstop;
11      private boolean blplay = true;
12      private boolean blstop = false;
13      private Intent it;
14      @Override
15      protected void onCreate(Bundle savedInstanceState) {
16          super.onCreate(savedInstanceState);
17          setContentView(R.layout.activity_my_mp3_service);
18          myInitGUI();
19      }
20      private void myInitGUI(){
21          imbplay = (ImageButton)findViewById(R.id.imbplay);
22          imbstop = (ImageButton)findViewById(R.id.imbstop);
23          imbplay.setImageResource(R.drawable.play1);
24          imbstop.setImageResource(R.drawable.stop2);
25          it = new Intent("myMP3ServIt");
26      }
```

第13行定义Intent对象it,第25行得到it对象。

```
27      public void myPlayMD(View v){
28          if(blplay){
29              startService(it);
30              blplay = false;
31              blstop = true;
32              imbstop.setImageResource(R.drawable.stop1);
33              imbplay.setImageResource(R.drawable.play2);
34          }
35      }
```

当单击"播放"按钮控件时,执行第27~35行的方法myPlayMD,第29行调用方法startService启动服务。

```
36      public void myStopMD(View v){
37          if(blstop){
38              stopService(it);
39              blstop = false;
40              blplay = true;
41              imbstop.setImageResource(R.drawable.stop2);
42              imbplay.setImageResource(R.drawable.play1);
43          }
44      }
```

当单击"停止"按钮控件时,执行第36~44行的方法myStopMD,第38行调用方法stopService停止服务。

```
45      public void myExitMD(View v){
46          this.finish();
47      }
48  }
```

当单击"关闭"按钮(图8-4(b)中圈住的按钮)时,关闭当前应用界面。

源文件 MyService.java 的内容如下：

```
1   package cn.edu.jxfue.zhangyong.mymp3serviceapp;
2
3   import android.app.Service;
4   import android.content.Intent;
5   import android.media.MediaPlayer;
6   import android.os.IBinder;
7
8   public class MyService extends Service {
9       private MediaPlayer mp;
10      @Override
11      public IBinder onBind(Intent arg0) {
12          return null;
13      }
14      @Override
15      public int onStartCommand(Intent intent, int flags, int startId){
16          int retVal = super.onStartCommand(intent,flags,startId);
17          mp = MediaPlayer.create(this, R.raw.ghsy);
18          mp.start();
19          return retVal;
20      }
```

当服务启动时自动调用 onStartCommand 方法，第 17 行从资源中取到音频媒体文件，第 18 行调用 start 方法播放音乐。

```
21      @Override
22      public void onDestroy(){
23          super.onDestroy();
24          mp.stop();
25      }
26  }
```

当服务停止时自动调用 onDestroy 方法，第 24 行调用 stop 方法停止播放音乐。

从例 8-2 可知，服务对外部调用提供的方法为 startService 和 stopService，当外部调用 startService 方法时，服务将自动执行 onStartCommand 方法；当外部调用 stopService 方法时，服务将自动执行 onDestroy 方法。所谓的"自动"是指 Android 系统管理了这些调用工作。

8.3 视频文件播放

视频文件是由音频和图像组合在一起的文件，连续地播放图像形成视频图像流，因此播放视频需要有较大的内存和快速的处理器。视频文件的压缩格式众多，Android 系统支持 MP4 和 3GP 等流媒体格式，由于 Dalvik 虚拟机是解释执行的，由 MediaPlayer 控件实现的视频播放在 Android 模拟器上运行，需要配置较高的计算机才能流畅播放。

由于播放视频和播放 MP3 都使用 MediaPlayer 类，因此，在例 8-1 的基础上添加视频图像播放的 SurfaceView 类即可完成视频播放功能。

例 8-3 视频(包括音频)播放器实例。

新建应用 MyMP4PlayerApp,应用名为 MyMP4PlayerApp,活动界面名为 MyMP4PlayerAct。应用 MyMP4PlayerApp 实现的功能如图 8-6 所示,应程序启动后的界面如图 8-6(a)所示,单击"wolf.MP4"文件后单击"播放"按钮,如图 8-6(b)所示;单击"bear.3GP"文件后单击"播放"按钮如图 8-6(c)所示;单击"gln_tt.mp3"音频文件后单击"播放"按钮如图 8-6(d)所示,即应用 MyMP4PlayerApp 可以播放音频和视频文件。

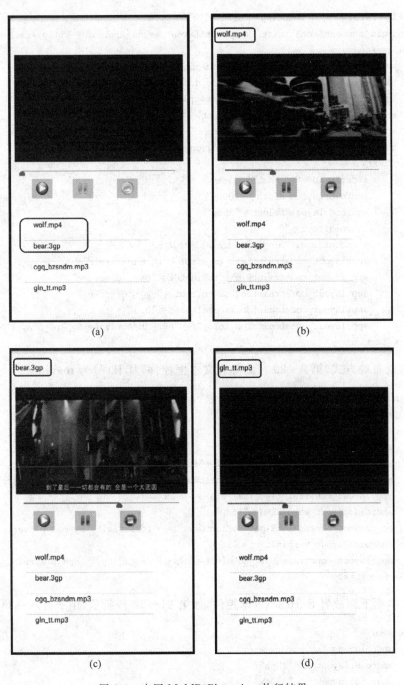

图 8-6 应用 MyMP4PlayerApp 执行结果

应用 MyMP4PlayerApp 包括源文件 MyMP4PlayerAct.java、布局文件 activity_my_mp4_player.xml、列表框布局文件 medialist.xml、颜色资源文件 myguicolor.xml 以及图像资源文件 pause1.png、pause2.png、pause3.png、play1.png、stop1.png 和 stop2.png 等。其中,6 个图像文件与例 8-1 中的同名文件相同,保存在 drawable 目录下;文件 myguicolor.xml 与例 8-1 中的同名文件内容相同,保存在 values 目录下。

例 8-3 的布局如图 8-6 所示,其布局文件 activity_my_mp4_player.xml 的内容如下:

```
1   <?xml version = "1.0" encoding = "utf-8"?>
2   <android.support.constraint.ConstraintLayout xmlns:android = "http://schemas.android.com/apk/res/android"
3       xmlns:app = "http://schemas.android.com/apk/res-auto"
4       xmlns:tools = "http://schemas.android.com/tools"
5       android:id = "@+id/activity_my_mp4_player"
6       android:layout_width = "match_parent"
7       android:layout_height = "match_parent"
8       tools:context = "cn.edu.jxfue.zhangyong.mymp4playerapp.MyMP4PlayerAct">
9       <TextView
10          android:id = "@+id/textview"
11          android:layout_width = "100dp"
12          android:layout_height = "25dp"
13          android:text = ""
14          android:textColor = "@drawable/black"
15          app:layout_constraintLeft_toLeftOf = "@+id/screen"
16          app:layout_constraintRight_toRightOf = "@+id/screen"
17          app:layout_constraintBottom_toTopOf = "@+id/screen"
18          app:layout_constraintHorizontal_bias = "0.0"
19          app:layout_constraintTop_toTopOf = "@+id/activity_my_mp4_player">
20      </TextView>
```

采用约束布局方式,第 9～20 行为静态文本框控件,其 ID 号为 textview。

```
21      <SurfaceView
22          android:id = "@+id/screen"
23          android:layout_width = "343dp"
24          android:layout_height = "209dp"
25          app:layout_constraintTop_toTopOf = "@+id/activity_my_mp4_player"
26          app:layout_constraintBottom_toBottomOf = "@+id/activity_my_mp4_player"
27          app:layout_constraintVertical_bias = "0.19"
28          android:layout_marginEnd = "16dp"
29          app:layout_constraintRight_toRightOf = "@+id/activity_my_mp4_player"
30          android:layout_marginStart = "16dp"
31          app:layout_constraintLeft_toLeftOf = "@+id/activity_my_mp4_player">
32      </SurfaceView>
```

静态文本框下面放置 SurfaceView 控件,如第 21～32 行所示,用于显示视频图像。

```
33      <SeekBar
34          android:id = "@+id/seekbar"
35          android:layout_x = "10px"
36          android:layout_y = "390px"
```

```
37          android:layout_height = "19dp"
38          android:layout_width = "344dp"
39          android:max = "100"
40          android:progress = "0"
41          app:layout_constraintLeft_toLeftOf = "@ + id/screen"
42          app:layout_constraintRight_toRightOf = "@ + id/screen"
43          app:layout_constraintHorizontal_bias = "1.0"
44          app:layout_constraintBottom_toBottomOf = "@ + id/activity_my_mp4_player"
45          app:layout_constraintTop_toBottomOf = "@ + id/screen"
46          app:layout_constraintVertical_bias = "0.03">
47      </SeekBar>
```

SurfaceView 控件下面放置进度条控件，如第 33～47 行所示。

```
48      < ImageButton
49          android:id = "@ + id/imbplay"
50          android:layout_width = "50dp"
51          android:layout_height = "50dp"
52          android:src = "@drawable/play1"
53          android:scaleType = "fitCenter"
54          android:onClick = "myPlayMD"
55          app:layout_constraintLeft_toLeftOf = "@ + id/seekbar"
56          android:layout_marginStart = "32dp"
57          app:layout_constraintTop_toBottomOf = "@ + id/seekbar"
58          app:layout_constraintBottom_toBottomOf = "@ + id/activity_my_mp4_player"
59          app:layout_constraintVertical_bias = "0.0">
60      </ImageButton>
61      < ImageButton
62          android:id = "@ + id/imbpause"
63          android:layout_width = "50dp"
64          android:layout_height = "50dp"
65          android:src = "@drawable/pause1"
66          android:scaleType = "fitCenter"
67          android:onClick = "myPauseMD"
68          app:layout_constraintBottom_toBottomOf = "@ + id/imbplay"
69          android:layout_marginStart = "8dp"
70          app:layout_constraintLeft_toRightOf = "@ + id/imbplay"
71          app:layout_constraintRight_toLeftOf = "@ + id/imbstop"
72          android:layout_marginEnd = "8dp"
73          app:layout_constraintHorizontal_bias = "0.45">
74      </ImageButton>
75      < ImageButton
76          android:id = "@ + id/imbstop"
77          android:layout_width = "50dp"
78          android:layout_height = "50dp"
79          android:src = "@drawable/stop1"
80          android:scaleType = "fitCenter"
81          android:onClick = "myStopMD"
82          app:layout_constraintBottom_toBottomOf = "@ + id/imbpause"
83          app:layout_constraintRight_toRightOf = "@ + id/seekbar"
84          android:layout_marginEnd = "88dp"
85      </ImageButton>
```

在进度条控制下面水平放置三个图像按钮,如第 48~85 行所示,为"播放"、"暂停"和"停止"按钮,其事件方法依次为 myPlayMD、myPauseMD 和 myStopMD。

```
86      <ListView
87          android:id = "@ + id/listview"
88          android:layout_width = "255dp"
89          android:layout_height = "151dp"
90          app:layout_constraintTop_toBottomOf = "@ + id/imbplay"
91          app:layout_constraintBottom_toBottomOf = "@ + id/activity_my_mp4_player"
92          app:layout_constraintLeft_toLeftOf = "@ + id/imbplay"
93          app:layout_constraintVertical_bias = "0.44"
94          app:layout_constraintRight_toRightOf = "@ + id/imbstop"
95          app:layout_constraintHorizontal_bias = "0.38">
96      </ListView>
97  </android.support.constraint.ConstraintLayout>
```

第 86~96 行的 ListView 控件用于显示 SD 卡上的音频和视频文件。

列表框布局文件 medialist.xml 的内容如下:

```
1   <?xml version = "1.0" encoding = "utf - 8"?>
2   <android.support.constraint.ConstraintLayout xmlns:app = "http://schemas.android.com/apk/res - auto"
3       xmlns:tools = "http://schemas.android.com/tools"
4       android:id = "@ + id/widget0"
5       android:layout_width = "fill_parent"
6       android:layout_height = "fill_parent"
7       xmlns:android = "http://schemas.android.com/apk/res/android">
8       <TextView
9           android:id = "@ + id/filename"
10          android:layout_width = "160dp"
11          android:layout_height = "40dp"
12          android:text = "TextView"
13          android:textColor = "@drawable/black"
14          android:layout_marginStart = "16dp"
15          app:layout_constraintLeft_toLeftOf = "@ + id/widget0"
16          android:layout_marginEnd = "16dp"
17          app:layout_constraintRight_toRightOf = "@ + id/widget0"
18          app:layout_constraintTop_toTopOf = "@ + id/widget0"
19          app:layout_constraintBottom_toBottomOf = "@ + id/widget0"
20          app:layout_constraintHorizontal_bias = "0.04"
21          app:layout_constraintVertical_bias = "0.05"></TextView>
22  </android.support.constraint.ConstraintLayout>
```

上述列表框布局中包含一个 TextView 控件,用于显示音频或视频文件的文件名。

源文件 MyMP4PlayerAct.java 的内容如下所示,重点介绍与例 8-1 中文件 MyMP3PlayerAct.java 不同的地方。

```
1   package cn.edu.jxfue.zhangyong.mymp4playerapp;
2
3   import android.support.v7.app.AppCompatActivity;
4   import java.io.File;
```

```
5   import java.io.FilenameFilter;
6   import java.io.IOException;
7   import java.util.ArrayList;
8   import java.util.HashMap;
9   import java.util.Timer;
10  import java.util.TimerTask;
11  import android.media.AudioManager;
12  import android.media.MediaPlayer;
13  import android.media.MediaPlayer.OnCompletionListener;
14  import android.os.Bundle;
15  import android.view.SurfaceHolder;
16  import android.view.SurfaceView;
17  import android.view.View;
18  import android.view.WindowManager;
19  import android.widget.AdapterView;
20  import android.widget.AdapterView.OnItemClickListener;
21  import android.widget.ImageButton;
22  import android.widget.ListView;
23  import android.widget.SeekBar;
24  import android.widget.SimpleAdapter;
25  import android.widget.TextView;
26
27  public class MyMP4PlayerAct extends AppCompatActivity
28          implements SeekBar.OnSeekBarChangeListener{
29      private SeekBar mySeekbar;
30      private TextView mytv;
31      private ImageButton imbplay, imbpause, imbstop;
32      private ListView lvMP4List;
33      private boolean blstop = false;
34      private int blpause = 0;
35      private MediaPlayer myPlayer;
36      private Timer timer;
37      private TimerTask timerTask;
38      private boolean stopTrack = false;
39      private SurfaceView screen;
40      private SurfaceHolder scrholder;
```

第 39 行定义 SurfaceView 类的对象 screen，第 40 行定义 SurfaceHolder 类的对象 scrholder。

```
41  @Override
42  protected void onCreate(Bundle savedInstanceState) {
43      super.onCreate(savedInstanceState);
44      if(getSupportActionBar()!= null){
45          getSupportActionBar().hide();
46      }
47      getWindow().setFlags(WindowManager.LayoutParams.FLAG_FULLSCREEN,
48              WindowManager.LayoutParams.FLAG_FULLSCREEN);
49      setContentView(R.layout.activity_my_mp4_player);
50      myInitGUI();
51  }
```

第44~46行设置窗口不显示标题；第47和第48行设置不显示状态栏，这里的状态栏是指窗口中最顶部的提示信号强度等手机状态的横栏。

```
52    private void myInitGUI(){
53        getWindow().addFlags(WindowManager.LayoutParams.FLAG_KEEP_SCREEN_ON);
54        mySeekbar = (SeekBar)findViewById(R.id.seekbar);
55        mySeekbar.setMax(100);
56        mySeekbar.setProgress(0);
57        mySeekbar.setOnSeekBarChangeListener(this);
58        mytv = (TextView)findViewById(R.id.textview);
59        imbplay = (ImageButton)findViewById(R.id.imbplay);
60        imbstop = (ImageButton)findViewById(R.id.imbstop);
61        imbpause = (ImageButton)findViewById(R.id.imbpause);
62        imbplay.setImageResource(R.drawable.play1);
```

第53行保持屏幕常亮；第62行借助setImageResource方法将资源中的ID号为R.drawable.play1的图形对象设置为该图像控件中显示的图标。第63、64行采用同样的方法给图像按钮添加图标。

```
63        imbstop.setImageResource(R.drawable.stop2);
64        imbpause.setImageResource(R.drawable.pause3);
65        lvMP4List = (ListView)findViewById(R.id.listview);
66        lvMP4List.setOnItemClickListener(new OnItemClickListener(){
67            @Override
68            public void onItemClick(AdapterView<?> arg0, View arg1, int arg2, long arg3) {
69                TextView tv_name = (TextView)arg1.findViewById(R.id.filename);
70                mytv.setText(tv_name.getText().toString());
71            }
72        });
73        MPPlayList();
74        myPlayer = new MediaPlayer();
```

第74行创建myPlayer对象。

```
75        timer = new Timer();
76        timerTask = new TimerTask(){
77            @Override
78            public void run() {
79                if(!stopTrack){
80                    if(myPlayer.isPlaying())
81                        mySeekbar.setProgress(myPlayer.getCurrentPosition());
82                }
83            }
84        };
85        timer.schedule(timerTask, 100,100);
86        screen = (SurfaceView)findViewById(R.id.screen);
87        scrholder = screen.getHolder();
88        scrholder.addCallback(new SurfaceHolder.Callback() {
89            @Override
90            public void surfaceDestroyed(SurfaceHolder holder) {
91                if(myPlayer!= null)
```

```
92              myPlayer.release();
93          }
94          @Override
95          public void surfaceCreated(SurfaceHolder holder) {
96          }
97          @Override
98          public void surfaceChanged(SurfaceHolder holder, int format, int width,
99                                    int height) {
100         }
101     });
102     //scrholder.setType(3);      //SurfaceHolder.SURFACE_TYPE_PUSH_BUFFERS);
103 }
```

第 86 行得到 screen 对象；第 87 行得到 scrholder 对象，通过该对象访问 screen 对象；第 88～101 行添加 scrholder 的回调函数；第 102 行设置 scrholder 的显示类型为 SURFACE_TYPE_PUSH_BUFFERS，该 Android 系统常量值为 3，Android 版本 4.4 以后系统将取消 setType 方法和显示类型常量，所以这里注释掉了。

```
104 @Override
105 public void onProgressChanged(SeekBar seekBar, int progress, boolean fromTouch){
106 }
107 @Override
108 public void onStartTrackingTouch(SeekBar seekBar) {
109     stopTrack = true;
110 }
111
112 @Override
113 public void onStopTrackingTouch(SeekBar seekBar) {
114     stopTrack = false;
115     myPlayer.seekTo(seekBar.getProgress());
116 }
117
118 public void myPlayMD(View v){
119     if(playMP3()){
120         blstop = true;
121         blpause = 1;
122         imbstop.setImageResource(R.drawable.stop1);
123         imbpause.setImageResource(R.drawable.pause1);
124     }
125 }
126
127 public void myPauseMD(View v){
128     switch(blpause){
129         case 1:
130             imbpause.setImageResource(R.drawable.pause2);
131             myPlayer.pause();
132             blpause = 2;
133             break;
134         case 2:
135             imbpause.setImageResource(R.drawable.pause1);
```

```
136                myPlayer.start();
137                blpause = 1;
138                break;
139          }
140    }
141
142    public void myStopMD(View v){
143        if(blstop){
144            blstop = false;
145            blpause = 0;
146            imbstop.setImageResource(R.drawable.stop2);
147            imbpause.setImageResource(R.drawable.pause3);
148            myPlayer.reset();
149            mySeekbar.setProgress(0);
150        }
151    }
152
153    private boolean playMP3(){
154        myPlayer.reset();
155        try {
156            File mfile = new File(getExternalFilesDir(null).toString() + "/" + mytv.getText().toString());
157            if(mfile.exists()){
158                myPlayer.setDataSource(mfile.getAbsolutePath());
159                if(mfile.getName().toLowerCase().endsWith(".mp4")
160                    || mfile.getName().toLowerCase().endsWith(".3gp")){
161                    myPlayer.setAudioStreamType(AudioManager.STREAM_MUSIC);
162                    myPlayer.setDisplay(scrholder);
163
164                }
165                myPlayer.prepare();
166                mySeekbar.setMax(myPlayer.getDuration());
167                myPlayer.start();
168                myPlayer.setOnCompletionListener(new OnCompletionListener(){
169                    @Override
170                    public void onCompletion(MediaPlayer mp) {
171                        myPlayer.reset();
172                        blstop = false;
173                        blpause = 0;
174                        mySeekbar.setProgress(0);
175                        imbstop.setImageResource(R.drawable.stop2);
176                        imbpause.setImageResource(R.drawable.pause3);
177                    }
178                });
179                return true;
180            }
```

第154行调用reset方法复位myPlayer；第156行取得文件名；第157行判断该文件是否存在,当文件存在时,执行第158～180行的代码；第159、160行判断文件类型是否为视频文件,即扩展名是否为.mp4或.3gp,如果为视频文件,则第161和第162行设置音频流类型和图像显示容器；第 165 行调用 prepare 方法准备播放；第 167 行调用 start 方法播放视频或音频文件。

```
181         } catch (IllegalArgumentException e) {
182             e.printStackTrace();
183         } catch (IllegalStateException e) {
184             e.printStackTrace();
185         } catch (IOException e) {
186             e.printStackTrace();
187         }
188         return false;
189     }
190
191     class MyFilter implements FilenameFilter{
192         @Override
193         public boolean accept(File dir, String filename) {
194             return filename.toLowerCase().endsWith(".mp4")
195                 || filename.toLowerCase().endsWith(".mp3")
196                 || filename.toLowerCase().endsWith(".3gp");
197         }
198     }
199
```

第 191～198 行的 MyFilter 类添加了对视频文件的过滤,结合第 204～210 行将把 SD 卡上的音频和视频文件名存入数组列表对象 mpNameList 中。

```
200     private void MPPlayList(){
201         ArrayList<HashMap<String, String>> mpNameList =
202             new ArrayList<HashMap<String,String>>();
203         File file = new File(getExternalFilesDir(null).toString());
204         if(file.listFiles(new MyFilter()).length>0){
205             for(File mfile:file.listFiles(new MyFilter())){
206                 HashMap<String, String> hashMap = new HashMap<String, String>();
207                 hashMap.put("mediaName", mfile.getName());
208                 mpNameList.add(hashMap);
209             }
210         }
211         SimpleAdapter listAdapter = new SimpleAdapter(this, mpNameList,
212             R.layout.medialist,
213             new String[]{"mediaName"},
214             new int[]{R.id.filename});
215         lvMP4List.setAdapter(listAdapter);
216     }
217 }
```

第 200～217 行的 MPPlayList 方法在第 73 行调用,即应用程序启动时将调用 MPPlayList 方法搜索 SD 卡上所有的音频和视频文件,将搜到的文件名显示在于列表框 lvMP4List 中,如果音视频文件较多,列表框自动添加垂直滚动条。

这里使用的计算机的配置为 Windows10、Intel Core I7-4720HQ CPU 和 8GB 内存,例 8-3 播放音频文件和视频文件的效果很好,如图 8-6(b)和图 8-6(c)所示,同时,例 8-3 在三星 N7100 智能手机（配置为 2GB 内存、1.6GHz 四核 Exynos4412 微处理器）上测试通过，可以非常流畅地播放视频文件。

8.4 本章小结

Android 系统集成了可以播放音频和视频文件的 MediaPlayer 类，该类集成了常用音频和视频的解码器，因此，在播放多媒体文件时，只需要调用 MediaPlayer 类的对象方法即可。MediaPlayer 类支持的音频文件类型有.wav、.ogg、.mid、.ota、.mp3、.3gp 等，支持的视频文件类型有.mp4 和.3gp 等，支持的音视频解码方式有 ACC LC/LTP、HE-AACv1、HE-AACv2、MP3、MIDI、AMR-NB、AMR-WB、Ogg Vorbis、H.263、H.264 AVC 和 MPEG-4 SP 等，其中只有 AAC LC/LTP、AMR-NB、AMR-WB、H.263 和 H.264 AVC 提供了编码功能。Android 系统推荐的高质量视频标准为：采用 H.264 编码方式、分辨率为 1280×720、帧率为 30fps、比特率为 500Kbps、语音编码为 AAC-LC，具有双声道的立体声、语音速率为 128Kbps。除了多媒体播放技术外，Android 系统还支持音视频录制，可借助于类 MediaRecorder 的对象方法实现。

第 9 章

通信应用技术

Andriod 系统主要应用于智能手机或移动设备上,广泛支持各种通信手段,例如 WiFi、USB、NFC(近距离无线通信)、蓝牙、红外线和 2G 至 4G 无线移动通信技术等,Android 系统为每类通信技术提供了协议栈或可直接调用的软件包或驱动程序,使得基于 Android 系统的通信应用设计变得相对简单。本章仅介绍基于 Android 系统进行短信息通信的应用设计,并介绍短信息加密与解密技术。

9.1 短信息发送

短信息发送由 android.telephony.SmsManager 类管理,该类的父类为 java.lang.object。短信息发送的方法为:① 定义 SmsManager 类的对象,并调用其静态方法 getDefault 初始化该对象;② 调用 sendTextMessage 方法发送短信。sendTextMessage 方法的原型如下:

```
1  void  sendTextMessage(String destinationAddress,
2                       String scAddress,
3                       String text,
4                       PendingIntent sentIntent,
5                       PendingIntent deliveryIntent)
```

其中,destinationAddress 表示要发送短信息的目的地手机号;scAddress 一般设为 null,表示使用本机的手机号;text 为要发送的短信息内容;sentIntent 以广播的形式返回短信息发送成功或失败的标志码,发送成功则返回 Activity.RESULT_OK,发送失败则返回 RESULT_ERROR_RADIO_OFF 等;deliveryIntent 以广播的形式返回对方接收成功后的通知。如果不关心发送成功或对方是否成功接收,则参数 sentIntent 和 deliveryIntent 均可以设为 null。

例 9-1 短信息发送实例。

新建应用 MySMSender,应用名为 MySMSender,活动界面名为 MainActivity。应用 MySMSender 包括源程序文件 MainActivity.java、字符串资源文件 strings.xml、布局文件 activity_main.xml 和 AndroidManifest.xml 文件等。其中,字符串资源文件 strings.xml 定义了应用中用到的字符串,其代码如下:

```xml
1  <?xml version = "1.0" encoding = "utf-8"?>
2  <resources>
3      <string name = "app_name">MySMSender</string>
4      <string name = "app_name_hz">短信息</string>
5      <string name = "phone_number">手机号:</string>
6      <string name = "sms_content">信息内容:</string>
7      <string name = "secret_key">密钥:</string>
8      <string name = "decipher_sms">解密信息</string>
9      <string name = "send_sms">发送</string>
10     <string name = "send_sec_sms">加密发送</string>
11     <string name = "sms_rec_cont">接收到的信息</string>
12     <string name = "clr_sms">删除短信</string>
13 </resources>
```

AndroidManifest.xml 文件的内容如下:

```xml
1  <?xml version = "1.0" encoding = "utf-8"?>
2  <manifest xmlns:android = "http://schemas.android.com/apk/res/android"
3      package = "cn.edu.jxufe.zhangyong.mysmsender">
4      <application
5          android:allowBackup = "true"
6          android:icon = "@mipmap/ic_launcher"
7          android:label = "@string/app_name"
8          android:supportsRtl = "true"
9          android:theme = "@style/AppTheme">
10         <activity android:name = ".MainActivity">
11             <intent-filter>
12                 <action android:name = "android.intent.action.MAIN" />
13                 <category android:name = "android.intent.category.LAUNCHER" />
14             </intent-filter>
15         </activity>
16     </application>
17     <uses-permission android:name = "android.permission.SEND_SMS" />
18 </manifest>
```

上述代码中,第 17 行添加了权限说明,即允许应用发送短信息。

应用 MySMSender 的布局如图 9-2 所示,布局文件 activity_main.xml 的内容如下:

```xml
1  <?xml version = "1.0" encoding = "utf-8"?>
2  <android.support.constraint.ConstraintLayout xmlns:android = "http://schemas.android.com/apk/res/android"
3      xmlns:app = "http://schemas.android.com/apk/res-auto"
4      xmlns:tools = "http://schemas.android.com/tools"
5      android:id = "@+id/activity_main"
```

```xml
6       android:layout_width = "match_parent"
7       android:layout_height = "match_parent"
8       tools:context = "cn.edu.jxufe.zhangyong.mysmsender.MainActivity">
9       <TextView
10          android:layout_width = "85dp"
11          android:layout_height = "40dp"
12          android:text = "@string/phone_number"
13          app:layout_constraintBottom_toBottomOf = "@ + id/activity_main"
14          app:layout_constraintLeft_toLeftOf = "@ + id/activity_main"
15          app:layout_constraintRight_toRightOf = "@ + id/activity_main"
16          app:layout_constraintTop_toTopOf = "@ + id/activity_main"
17          app:layout_constraintHorizontal_bias = "0.03"
18          app:layout_constraintVertical_bias = "0.04000002"
19          android:id = "@ + id/textView2" />
```

第 9～19 行为静态文本框,显示"手机号:"。

```xml
20      <TextView
21          android:text = "@string/sms_content"
22          android:layout_width = "85dp"
23          android:layout_height = "40dp"
24          android:id = "@ + id/textView"
25          app:layout_constraintLeft_toLeftOf = "@ + id/textView2"
26          app:layout_constraintTop_toBottomOf = "@ + id/textView2"
27          app:layout_constraintBottom_toBottomOf = "@ + id/activity_main"
28          app:layout_constraintVertical_bias = "0.03" />
```

第 20～28 行为静态文本框,显示"信息内容:"。

```xml
29      <EditText
30          android:layout_width = "200dp"
31          android:layout_height = "45dp"
32          android:inputType = "textPersonName"
33          android:ems = "10"
34          android:id = "@ + id/etnumber"
35          app:layout_constraintBaseline_toBaselineOf = "@ + id/textView2"
36          app:layout_constraintLeft_toRightOf = "@ + id/textView2"
37          app:layout_constraintRight_toRightOf = "@ + id/activity_main"
38          app:layout_constraintHorizontal_bias = "0.06" />
```

第 29～38 行为编辑框,在其中可输入手机号码。

```xml
39      <EditText
40          android:layout_width = "200dp"
41          android:layout_height = "45dp"
42          android:inputType = "text"
43          android:ems = "10"
44          android:id = "@ + id/etmessage"
45          app:layout_constraintLeft_toRightOf = "@ + id/textView"
46          app:layout_constraintBaseline_toBaselineOf = "@ + id/textView"
47          app:layout_constraintRight_toRightOf = "@ + id/activity_main"
48          app:layout_constraintHorizontal_bias = "0.1" />
```

第 39～48 行为编辑框，在其中可输入信息内容。

```
49    < Button
50        android:text = "@string/send_sms"
51        android:layout_width = "wrap_content"
52        android:layout_height = "wrap_content"
53        android:id = "@ + id/btsmssend"
54        app:layout_constraintTop_toBottomOf = "@ + id/etmessage"
55        app:layout_constraintRight_toRightOf = "@ + id/etmessage"
56        app:layout_constraintBottom_toBottomOf = "@ + id/activity_main"
57        app:layout_constraintVertical_bias = "0.09"
58        android:onClick = "mySmSendMD" />
59  </android.support.constraint.ConstraintLayout >
```

第 49～58 行为"发送"信息命令按钮，其单击方法为 mySmSendMD（第 58 行）。

应用 MySMSender 需在智能手机上运行，当执行该应用时，弹出图 9-1 所示的界面，在其中，选择"Samsung GT-N7100（Android 4.4.4，API 19）"。可以使用任何 Android 系统版本在 4.4.4 以上的任何智能手机。

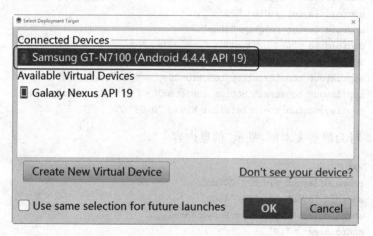

图 9-1　选择智能手机作为目标设备

图 9-2(a)为应用 MySMSender 的布局，在图 9-2(b)中，输入要发送短信息的目的地手机号和发送的信息内容"今晚上映印度电影。"，单击"发送"按钮，则短信息将发送给对方手机。

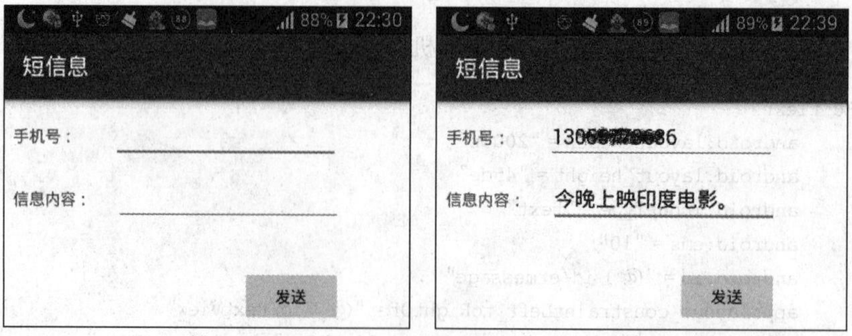

图 9-2　应用 MySMSender

源程序文件 MainActivity.java 的代码如下：

```
1   package cn.edu.jxufe.zhangyong.mysmsender;
2
3   import android.app.PendingIntent;
4   import android.content.Intent;
5   import android.support.v7.app.AppCompatActivity;
6   import android.os.Bundle;
7   import android.telephony.SmsManager;
8   import android.view.View;
9   import android.widget.EditText;
10  import android.widget.Toast;
11  import java.util.regex.Matcher;
12  import java.util.regex.Pattern;
13
14  public class MainActivity extends AppCompatActivity {
15      private EditText etPhoneNumber;
16      private EditText etSmContent;
17      @Override
18      protected void onCreate(Bundle savedInstanceState) {
19          super.onCreate(savedInstanceState);
20          this.setTitle(R.string.app_name_hz);
21          setContentView(R.layout.activity_main);
22
23          myInitGUI();
24      }
```

第 20 行设置应用的标题为"短信息"，即字符串 app_name_hz。

```
25  public void myInitGUI(){
26      etPhoneNumber = (EditText)findViewById(R.id.etnumber);
27      etSmContent = (EditText)findViewById(R.id.etmessage);
28  }
29  public void mySmSendMD(View v){
30      String strAddr = etPhoneNumber.getText().toString();
31      String strMsg = etSmContent.getText().toString();
32      SmsManager smsManager = SmsManager.getDefault();
33      if(checkPhoneNumber(strAddr) && (strMsg.length()<=70)){
34          PendingIntent pendingIntent = PendingIntent.getBroadcast(MainActivity.this,
35              0,new Intent(),0);
36          smsManager.sendTextMessage(strAddr,null,strMsg,pendingIntent,null);
37          Toast.makeText(this,"OK!",Toast.LENGTH_LONG).show();
38      }
39  }
```

第 29~39 行为发送短信息的函数 mySmSendMD，当单击图 9-2(b)中的"发送"按钮时该函数将得到执行。第 30 行获得目的地手机号，第 31 行获得发送的信息内容，第 32 行得到 SmsManager 类的对象 smsManager，第 33 行用于判断当手机号合法且短信息长度小于

70时,将执行第34～37行。第34和第35行创建一个pendingIntent接口对象,用于信息发送成功后进行广播(这里没有对广播进一步处理);第36行调用sendTextMessage发送短信,这里的pendingIntent参数可以使用null替换;第37行用Toast方法显示"OK"表示发送动作完成。

```
40    private boolean checkPhoneNumber(String phoneNumber){
41        boolean res = false;
42        Pattern pattern = Pattern.compile("1[0-9]{10}");
43        Matcher matcher = pattern.matcher(phoneNumber);
44        if(matcher.matches())
45            res = true;
46        return res;
47    }
48 }
```

第40～47行的函数checkPhoneNumber用于检查手机号是否合法,第42行表示手机号匹配的模式为:第一个数字为1,然后是10个0～9的数字。这里的{10}表示重复10次,[0-9]表示数字0～9。如果手机号满足上述的条件,则第44行的matcher.matches返回真。

9.2 短信息接收

Android系统管理了短信息的发送和接收,第9.1节介绍了短信息的发送,这里介绍短信息的接收。与短信息的发送不同,短信息的接收是被动事件,因此,Android系统使用服务和广播进行处理。广播类似于Windows CE系统中的消息与事件处理,Android系统提供了接收短信息的服务,始终处于后台工作中,当收到短信息后,该服务将短信息发送(称为"广播")到Android内核应用程序中,该应用程序将短信息保存在内部的短信息数据库中,键名为"pdus"。因此,短信息接收应用程序需要做的工作就是当收到短信息广播后,从短信息数据库中读出短信息。

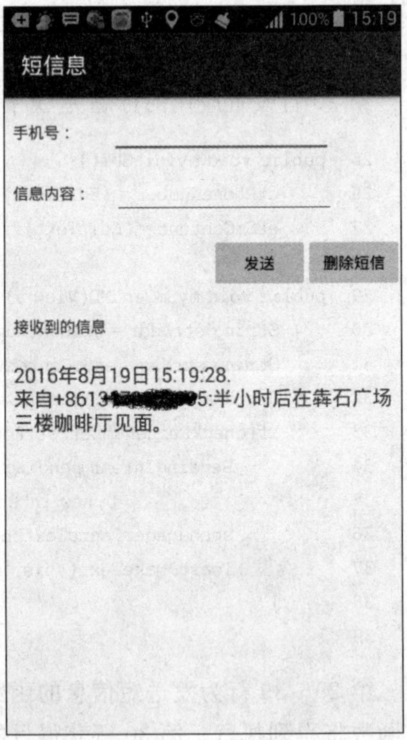

例9-2 短信息接收实例。

在例9-1的基础上,新建应用MySMSendRev,应用名为MySMSendRev,活动界面名为MainActivity。应用MySMSendRev包括源文件MainActivity.java、短信接收源文件MySmsRev.java、字符串资源文件strings.xml、布局文件activity_main.xml和AndroidManifest.xml等。其中,strings.xml与例9-1中的同名文件内容相同。

图9-3中为收到一条短信息后的显示情况。应用MySMSendRev布局如图9-3所示,布局文件activity_main.xml的内容如下所示,仅介绍了相对

图9-3 应用MySMSendRev执行情况

于例 9-1 中的同名文件不同的地方：

```xml
1   <?xml version = "1.0" encoding = "utf-8"?>
2   <android.support.constraint.ConstraintLayout xmlns:android = "http://schemas.android.com/apk/res/android"
3       xmlns:app = "http://schemas.android.com/apk/res-auto"
4       xmlns:tools = "http://schemas.android.com/tools"
5       android:id = "@+id/activity_main"
6       android:layout_width = "match_parent"
7       android:layout_height = "match_parent"
8       tools:context = "cn.edu.jxufe.zhangyong.mysmssendrev.MainActivity">
9       <TextView
10          android:layout_width = "85dp"
11          android:layout_height = "40dp"
12          android:text = "@string/phone_number"
13          app:layout_constraintBottom_toBottomOf = "@+id/activity_main"
14          app:layout_constraintLeft_toLeftOf = "@+id/activity_main"
15          app:layout_constraintRight_toRightOf = "@+id/activity_main"
16          app:layout_constraintTop_toTopOf = "@+id/activity_main"
17          app:layout_constraintHorizontal_bias = "0.03"
18          app:layout_constraintVertical_bias = "0.04000002"
19          android:id = "@+id/textView2" />
20      <TextView
21          android:text = "@string/sms_content"
22          android:layout_width = "85dp"
23          android:layout_height = "40dp"
24          android:id = "@+id/textView"
25          app:layout_constraintLeft_toLeftOf = "@+id/textView2"
26          app:layout_constraintTop_toBottomOf = "@+id/textView2"
27          app:layout_constraintBottom_toBottomOf = "@+id/activity_main"
28          app:layout_constraintVertical_bias = "0.03" />
29      <EditText
30          android:layout_width = "200dp"
31          android:layout_height = "45dp"
32          android:inputType = "textPersonName"
33          android:ems = "10"
34          android:id = "@+id/etnumber"
35          app:layout_constraintBaseline_toBaselineOf = "@+id/textView2"
36          app:layout_constraintLeft_toRightOf = "@+id/textView2"
37          app:layout_constraintRight_toRightOf = "@+id/activity_main"
38          app:layout_constraintHorizontal_bias = "0.06" />
39      <EditText
40          android:layout_width = "200dp"
41          android:layout_height = "45dp"
42          android:inputType = "text"
43          android:ems = "10"
44          android:id = "@+id/etmessage"
45          app:layout_constraintLeft_toRightOf = "@+id/textView"
46          app:layout_constraintBaseline_toBaselineOf = "@+id/textView"
47          app:layout_constraintRight_toRightOf = "@+id/activity_main"
48          app:layout_constraintHorizontal_bias = "0.1" />
```

```
49      <Button
50          android:text = "@string/send_sms"
51          android:layout_width = "wrap_content"
52          android:layout_height = "wrap_content"
53          android:id = "@+id/btsmssend"
54          app:layout_constraintTop_toBottomOf = "@+id/etmessage"
55          app:layout_constraintRight_toRightOf = "@+id/etmessage"
56          app:layout_constraintBottom_toBottomOf = "@+id/activity_main"
57          app:layout_constraintVertical_bias = "0.04000002"
58          android:onClick = "mySmSendMD"
59          android:layout_marginEnd = "24dp" />
60      <TextView
61          android:text = "@string/sms_rec_cont"
62          android:layout_width = "140dp"
63          android:layout_height = "40dp"
64          android:id = "@+id/textView3"
65          app:layout_constraintLeft_toLeftOf = "@+id/textView"
66          app:layout_constraintRight_toLeftOf = "@+id/btsmssend"
67          android:layout_marginEnd = "8dp"
68          app:layout_constraintHorizontal_bias = "0.0"
69          app:layout_constraintTop_toBottomOf = "@+id/textView"
70          app:layout_constraintBottom_toBottomOf = "@+id/activity_main"
71          app:layout_constraintVertical_bias = "0.19" />
```

第60~71行为静态文本框,显示"接收到的信息"提示信息。

```
72      <TextView
73          android:layout_width = "349dp"
74          android:layout_height = "105dp"
75          android:id = "@+id/tvsmsrev"
76          app:layout_constraintTop_toBottomOf = "@+id/textView3"
77          app:layout_constraintBottom_toBottomOf = "@+id/activity_main"
78          app:layout_constraintLeft_toLeftOf = "@+id/textView3"
79          app:layout_constraintVertical_bias = "0.0"
80          android:textSize = "18sp"
81          app:layout_constraintRight_toRightOf = "@+id/activity_main"
82          app:layout_constraintHorizontal_bias = "0.0"
83          android:layout_marginEnd = "16dp" />
```

第72~83行为静态文本框,ID号为tvsmsrev,接收到的短信息将在该控件显示出来。

```
84      <Button
85          android:text = "@string/clr_sms"
86          android:layout_width = "wrap_content"
87          android:layout_height = "wrap_content"
88          android:id = "@+id/btsmsclr"
89          app:layout_constraintBaseline_toBaselineOf = "@+id/btsmssend"
90          app:layout_constraintLeft_toRightOf = "@+id/btsmssend"
91          app:layout_constraintRight_toRightOf = "@+id/activity_main"
92          app:layout_constraintHorizontal_bias = "0.75"
93          android:onClick = "mySmClrMD" />
94  </android.support.constraint.ConstraintLayout>
```

第 84～93 行为"删除信息"命令控件。

应用配置 AndroidManifest.xml 文件的内容如下：

```
1   <?xml version = "1.0" encoding = "utf-8"?>
2   <manifest xmlns:android = "http://schemas.android.com/apk/res/android"
3       package = "cn.edu.jxufe.zhangyong.mysmsendrev">
4       <application
5           android:allowBackup = "true"
6           android:icon = "@mipmap/ic_launcher"
7           android:label = "@string/app_name"
8           android:supportsRtl = "true"
9           android:theme = "@style/AppTheme">
10          <activity android:name = ".MainActivity">
11              <intent-filter>
12                  <action android:name = "android.intent.action.MAIN" />
13                  <category android:name = "android.intent.category.LAUNCHER" />
14              </intent-filter>
15          </activity>
16          <receiver android:name = "MySmsRev">
17          <intent-filter>
18              <action android:name = "android.provider.Telephony.SMS_RECEIVED" />
19          </intent-filter>
20          </receiver>
21      </application>
22      <uses-permission android:name = "android.permission.SEND_SMS" />
23      <uses-permission android:name = "android.permission.RECEIVE_SMS" />
24      <uses-permission android:name = "android.permission.READ_SMS" />
25  </manifest>
```

第 16～20 行为类 MySmsRev 的注册，此时，该应用可接收短信息的广播。第 22～24 行为允许应用发送短信息、接收短信息和读取短信息。

短信接收源文件 MySmsRev.java 的内容如下：

```
1   package cn.edu.jxufe.zhangyong.mysmsendrev;
2
3   import android.content.BroadcastReceiver;
4   import android.content.Context;
5   import android.content.Intent;
6   import android.os.Bundle;
7   import android.telephony.SmsMessage;
8   import android.widget.Toast;
9
10  public class MySmsRev extends BroadcastReceiver {
11      private final String strItEv = "android.provider.Telephony.SMS_RECEIVED";
12      public void onReceive(Context context, Intent intent){
13          String strItAct = intent.getAction();
14          if(strItEv.equals(strItAct)){
```

第 10 行自定义类 MySmsRev 继承自类 BroadcastReceiver，用于实现短信息的接收。第 12 行在 onReceive 方法中实现短信息的接收处理。当发生动作"android.provider.

Telephony.SMS_RECEIVED"时(实际上是一条服务消息),即第 14 行为真时,表明收到了短信息。

```
15    StringBuilder stringBuilder = new StringBuilder();
16    Bundle bundle = intent.getExtras();
17    if(bundle!= null){
18        Object[] objects = (Object[]) bundle.get("pdus");
19        SmsMessage[] messages = new SmsMessage[objects.length];
20        for(int i = 0;i < objects.length;i++){
21            messages[i] = SmsMessage.createFromPdu((byte[]) objects[i]);
22        }
```

短信息在短信息数据库中的键名为"pdus",第 18 行获取短信息,第 19 行创建 messages 对象,第 20~22 行读取短信息(可读取多条短信息)。这里的 createFromPdu 函数在 Andriod API 版本 23 以后,其参数由一个 byte[]变为两个参数,即 byte[]和 String 类型的 format,后者表示 GSM 制式还是 CDMA 制式的短信息。因为这里智能手机 N7100 用的是 Android API 19 版,所以仍然使用了 createFromPdu 单参数版本。

```
23        for(SmsMessage message:messages){
24            stringBuilder.append("来自");
25            stringBuilder.append(message.getDisplayOriginatingAddress());
26            stringBuilder.append(":");
27            stringBuilder.append(message.getDisplayMessageBody());
28        }
```

第 23~28 行将短信息保存在字符串变量 stringBuilder 中。

```
29                //Toast.makeText(context,stringBuilder.toString(),Toast.LENGTH_LONG).show();
30                Intent it = new Intent(context,MainActivity.class);
31                Bundle bd = new Bundle();
32                bd.putString("SMS",stringBuilder.toString());
33                it.putExtras(bd);
34                it.addFlags(Intent.FLAG_ACTIVITY_NEW_TASK);
35                context.startActivity(it);
36            }
37        }
38    }
39 }
```

第 31、32 行将短信息字符串保存在 Bundle 对象 bd 中,第 30、34 和第 35 行打开一个新的 MainActivity 界面(实际上是启动了一个新的线程,在该线程打开 MainActivity)。

源文件 MainActivity.java 的内容如下:

```
1  package cn.edu.jxufe.zhangyong.mysmssendrev;
2
3  import android.app.PendingIntent;
4  import android.content.Intent;
5  import android.support.v7.app.AppCompatActivity;
6  import android.os.Bundle;
7  import android.telephony.SmsManager;
```

```
8    import android.view.View;
9    import android.widget.EditText;
10   import android.widget.TextView;
11   import android.widget.Toast;
12
13   import java.util.Calendar;
14   import java.util.regex.Matcher;
15   import java.util.regex.Pattern;
16
17   public class MainActivity extends AppCompatActivity {
18       private EditText etPhoneNumber;
19       private EditText etSmContent;
20       public TextView tvSmRev;
21       @Override
22       protected void onCreate(Bundle savedInstanceState) {
23           super.onCreate(savedInstanceState);
24           this.setTitle(R.string.app_name_hz);
25           setContentView(R.layout.activity_main);
26
27           myInitGUI();
28
29           Bundle bd = getIntent().getExtras();
30           if(bd!= null){
31               String sms = bd.getString("SMS");
32               Calendar calendar = Calendar.getInstance();
33               int year = calendar.get(Calendar.YEAR);
34               int month = calendar.get(Calendar.MONTH);
35               int day = calendar.get(Calendar.DAY_OF_MONTH);
36               int hour = calendar.get(Calendar.HOUR_OF_DAY);
37               int minute = calendar.get(Calendar.MINUTE);
38               int second = calendar.get(Calendar.SECOND);
39               tvSmRev.setText(Integer.toString(year) + "年" + Integer.toString(month + 1)
40                       + "月" + Integer.toString(day) + "日" + Integer.toString(hour) + ":"
41                       + Integer.toString(minute) + ":" + Integer.toString(second) + ".\n"
42                       + sms);
43           }
44       }
```

第29～31行通过Bundle对象bd获得短信息的内容,保存在字符串变量sms中。第32～38行由Calendar类的对象calendar获得系统时间,第39～42行在tvSmRev静态文本框中显示接收到的信息,如图9-3所示。

```
45       public void myInitGUI(){
46           etPhoneNumber = (EditText)findViewById(R.id.etnumber);
47           etSmContent = (EditText)findViewById(R.id.etmessage);
48           tvSmRev = (TextView)findViewById(R.id.tvsmsrev);
49       }
50       public void mySmSendMD(View v){
51           String strAddr = etPhoneNumber.getText().toString();
52           String strMsg = etSmContent.getText().toString();
53           SmsManager smsManager = SmsManager.getDefault();
```

```
54          if(checkPhoneNumber(strAddr) && (strMsg.length()<=70)){
55              PendingIntent pendingIntent = PendingIntent.getBroadcast(MainActivity.this,
56                      0,new Intent(),0);
57              smsManager.sendTextMessage(strAddr,null,strMsg,pendingIntent,null);
58              Toast.makeText(this,"OK!",Toast.LENGTH_LONG).show();
59          }
60      }
61      private boolean checkPhoneNumber(String phoneNumber){
62          boolean res = false;
63          Pattern pattern = Pattern.compile("1[0-9]{10}");
64          Matcher matcher = pattern.matcher(phoneNumber);
65          if(matcher.matches())
66              res = true;
67          return res;
68      }
69      public void mySmClrMD(View v){
70          tvSmRev.setText("");
71      }
72  }
```

第 50~60 行为发送信息的函数 mySmSendMD，第 61~68 行为检查手机号是否合法的函数 checkPhoneNumber，其内容与例 9-1 中的同名文件中的同名函数内容相同。第 69~71 行为"删除短信"按钮的响应函数，第 70 行执行时清空 tvSmRev 静态文本框。

9.3 短信息加密

短信息是一种方便实时的通信方式，但是短信息的安全性较低。本节通过实例介绍短信息加密处理的方法。

例 9-3 短信息加密实例。

新建应用 MySMSecret，应用名为 MySMSecret，活动界面名为 MainActivity。应用 MySMSecret 包括源文件 MainActivity.java、短信接收源文件 MySmsRev.java、加密与解密源文件 MySecretAlg.java、布局文件 activity_main.xml、字符串资源文件 strings.xml 和应用配置文件 AndroidManifest.xml。其中，文件 strings.xml 和 AndroidManifest.xml 与例 9-2 中的同名文件内容相同。应用 MySMSecret 的布局如图 9-4(a)所示，其布局文件的代码如下所示，重点介绍了与例 9-2 中同名文件内容不同的部分：

```
1   <?xml version="1.0" encoding="utf-8"?>
2   <android.support.constraint.ConstraintLayout xmlns:android="http://schemas.android.
    com/apk/res/android"
3       xmlns:app="http://schemas.android.com/apk/res-auto"
4       xmlns:tools="http://schemas.android.com/tools"
5       android:id="@+id/activity_main"
6       android:layout_width="match_parent"
7       android:layout_height="match_parent"
8       tools:context="cn.edu.jxufe.zhangyong.mysmsecret.MainActivity">
9       <TextView
10          android:layout_width="85dp"
```

```xml
11          android:layout_height = "40dp"
12          android:text = "@string/phone_number"
13          app:layout_constraintBottom_toBottomOf = "@+id/activity_main"
14          app:layout_constraintLeft_toLeftOf = "@+id/activity_main"
15          app:layout_constraintRight_toRightOf = "@+id/activity_main"
16          app:layout_constraintTop_toTopOf = "@+id/activity_main"
17          app:layout_constraintHorizontal_bias = "0.03"
18          app:layout_constraintVertical_bias = "0.04000002"
19          android:id = "@+id/textView2" />
20      <TextView
21          android:text = "@string/sms_content"
22          android:layout_width = "85dp"
23          android:layout_height = "40dp"
24          android:id = "@+id/textView"
25          app:layout_constraintLeft_toLeftOf = "@+id/textView2"
26          app:layout_constraintTop_toBottomOf = "@+id/textView2"
27          app:layout_constraintBottom_toBottomOf = "@+id/activity_main"
28          app:layout_constraintVertical_bias = "0.03" />
29      <EditText
30          android:layout_width = "200dp"
31          android:layout_height = "45dp"
32          android:inputType = "textPersonName"
33          android:ems = "10"
34          android:id = "@+id/etnumber"
35          app:layout_constraintBaseline_toBaselineOf = "@+id/textView2"
36          app:layout_constraintLeft_toRightOf = "@+id/textView2"
37          app:layout_constraintRight_toRightOf = "@+id/activity_main"
38          app:layout_constraintHorizontal_bias = "0.06" />
39      <EditText
40          android:layout_width = "200dp"
41          android:layout_height = "45dp"
42          android:inputType = "text"
43          android:ems = "10"
44          android:id = "@+id/etmessage"
45          app:layout_constraintLeft_toRightOf = "@+id/textView"
46          app:layout_constraintBaseline_toBaselineOf = "@+id/textView"
47          app:layout_constraintRight_toRightOf = "@+id/activity_main"
48          app:layout_constraintHorizontal_bias = "0.1" />
49      <Button
50          android:text = "@string/send_sms"
51          android:layout_width = "wrap_content"
52          android:layout_height = "wrap_content"
53          android:id = "@+id/btsmssend"
54          app:layout_constraintTop_toBottomOf = "@+id/etmessage"
55          app:layout_constraintRight_toRightOf = "@+id/etmessage"
56          app:layout_constraintBottom_toBottomOf = "@+id/activity_main"
57          app:layout_constraintVertical_bias = "0.04000002"
58          android:onClick = "mySmSendMD"
59          android:layout_marginEnd = "24dp" />
60      <TextView
61          android:text = "@string/sms_rec_cont"
```

```
62          android:layout_width = "104dp"
63          android:layout_height = "40dp"
64          android:id = "@ + id/textView3"
65          app:layout_constraintLeft_toLeftOf = "@ + id/textView"
66          app:layout_constraintRight_toLeftOf = "@ + id/btsmssend"
67          android:layout_marginEnd = "8dp"
68          app:layout_constraintHorizontal_bias = "0.0"
69          app:layout_constraintTop_toBottomOf = "@ + id/textView"
70          app:layout_constraintBottom_toBottomOf = "@ + id/activity_main"
71          app:layout_constraintVertical_bias = "0.19" />
72      < TextView
73          android:layout_width = "349dp"
74          android:layout_height = "105dp"
75          android:id = "@ + id/tvsmsrev"
76          app:layout_constraintTop_toBottomOf = "@ + id/textView3"
77          app:layout_constraintBottom_toBottomOf = "@ + id/activity_main"
78          app:layout_constraintLeft_toLeftOf = "@ + id/textView3"
79          app:layout_constraintVertical_bias = "0.0"
80          android:textSize = "18sp"
81          app:layout_constraintRight_toRightOf = "@ + id/activity_main"
82          app:layout_constraintHorizontal_bias = "0.0"
83          android:layout_marginEnd = "16dp" />
84      < Button
85          android:text = "@string/clr_sms"
86          android:layout_width = "wrap_content"
87          android:layout_height = "wrap_content"
88          android:id = "@ + id/btsmsclr"
89          app:layout_constraintBaseline_toBaselineOf = "@ + id/btsmssend"
90          app:layout_constraintLeft_toRightOf = "@ + id/btsmssend"
91          app:layout_constraintRight_toRightOf = "@ + id/activity_main"
92          app:layout_constraintHorizontal_bias = "0.75"
93          android:onClick = "mySmClrMD" />
94      < TextView
95          android:text = "@string/secret_key"
96          android:layout_width = "52dp"
97          android:layout_height = "32dp"
98          android:id = "@ + id/textView4"
99          app:layout_constraintBaseline_toBaselineOf = "@ + id/textView3"
100         app:layout_constraintLeft_toRightOf = "@ + id/textView3"
101         app:layout_constraintRight_toRightOf = "@ + id/btsmssectsend"
102         app:layout_constraintHorizontal_bias = "0.29" />
```

第 94～102 行为静态文本框,显示"密钥"提示信息。

```
103     < EditText
104         android:layout_width = "204dp"
105         android:layout_height = "48dp"
106         android:inputType = "textPassword"
107         android:ems = "10"
108         android:id = "@ + id/etKeys"
109         app:layout_constraintBaseline_toBaselineOf = "@ + id/textView4"
```

```
110           app:layout_constraintRight_toRightOf = "@ + id/activity_main"
111           app:layout_constraintLeft_toRightOf = "@ + id/textView4" />
```

第103~111行为编辑框,用于输入密钥,密钥长度为128位,即32个四位一组的十六进制字符0~9、A~F(或a~f),因此,输入密钥长度为32个字符,每个字符只能取0~9或A~F(或a~f)。

```
112     < Button
113           android:text = "@string/send_sec_sms"
114           android:layout_width = "wrap_content"
115           android:layout_height = "wrap_content"
116           android:id = "@ + id/btsmssectsend"
117           app:layout_constraintBaseline_toBaselineOf = "@ + id/btsmssend"
118           app:layout_constraintLeft_toRightOf = "@ + id/textView"
119           android:onClick = "mySecSendMD" />
```

第112~119行为"加密发送"命令按钮,该命令按钮的方法为mySecSendMD(第119行)。

```
120     < TextView
121           android:layout_width = "349dp"
122           android:layout_height = "105dp"
123           android:id = "@ + id/tvplain"
124           app:layout_constraintTop_toBottomOf = "@ + id/tvsmsrev"
125           app:layout_constraintBottom_toBottomOf = "@ + id/activity_main"
126           app:layout_constraintVertical_bias = "0.75"
127           app:layout_constraintLeft_toLeftOf = "@ + id/btdecipher" />
```

第120~127行为静态文本框tvplain,用于显示解密后的短信息。

```
128     < Button
129           android:text = "@string/decipher_sms"
130           android:layout_width = "wrap_content"
131           android:layout_height = "wrap_content"
132           android:id = "@ + id/btdecipher"
133           app:layout_constraintLeft_toLeftOf = "@ + id/tvsmsrev"
134           app:layout_constraintTop_toBottomOf = "@ + id/tvsmsrev"
135           app:layout_constraintBottom_toTopOf = "@ + id/tvplain"
136           android:onClick = "myDeciphMD" />
137     </android.support.constraint.ConstraintLayout >
```

第128~136行为"解密信息"命令按钮,其单击事件方法为"myDeciphMD"(第136行)。

应用MySMSecret的执行情况如图9-4和图9-5所示。在两部智能手机上安装好应用MySMSecrct,图9-3为发送方智能手机的情况,输入要发送的短信息和密钥,密钥为32个字符,每个字符只能为0~9、a~f或A~F(例如0123456789ABCDEF0123456789ABCDEF,这是默认的密钥,密钥可手动修改),然后,单击"加密发送"按钮,将发送加密后的短信息。图9-5(a)和图9-5(b)为接收方智能手机的情况,接收到的秘密信息如图9-5(a)所示,是完全不可读的无意义的密文信息;在图9-5(b)中,输入与发送方相同的合法密钥,然后,单击"解密信息"按钮,可以得到发送方发送的明文信息,即"半小时后在咖啡厅见。"。图9-5(c)显示了窃听方拦截到的短信息。应用MySMSecret没有使用AES(高级加密标准)或DES

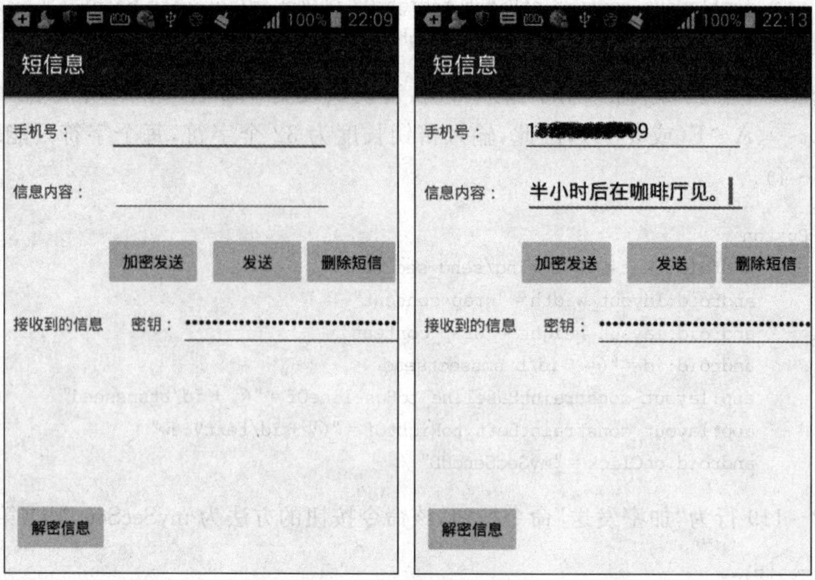

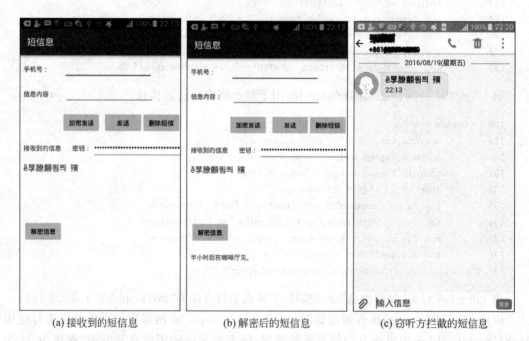

图 9-4 应用 MySMSecret 执行情况——发送方

(a) 接收到的短信　　(b) 解密后的短信　　(c) 窃听方拦截的短信

图 9-5 应用 MySMSecret 执行情况——接收方与窃听方

(数据加密标准)这类加密算法,而是使用一种基于分段线性映射的简单实用的加密算法,如果借助于计算机(配置 CPU 为 Intel i7-4720),使用穷举法进行攻击,100 年内无法解密该信息。限于本书主要介绍 Android 系统的应用设计,这里没有展开讲述该加密算法的性能分析。

短信息接收源文件 MySmsRev.java 相对于例 9-2 中的同名文件,去掉以下这三行:

```
1   //stringBuilder.append("来自");
2   //stringBuilder.append(message.getDisplayOriginatingAddress());
3   //stringBuilder.append(":");
```

即在这三条语句的前面添加"//"将其注释掉,表示不获取短信的发送手机号。

源文件 MainActivity.java 的内容如下:

```
1   package cn.edu.jxufe.zhangyong.mysmsecret;
2
3   import android.app.PendingIntent;
4   import android.content.Intent;
5   import android.support.v7.app.AppCompatActivity;
6   import android.os.Bundle;
7   import android.telephony.SmsManager;
8   import android.view.View;
9   import android.widget.EditText;
10  import android.widget.TextView;
11  import android.widget.Toast;
12  import java.util.regex.Matcher;
13  import java.util.regex.Pattern;
14
15  public class MainActivity extends AppCompatActivity {
16      private EditText etPhoneNumber;
17      private EditText etSmContent;
18      private EditText etKeys;
19      public TextView tvSmRev;
20      public TextView tvPlain;
21
22      @Override
23      protected void onCreate(Bundle savedInstanceState) {
24          super.onCreate(savedInstanceState);
25          this.setTitle(R.string.app_name_hz);
26          setContentView(R.layout.activity_main);
27
28          myInitGUI();
29
30          Bundle bd = getIntent().getExtras();
31          if(bd!= null){
32              String sms = bd.getString("SMS");
33              tvSmRev.setText(sms);
34          }
35      }
```

第30~34行通过Bundle对象从文件MySmsRev.java中得到短信息sms。如果是加密的短信息,则sms为密文信息,该短信息显示在tvSmRev中,如图9-5(a)所示。

```
36  public void myInitGUI(){
37      etPhoneNumber = (EditText)findViewById(R.id.etnumber);
38      etSmContent = (EditText)findViewById(R.id.etmessage);
39      tvSmRev = (TextView)findViewById(R.id.tvsmsrev);
40      tvPlain = (TextView)findViewById(R.id.tvplain);
```

```
41      etKeys = (EditText)findViewById(R.id.etKeys);
42      etKeys.setText("0123456789abcdef0123456789abcdef");
43  }
44  public void mySmSendMD(View v){
45      String strAddr = etPhoneNumber.getText().toString();
46      String strMsg = etSmContent.getText().toString();
47      SmsManager smsManager = SmsManager.getDefault();
48      if(checkPhoneNumber(strAddr) && (strMsg.length()<=70)){
49          PendingIntent pendingIntent = PendingIntent.getBroadcast(MainActivity.this,
50              0,new Intent(),0);
51          smsManager.sendTextMessage(strAddr,null,strMsg,pendingIntent,null);
52          Toast.makeText(this,"OK!",Toast.LENGTH_LONG).show();
53      }
54  }
```

第 44~54 行为不加密信息进行短信息发送的函数 mySmSendMD, 即单击图 9-5(a)中"发送"按钮的事件响应方法。

```
55  private boolean checkPhoneNumber(String phoneNumber){
56      boolean res = false;
57      Pattern pattern = Pattern.compile("1[0-9]{10}");
58      Matcher matcher = pattern.matcher(phoneNumber);
59      if(matcher.matches())
60          res = true;
61      return res;
62  }
63  public void mySmClrMD(View v){
64      tvSmRev.setText("");
65  }
66  public void mySecSendMD(View v){
67      String strkey = etKeys.getText().toString();
68      char[] chkey = strkey.toCharArray();
69      MySecretAlg mySec = new MySecretAlg(chkey);
70      String strMsg = mySec.enCipher(etSmContent.getText().toString());
71      String strAddr = etPhoneNumber.getText().toString();
72      SmsManager smsManager = SmsManager.getDefault();
73      if(checkPhoneNumber(strAddr) && (strMsg.length()<=70)){
74          PendingIntent pendingIntent = PendingIntent.getBroadcast(MainActivity.this,
75              0,new Intent(),0);
76          smsManager.sendTextMessage(strAddr,null,strMsg,pendingIntent,null);
77          Toast.makeText(this,"OK!",Toast.LENGTH_LONG).show();
78      }
79  }
```

第 66~79 行为发送加密后的短信息的函数 mySecSendMD, 即单击图 9-5(a)中"加密发送"按钮的事件响应方法。这里,第 67 行得到密钥;第 68 行将密钥字符串转化为字符数组;第 69 行创建类 MySecretAlg 的对象 mySec;第 70 行调用 enCipher 方法加密编辑框 etSmContent 中输入的明文字符串,得到加密后的短信息字符串 strMsg;第 71 行获得发送短信息的目的地手机号;第 72~78 行发送加密后的短信息 strMsg。

```
80    public void myDeciphMD(View v){
81        String strkey = etKeys.getText().toString();
82        char[] chkey = strkey.toCharArray();
83        MySecretAlg mySec = new MySecretAlg(chkey);
84        String strMsg = mySec.deCipher(tvSmRev.getText().toString());
85        tvPlain.setText(strMsg);
86    }
87  }
```

第 80～86 行为"解密信息"按钮的事件响应方法。第 81 行得到密钥字符串 strkey，第 82 行将密钥字符串转化为字符数组 chkey。第 83 行创建 MySecretAlg 类的对象 mySec，第 84 行调用方法 deCipher 解密静态文本框 tvSmRev 中的字符串（即接收到的密文字符串）。第 85 行将解密后的字符串显示在 tvPlain 静态文本框中。

加密与解密算法源文件 MySecretAlg.java 的内容如下：

```
1  package cn.edu.jxufe.zhangyong.mysmsecret;
2
3  public class MySecretAlg {
4      private int[] keys = new int[32];
5      private double x0,p;
6      private int[] secCode = new int[512];
7      MySecretAlg(){}
8      MySecretAlg(char[] chs){
9          if(chs.length == 32) {
10             for (int i = 0; i < 32; i++) {
11                 switch (chs[i]) {
12                     case '0':
13                     case '1':
14                     case '2':
15                     case '3':
16                     case '4':
17                     case '5':
18                     case '6':
19                     case '7':
20                     case '8':
21                     case '9':
22                         keys[i] = chs[i] - '0';
23                         break;
24                     case 'A':
25                     case 'B':
26                     case 'C':
27                     case 'D':
28                     case 'E':
29                     case 'F':
30                         keys[i] = chs[i] - 'A' + 10;
31                         break;
32                     case 'a':
33                     case 'b':
34                     case 'c':
35                     case 'd':
```

```
36                        case 'e':
37                        case 'f':
38                            keys[i] = chs[i] - 'a' + 10;
39                    }
40                }
41            }
42            else{
43                for(int i = 0;i < 32;i++){
44                    keys[i] = i % 16;
45                }
46            }
47            calcParam();
48            calcSecCode();
49        }
```

第 8~49 行为类 MySecretAlg 的构造函数，参数为字符数组形式的密钥 chs。第 9 行判断字符数组的长度是否为 32，如果为真，则第 10~40 行得到执行，将字符数组 chs 的值赋给密钥 keys；否则第 43~45 行执行，即密钥固定配置为"0123456789ABCDEF0123456789ABCDEF"。第 47 行调用 calcParam 函数（第 50~72 行），由密钥 keys 得到分段线性映射的初始值 x0 和参数 p。第 48 行调用 calcSecCode 函数（第 83~88 行）得到加密用的伪随机数密码。

```
50        private void calcParam(){
51            double x00,x01,p0,p1;
52            x00 = 0;p0 = 0;
53            for(int i = 0;i < 8;i++){
54                x00 = x00 + keys[i]/Math.pow(2,4 * (i + 1));
55                p0 = p0 + keys[i + 8]/Math.pow(2,4 * (i + 1));
56            }
57            x00 = 0.8 * x00 + 0.1;
58            p0 = 0.4 * p0 + 0.1;
59            p = p0;x0 = x00;
60            for(int i = 0;i < 64;i++){
61                x0 = PWLCM(x0);
62            }
63            x01 = 0;p1 = 0;
64            for(int i = 0;i < 8;i++){
65                x01 = x01 + keys[i + 16]/Math.pow(2,4 * (i + 1));
66                p1 = p1 + keys[i + 24]/Math.pow(2,4 * (i + 1));
67            }
68            x01 = 0.8 * x01 + 0.1;
69            p1 = 0.4 * p1 + 0.1;
70            p = 0.382 * p1 + 0.618 * p0;
71            x0 = 0.382 * x01 + 0.618 * x00;
72        }
```

第 50~72 行的函数 calcParam 根据密钥 keys 计算分段线性映射的初始值 x0 和参数 p。

```
73        private double PWLCM(double x){
74            double res;
```

```
75      if(x>=0 && x<p)
76          res = x/p;
77      else if(x<0.5)
78          res = (x-p)/(0.5-p);
79      else
80          res = PWLCM(1.0-x);
81      return res;
82  }
```

第 73~82 行为分段线性映射函数 PWLCM,输入参数 x 得到它迭代一次的值。

```
83  private void calcSecCode(){
84      for(int i=0;i<512;i++){
85          x0 = PWLCM(x0);
86          secCode[i] = (int)(x0 * Math.pow(10,11) % Math.pow(2,16));
87      }
88  }
```

第 83~88 行的函数 calcSecCode 由初始值 x0 和参数 p 不断迭代分段线性映射的各个状态,并将该状态值转化为 0~65535 间的整型数,赋给 secCode 数组,用作加密的密码,密码的长度为 512。需要说明的是:密码的长度越长,则加密性能越好,相应的加密时间也越长。

```
89  public String enCipher(String strp){
90      int len = strp.length();
91      char[] seqs = strp.toCharArray();
92      int[] seq1 = new int[512+70];       //Secret Code
93      int[] seq2 = new int[100];           //Plaintext
94      int[] seq3 = new int[100];           //Cipher text
95      int k = (int)(512/len);
96      int m = 512 % len;
97      if(m>0){
98          k = k+1;
99          for(int i=0;i<512+len-m;i++){
100             if(i<512){
101                 seq1[i] = secCode[i];
102             }
103             else{
104                 seq1[i] = secCode[i-512];
105             }
106         }
107     }
108     for(int i=0;i<len;i++){
109         seq2[i] = seqs[i];
110     }
111     for(int i=0;i<k;i++){
112         for(int j=0;j<len;j++){
113             if(j==0){
114                 seq3[j] = (seq2[j]+seq1[j+i*len]) % 65536;
115             }
116             else{
```

```
117                    seq3[j] = (seq2[j] + seq1[j + i * len] + seq3[j-1]) % 65536;
118                }
119            }
120            for(int j = 0;j < len;j++){
121                seq2[j] = seq3[j];
122            }
123        }
124        for(int i = 0;i < len;i++){
125            seqs[i] = (char)seq3[i];
126        }
127        return String.valueOf(seqs);
128    }
```

第 89~128 行为 enCipher 加密函数，该函数为 public 型，可供类 MySecretAlg 外部的函数调用。enCipher 加密函数实现的基于加密原理为：密码的长度为 512，明文消息不会超过 70 个汉字或 160 个字符，所以，将密码按明文消息的长度进行分段，每一段密码的长度与明文消息的长度相同，最后一段密码的长度不是 0，就用密码的首部的字符去填充，保最后一段密码的长度也是明文消息的长度；然后，依次用每一段密码加密明文消息，直到用尽全部密码；为了简化加密过程，这里每一段密码加密明文消息的算法相同，即对于每个明文字符，使用相应位置上的密码字符加上明文字符，再加上前一个位置上的密文字符，其和对 65536 取模，是典型的"加取模"操作。

```
129    public String deCipher(String strc){
130        int len = strc.length();
131        char[] seqs = strc.toCharArray();
132        int[] seq1 = new int[512 + 70];//Secret Code
133        int[] seq2 = new int[100];      //Cipher-text
134        int[] seq3 = new int[100];      //Plaintext
135        int k = (int)(512/len);
136        int m = 512 % len;
137        if(m > 0){
138            k = k + 1;
139            for(int i = 0;i < 512 + len - m;i++){
140                if(i < 512){
141                    seq1[i] = secCode[i];
142                }
143                else{
144                    seq1[i] = secCode[i - 512];
145                }
146            }
147        }
148        for(int i = 0;i < len;i++){
149            seq2[i] = seqs[i];
150        }
151        for(int i = k - 1;i >= 0;i--){
152            for(int j = 0;j < len;j++){
153                if(j == 0){
154                    seq3[j] = (65536 + seq2[j] - seq1[j + i * len]) % 65536;
155                }
```

```
156                  else{
157                      seq3[j] = (65536 * 2 + seq2[j] - seq1[j + i * len] - seq2[j - 1])
                             % 65536;
158                  }
159              }
160              for(int j = 0;j < len;j++){
161                  seq2[j] = seq3[j];
162              }
163          }
164          for(int i = 0;i < len;i++){
165              seqs[i] = (char)seq3[i];
166          }
167          return String.valueOf(seqs);
168      }
169  }
```

第 129～168 行为 deCipher 解密函数,是 enCipher 加密函数的逆过程。输入为密文字符串 strc,输出(即返回)为解密后的字符串。如果解密密钥是与加密密钥相同的合法密钥,则返回原始的明文消息。

9.4 本章小结

本章详细介绍了基于 Andriod 系统智能手机的短信发送和短信接收实现方法,介绍了短信息发送和接收的函数。详细的短信息发送和接收底层机制可参考 Android 用户开发手册。在阐述了短信息发送与接收实例的基础上,研究了短信息的加密技术,通过设计一个基于分段线性映射(PWLCM)的字符加密算法,用实例的方式详细阐述了 Android 系统智能手机短信息加密技术的可行性,同时,这里提出的短信息加密与解密技术具有加密与解密速度快、密码长、保密性好、实用性强等优点。

参 考 文 献

[1] 张勇,夏家莉,陈滨,等. Google Android 开发技术. 西安:西安电子科技大学出版社,2011.
[2] Gerber A,Craig C. Learn Android Studio—Build Android Apps Quickly and Effectively. Berkeley: Apress,2015.
[3] http://www.android-studio.org/. (Android Studio 中文社区)
[4] http://android.xsoftlab.net/reference/packages.html. (Android 开发者手册)
[5] 梁勇. Java 语言程序设计(基础篇). 北京:机械工业出版社,2015.